Advances in Dielectrics

Series Editor

Friedrich Kremer, Fakultät für Physik und Geowissenschaften, Peter-Debye-Institut für Physik der weichen Materie, Universität Leipzig, Leipzig, Germany

Aims and Scope

Broadband Dielectric Spectroscopy (BDS) has developed tremendously in the last decades. For dielectric measurements it is now state of the art to cover typically 8-10 decades in frequency and to carry out the experiments in a wide temperature and pressure range. In this way a wealth of fundamental studies in molecular physics became possible, e.g. the scaling of relaxation processes, the interplay between rotational and translational diffusion, charge transport in disordered systems, and molecular dynamics in the geometrical confinement of different dimensionality – to name but a few. BDS has also proven to be an indispensable tool in modern material science; it plays e.g. an essential role in the characterization of Liquid Crystals or Ionic Liquids and the design of low-loss dielectric materials.

It is the aim of "Advances in Dielectrics" to reflect this rapid progress with a series of monographs devoted to specialized topics.

Target Group

Solid state physicists, molecular physicists, material scientists, ferroelectric scientists, soft matter scientists, polymer scientists, electronic and electrical engineers.

Silvina Cerveny · Jan Swenson
Editors

Dynamics of Biomolecules Studied by Dielectric Spectroscopy

Springer

Editors
Silvina Cerveny
Centro de Fisica de Materiales (CFM)
Consejo Superior de Investigaciones
Cientificas (CSIC)
Donostia-San Sebastián, Guipúzcoa, Spain

Jan Swenson
Department of Physics and Astronomy
Chalmers University of Technology
Gothenburg, Sweden

ISSN 2190-930X ISSN 2190-9318 (electronic)
Advances in Dielectrics
ISBN 978-3-032-30587-9 ISBN 978-3-032-30588-6 (eBook)
https://doi.org/10.1007/978-3-032-30588-6

This Springer imprint is published by the registered company Springer Nature Switzerland AG
The registered company address is: Gewerbestrasse 11, 6330 Cham, Switzerland

Preface

Understanding how proteins function requires understanding of how they move. Protein biological activity (from enzyme catalysis to molecular recognition) depends critically on dynamics that extend from femtoseconds to hundreds of seconds. These motions not only arise from the protein but are also intimately coupled to the surrounding hydration water that provides the entropy fluctuations necessary for proteins to transition between functional states. Broadband dielectric spectroscopy (BDS) and terahertz spectroscopy (THz) are increasingly used experimental techniques for the study of bio-molecular dynamics. This book focuses on the dynamics of proteins and their smaller building units, peptides, and amino acids. The aim is to present the current knowledge in this field and to show what kind of insights can be reached by these dielectric techniques. The different chapters contain different materials and different aspects of the dynamics, with the aim to complement each other and provide a complete overview of the different types of dynamics, but also present slightly different viewpoints regarding the exact interpretations of the observed relaxation processes. Most chapters demonstrate that hydration water critically governs biomolecular dynamics and exhibits distinct relaxation processes. However, it should be noted that unfortunately there is no universal names for the water relaxations nor for the different relaxations in the biomolecules. The reason for this is that the terminology used for the observed dielectric relaxation processes is influenced by historical conventions and depend on the specific research field. Therefore, below we are giving short descriptions of the contents in the different chapters, where we mention what the different relaxation processes are called.

In Chapter "Broadband Dielectric and THz Spectroscopy on Bio-related Matter: Water, Amino Acids, Proteins, and Blood" (*Lunkenheimer, Emmert, Wolf, and Loidl*), the interpretation of BDS and THz spectroscopy data on biological materials is discussed, with a particular focus on water, amino acids, proteins, and blood. The authors emphasize that many spectral features are influenced by ionic conductivity and polarization effects (e.g., at the electrodes), which can obscure intrinsic dynamics. It should also be noted that in this chapter the authors are focusing on the dynamics of biomolecules in diluted solutions, which, for instance in the case of proteins, means that not only intramolecular relaxation processes are observed (as in the case of hydrated protein powders), but also rotations and translational motions of the whole protein molecules. This further implies that a different terminology is used compared to that for hydrated biomolecules. In the biophysics nomenclature used in this chapter, the Greek letters α, β, γ, and δ are consecutively assigned to relaxation processes from lowest to highest frequency (or slowest to fastest), with the exception of the δ-relaxation, which was observed later and has a relaxation time between the β and γ processes. However, in contrast to the glass-physics community (and most other chapters in this book), where the viscosity and glass transition (T_g) related structural relaxation is denoted α, the α-relaxation is here due to motions of "counterions" on the surface of the biomolecules. An equally big difference occurs for the β-relaxation, which is here attributed to reorientational motions of the whole biomolecule, in contrast to the local secondary relaxation called β by the glass-physics community. The water-related relaxation observed around 20 GHz at room temperature is due to reorientational motions of mainly unbound water molecules and is here called the γ-relaxation. The δ-relaxation, as mentioned above, lies between β and γ at a frequency around 100 MHz. However, its assignment is less clear, and everything from bound water molecules to fluctuating polar side groups of the biomolecules, including coupled biomolecular-water motions, has been proposed in the literature.

In Chapter "The Glass "Transition" Temperature and the Dynamics of Polypeptides" (*Floudas and Spyridakou*), the authors challenge the idea that glass formation in proteins and polypeptides is driven by solvent dynamics, providing experimental evidence that it is also an intrinsic property of the polypeptide chain itself. This chapter contributes to our understanding of how polypeptide dynamics are influenced by their structural motifs, an area that was less explored before. The chapter first reviews early key studies on hydrated and nearly dry proteins (including myoglobin, silk fibroin, and lysozyme) showing how hydration influences protein dynamics and how systems with minimal solvent still display a T_g. Subsequently, the chapter examines poly(γ-benzyl-L-glutamate) (PBLG) as a model synthetic polypeptide, studied through a combination of temperature-modulated differential scanning calorimetry (TM-DSC), wide-angle X-ray scattering (WAXS), solid-state ^{13}C nuclear magnetic resonance (NMR), BDS, high-pressure BDS, and rheology. By systematically varying chain length and end-group chemistry over a wide range of molar masses, the authors distinguish between local and global dynamics associated with α-helical and β-sheet assemblies. They identify two glass transition temperatures in oligopeptides that stabilize both secondary structures: a lower T_g (the α segmental process) linked to amorphous segments disrupting the α-helices, and a higher T_g (the α* segmental process) caused by disordered segments within β-sheet domains. The influence of secondary structure type on fragility, pressure, and viscoelastic response shows that α-helical and β-sheet polypeptides exhibit different dynamics: β-sheets are notably more cooperative and stiffer, and become dynamically arrested at higher temperatures than α-helices. Finally, the impact of nanoconfinement in aluminum oxide nano-pores on PBLG is discussed, highlighting an acceleration of segmental dynamics and a partial destabilization of the secondary structure, similar to that observed in intrinsically disordered proteins. Overall, the chapter establishes a direct link between secondary structure organization and glass-forming behavior in polypeptides, complementing the solvent-slaved dynamics covered in other chapters by showing that intrinsic polypeptide structure and dynamics also influence the glass transition.

In Chapter "Dynamics of Peptides at Low Temperatures: From Monomer to Polymer and the Universal Nature of Slaving Behavior" (*Cerveny*), the author conducts a systematic investigation using dielectric spectroscopy and complementary techniques to analyze the dynamics of a series of lysine-based model systems (ranging from the amino acid L-lysine to oligomers and the polypeptide ε-polylysine), revealing that peptide motions are closely coupled to those of the surrounding water. Frauenfelder previously observed this behavior in studies of protein dynamics, and the phenomenon was termed "slaving": large-scale conformational fluctuations of the peptide exhibit the same non-Arrhenius temperature dependence as the cooperative water α-relaxation (also called "slow water relaxation" to distinguish it from the more local and universal "fast water relaxation" in this chapter), differing only by a constant time offset. The author demonstrates that this slaving behavior is not unique to proteins but represents a universal physical principle observable across both biological and non-biological polymer-water systems. The chapter also shows that slaving can be distinguished from simple plasticization. Slaving behavior is observed in solutions where the glass transition temperature decreases sharply and nonlinearly with increasing water concentration: in these systems, the α-relaxation of the solute and the slow water relaxation exhibit the same temperature dependence, indicating that the solute dynamics are slaved to those of the surrounding water. Conversely, slaving is not observed in solutions where the glass transition temperature decreases roughly linearly with increasing water concentration: water acts as a plasticizer, reducing the energy barriers to molecular motion without directly controlling the polymer dynamics, and the solute and water dynamics remain independent. The chapter further establishes that the onset of slaving behavior at hydration levels around 15–20 wt% water coincides with the minimum water content required for protein biological activity, thereby providing a fundamental physical explanation for why proteins absolutely require hydration to function and how water molecules provide the entropy contributions necessary for proteins to overcome energy barriers and transition between conformational states.

In Chapter "Atomistic Mechanisms of Protein Dynamics and the Essential Role of Hydration Water" (*Cerveny, Swenson and Sokolov*), the authors provide a comprehensive review of protein dynamics from picoseconds to microseconds, using a soft-matter physics framework to organize the hierarchy of biomolecular motions. Combining quasielastic neutron scattering (QENS), BDS, and molecular dynamics (MD) simulations, the chapter identifies five distinct classes of protein dynamics. The fast picosecond dynamics including the boson peak and cage rattling motions, the methyl group rotations, the sub-nanosecond coupled protein-solvent relaxation ("localized diffusion"), the nanosecond conformational jumps involving side-chain and domain motions, and the microsecond backbone relaxation, identified as the primary structural relaxation directly linked to the glass transition of hydrated proteins (relaxation time ~100 s at $T_g \approx 170$–220 K). The chapter demonstrates that hydration water plays an essential active role rather than merely acting as a passive solvent: water motions control protein dynamics through coupled relaxation processes, and artificially restricting water motion in MD simulations suppresses protein relaxation. Thus, in this chapter there is no specific water relaxation, but only the coupled protein-solvent relaxation where the solvent and protein move collectively on the same time-scale. However, it is the same relaxation process as the main water relaxation in the other chapters. In the studied temperature range it can be considered as a protein modified α-relaxation of the hydration water. Also the different protein relaxations have not been given any particular designations but are rather described with names based on their origins.

In Chapter "Studies of the Protein Glass Transition and Its Related Dynamics—Comparison with Various Aqueous Solutions" (*Hussan, Yagihara, Kita, Sasaki, and Shinyashiki*), the authors discuss the dynamics of unfrozen and frozen water and how it interacts with both smaller molecules, such as alcohol and sugar, and larger molecules, such as protein. They show that water exhibits universal relaxation behavior across a wide range of aqueous mixtures (whether with small molecules or proteins), as other authors also note in their Chapters ("Dynamics of Peptides at Low Temperatures: From Monomer to Polymer and the Universal Nature of Slaving Behavior", "Protein Dynamics Slaved by Solvent Dynamics as Revealed by Dielectric Spectroscopy", and "Combined Broadband Dielectric Spectroscopy and Nuclear Magnetic Resonance Studies on the Dynamics of Cooled Protein–Solvent Mixtures"). In particular, their results show that the smaller solute molecules exhibit a cooperative relaxation process together with the water molecules, whereas the main protein relaxation (here denoted as α-relaxation or process III) is slower than the water relaxation (mostly called ν-relaxation, but also β-process or process I in this chapter). On an intermediate time-scale between these two relaxations, an ice relaxation (or process II) is observed in the low temperature regime discussed in this chapter. For polymer-water systems such as PVP-water mixtures, a third water-related process, called the m-process, was also identified between the α- and ν-processes. It should be noted here that the ν-relaxation is the same relaxation as the one called fast water relaxation in Chapter "Dynamics of Peptides at Low Temperatures: From Monomer to Polymer and the Universal Nature of Slaving Behavior", and the m-process is the same relaxation as the one called slow water relaxation in Chapter "Dynamics of Peptides at Low Temperatures: From Monomer to Polymer and the Universal Nature of Slaving Behavior". In addition to discussing the observed relaxation processes, the authors emphasize the crucial role of water for the flexibility, stability, and functionality of proteins and other macromolecules.

In Chapter "Protein Dynamics Slaved by Solvent Dynamics as Revealed by Dielectric Spectroscopy" (*Swenson and Jansson*), the authors investigate the fundamental relationship between protein dynamics and solvent dynamics. By applying BDS over a broad temperature range, they show that the relaxation processes of protein and solvent exhibit similar temperature dependencies, in agreement with the slaving model proposed by Frauenfelder. This model implies that protein dynamics are governed (or "slaved") by solvent motions, and that protein dynamics and related biological functions are therefore critically dependent on solvent mobility. The results also show that both the α-relaxation in the solvent and various protein relaxations contribute to the calorimetric glass transition. Since these relaxation processes occur on widely different time-scales the calorimetric glass transition becomes anomalously broad. This chapter connects closely with Chapter "Dynamics of Peptides at Low Temperatures: From Monomer to Polymer and the Universal Nature of Slaving Behavior", where it is shown in more detail that the presented "slaving" phenomenon is not unique for proteins or even biomolecules. Chapter "Protein Dynamics Slaved by Solvent Dynamics as Revealed by Dielectric Spectroscopy" also proposes that "slaving" of proteins does not necessarily need a surrounding solvent since it seems to be sufficient that proteins are surrounded by molecules which are able to maintain the native protein structure and exhibit a viscosity related α-relaxation. In Chapter "Protein Dynamics Slaved by Solvent Dynamics as Revealed by Dielectric Spectroscopy", the main water relaxation is called Process IIa/IIb, representing the primary solvent α-relaxation (IIa) that undergoes a crossover to a local β-relaxation (IIb) below approximately 170 K. Below the crossover temperature, this process is described as "most likely of Johari-Goldstein type, also known as ν-relaxation". Process IIb is the same relaxation as the "fast water relaxation" in Chapter "Dynamics of Peptides at Low Temperatures: From Monomer to Polymer and the Universal Nature of Slaving Behavior" or ν-relaxation in Chapter "Studies of the Protein Glass Transition and Its Related Dynamics—Comparison with Various Aqueous Solutions", whereas Process IIa is the same relaxation as the one called slow water relaxation in Chapter "Dynamics of Peptides at Low Temperatures: From Monomer to Polymer and the Universal Nature of Slaving Behavior" and m-process in Chapter "Studies of the Protein Glass Transition and Its Related Dynamics—Comparison with Various Aqueous Solutions". The two slowest protein relaxations mentioned in Chapter "Atomistic Mechanisms of Protein Dynamics and the Essential Role of Hydration Water", "conformational jumps" (of e.g. side groups) and "main structural relaxation" correspond to Processes III and IV, respectively, in this chapter.

In Chapter "Combined Broadband Dielectric Spectroscopy and Nuclear Magnetic Resonance Studies on the Dynamics of Cooled Protein–Solvent Mixtures" (*Vogel*), the interplay between protein and solvent dynamics is investigated using a multi-technique approach. BDS measurements are combined with NMR, QENS and MD simulations to achieve component-selective characterization of water and protein dynamics across globular proteins (lysozyme, myoglobin), fibrous proteins (collagen, elastin), and the peptide ε-polylysine and to obtain further insights into the dynamical nature of the different relaxation processes, which both latter techniques can provide. The results confirm that reorientations in protein hydration water are slower than in bulk water, but also show that it behaves somewhat differently from a normal β-relaxation, exhibiting quasi-isotropic character coupled to translational diffusion down to ~200 K, which is more a signature of an α-relaxation. Therefore, the author denotes this process as a β-relaxation (also called P1 in the BDS data), while emphasizing its α-like character. This process is the same as the "fast water relaxation" in Chapter "Dynamics of Peptides at Low Temperatures: From Monomer to Polymer and the Universal Nature of Slaving Behavior", the ν-relaxation in Chapter "Studies of the Protein Glass Transition and Its Related Dynamics—Comparison with Various Aqueous Solutions", and Process IIb in Chapter "Protein Dynamics Slaved by Solvent Dynamics as Revealed by Dielectric Spectroscopy". Two slower relaxation processes, P2 and P3, are attributed to the α-relaxation of the solvent and conformational fluctuations of the protein (analogous to Process IV in Chapter "Protein Dynamics Slaved by Solvent Dynamics as Revealed by Dielectric Spectroscopy"), respectively. Component-selective NMR and QENS analyses reveal that proteins exhibit only highly restricted localized motions rather than large-amplitude quasi-isotropic reorientation, challenging interpretations that assign slower BDS processes (P3) to protein α-relaxation. Moreover, ^{1}H and ^{2}H NMR studies of mixtures of ε-PLL with H_2O or D_2O do not provide evidence for the existence of two distinct water fractions with different interactions with ε-PLL. Instead, the NMR findings suggested that the second-slowest BDS process was due to a reorientation of polar moieties in the peptide backbone. Therefore, the NMR studies did not support the idea of two separate water fractions initially suggested by the BDS analysis. From these studies it was also clear that the relaxation times of both the solvent and the protein are affected by the structure and chemical nature of the protein.

In Chapter "Terahertz Spectroscopy of Water and Proteins: Methodology and Insights" (*Novelli and Adams*), the application of THz spectroscopy to investigate water and protein dynamics at frequencies complementary to BDS (0.1–10 THz vs. mHz–GHz) is discussed. THz radiation probes the collective intermolecular motions of water, including restricted translations (~5 THz) and librations (~18 THz), while the lower-frequency dielectric response is described by two Debye relaxation processes with timescales $\tau_1 \approx 9.4$ ps and $\tau_2 \approx 0.3$ ps at room temperature. Particular focus is on how the interactions with the protein surface affect the dielectric properties and dynamics of hydration water. A key framework presented is the "water plus void" model, which considers that the solute is creating a cavity or void in water. If the experimental absorption is higher than model prediction, the protein has a detectable hydration shell with faster reorientation time-scales than bulk water, indicating a more disordered or weaker hydrogen-bond network (structure breaker). On the contrary, if the absorption is lower, it suggests a structure maker protein with a more ordered hydration shell. The authors also show that water structure and dynamics depend on the surface electrostatic potential and hydrophobicity of the protein, providing insights into how hydration water behaves differently from bulk water. Such knowledge helps us to understand the complex interplay between protein surfaces and their hydration layers.

In Chapter "Disorder at the Interface: Hydration Water and the Physics of Intrinsically Disordered Proteins" (*Leitner, Reid, Singh, and Vinh*), the dynamics of hydration water around intrinsically disordered proteins (IDPs) and intrinsically disordered regions (IDRs) is discussed. This topic has historically received less attention than structured proteins. By applying MHz-THz dielectric spectroscopy and

MD simulations, it was found that the concentration of charged groups, particularly anionic residues, significantly influences the dynamics of nearby water molecules. The authors show that water around IDPs exhibits slower and more heterogeneous dynamics and a higher density of tightly bound water than in structured proteins, providing new insights into the role of charged residues in modulating hydration dynamics. This chapter enhances our understanding of the molecular interactions at the protein-water interface, particularly in disordered systems.

Thus, from the summaries given above, it is clear that there is no universal terminology for the different relaxation processes. This causes authors to call processes differently even if they have the same physical origin. For instance, in the case of the Arrhenius-type relaxation of water molecules in aqueous mixtures below the glass transition temperature, it is called γ-relaxation in Chapter "Broadband Dielectric and THz Spectroscopy on Bio-related Matter: Water, Amino Acids, Proteins, and Blood", fast water relaxation in Chapter "Dynamics of Peptides at Low Temperatures: From Monomer to Polymer and the Universal Nature of Slaving Behavior", ν-relaxation in Chapter "Studies of the Protein Glass Transition and Its Related Dynamics—Comparison with Various Aqueous Solutions", Process IIb in Chapter "Protein Dynamics Slaved by Solvent Dynamics as Revealed by Dielectric Spectroscopy" and P1 in Chapter "Combined Broadband Dielectric Spectroscopy and Nuclear Magnetic Resonance Studies on the Dynamics of Cooled Protein–Solvent Mixtures". However, in all these chapters except Chapter "Broadband Dielectric and THz Spectroscopy on Bio-related Matter: Water, Amino Acids, Proteins, and Blood" the authors also refer to it as a β-relaxation. The cooperative water relaxation at higher temperatures that couples to solute dynamics is called slow water relaxation in Chapter "Dynamics of Peptides at Low Temperatures: From Monomer to Polymer and the Universal Nature of Slaving Behavior", α-relaxation and m-process in Chapter "Studies of the Protein Glass Transition and Its Related Dynamics—Comparison with Various Aqueous Solutions", Process IIa and α-relaxation in Chapter "Protein Dynamics Slaved by Solvent Dynamics as Revealed by Dielectric Spectroscopy" and P2 and α-relaxation in Chapter "Combined Broadband Dielectric Spectroscopy and Nuclear Magnetic Resonance Studies on the Dynamics of Cooled Protein–Solvent Mixtures". Understanding these equivalent designations is essential for readers navigating between chapters and interpreting the rich body of literature on water dynamics in biological systems.

As noted above, the terminology for water and biomolecular relaxation processes varies significantly across chapters. To facilitate reading between chapters, Table 1 lists the equivalent relaxation processes across all chapters, showing how the same physical phenomenon is designated by different names in different contexts. This table is essential for readers seeking to understand the relationships among the processes discussed across chapters and to connect results across the literature. Readers should refer to this table when encountering unfamiliar nomenclature in any chapter.

It should also be noted that irrespective of what a certain relaxation process is called, different authors make somewhat different interpretations of its physical

Table 1 Terminology for identifying the water and biomolecular relaxation processes by the different authors

Physical processes	Chapter "Dynamics of Peptides at Low Temperatures: From Monomer to Polymer and the Universal Nature of Slaving Behavior"	Chapter "Studies of the Protein Glass Transition and Its Related Dynamics—Comparison with Various Aqueous Solutions"	Chapter "Protein Dynamics Slaved by Solvent Dynamics as Revealed by Dielectric Spectroscopy"	Chapter "Combined Broadband Dielectric Spectroscopy and Nuclear Magnetic Resonance Studies on the Dynamics of Cooled Protein–Solvent Mixtures"	Alternative names
Universal low temperature water relaxation	Fast-water	ν-, β- or I-relaxation	IIb	P1	Water relaxation, JG relaxation, β-relaxation
Slow cooperative water relaxation	Slow-water	m-process	IIa	P2	α-relaxation of water
Protein structural relaxation	α-relaxation	α-relaxation or III	IV	P3	Large-scale conformational fluctuations

origin, reflecting ongoing debates about the microscopic mechanisms of water-biomolecule coupling. Rather than a limitation, this diversity of perspectives highlights the vitality of the field and the depth of insight achievable through multiple experimental and theoretical approaches. We hope that this book inspires readers to contribute discoveries in this fascinating and important area of research.

This book is intended for graduate students, postdoctoral researchers, and established scientists seeking a comprehensive overview of experimental approaches to protein-water dynamics, from fundamental water physics to biological function.

Finally, we would like to express our gratitude to Prof. Friedrich Kremer for inviting us to edit this book and for all the support we have received from him and Springer Nature in completing the book. We are also thankful to all the contributing authors. It has been a pleasure to work with all of them and read their excellent contributions to the book. Without their expertise and knowledge, this book would not have been possible to complete.

Donostia-San Sebastián, Spain Silvina Cerveny
Gothenburg, Sweden Jan Swenson

Contents

Abbreviations

3PG	Tripropylene glycol
ASN	L-asparagine
BDS	Broadband Dielectric Spectroscopy
BP	Boson Peak
BSA	Bovine Serum Albumin
CC	Cole–Cole (function)
CD	Cole–Davidson (function); Circular Dichroism
C_f	Concentration of fructose
COL	Collagen
CPC	C-phycocyanin
CS	Conformational Substates
c_w	Water concentration (weight fraction)
Cys	L-cysteine
D	Strength parameter (VFT equation)
DF	D-fibrinogen
DM	D-myocin
DRS	Dielectric Relaxation Spectroscopy
DS	Dielectric Spectroscopy
DSC	Differential Scanning Calorimetry
Ea	Activation Energy
EG	Ethylene glycol
EGO	Ethylene glycol oligomer
ELA	Elastin
EP	Electrode Polarization
FIR	Far-Infrared
FST	Fragile-to-strong transition
FTIR	Fourier Transform Infrared Spectroscopy
G'	Storage modulus
G''	Loss modulus
GL	Glycerol
h.b.	Hydrogen bonds

HAM	Highly Asymmetric Mixtures
HAPB	Highly Asymmetric Polymer Blends
HB	Hydrogen bond
hw	Hydration fraction (water content)
IDP	Intrinsically disordered proteins
IDR	Intrinsically disordered regions
INS	Inelastic Neutron Scattering
IR	Infrared (spectroscopy)
JG	Johari–Goldstein (relaxation)
KWW	Kohlrausch–Williams–Watts
LYS	Lysozyme
m	Fragility index
MD	Molecular dynamics
MSD	Mean-squared displacement
MYO	Myoglobin
n	Degree of polymerization
NMR	Nuclear Magnetic Resonance
NP4	Nitrophorin 4
NS	Neutron scattering
PBLG	Poly(γ-benzyl-L-glutamate)
PDP	Protein Data Bank
PEI	Polyethylenimine
PGly	Polyglycine
PNIPAM	Poly(N-isopropylacrylamide)
PVME	Poly(vinyl methyl ether)
PVP	Poly(vinyl pyrrolidone)/Polyvinylpyrrolidone
PZLL	Poly-(ε-carbobenzoxy-L-lysine)
q	Modulus of the scattering vector
QENS	Quasielastic neutron scattering
RBC	Red blood cells/erythrocytes
RSA	Relative accessible surface area
S	Dynamic order parameter
SASA	Solvent accessible surface area
SAXS	Small-Angle X-ray Scattering
Ser	L-serine
SUMO-1	Small ubiquitin-related modifier protein-1
TC	Critical temperature
TDDS	Time-domain dielectric spectroscopy
T_g	Glass transition temperature
THz	Terahertz (spectroscopy)
THz-TDS	Terahertz Time-Domain Spectroscopy
T_m	Melting temperature
T_S	Separation temperature
TSDC	Thermally Stimulated Depolarization Current
UBC9	Ubiquitin-conjugating enzyme

VFT	Vogel–Fulcher–Tammann (equation)
vGP	van Gurp–Palmen
VSHBβ	Vectorial summation of hydrogen-bonded β-strands
WAXS	Wide-Angle X-ray Scattering
δ	Phase angle
$\Delta\varepsilon$	Dielectric strength
$\Delta\varepsilon\alpha$	Dielectric strength of α-process
$\Delta\varepsilon\nu$	Dielectric strength of ν-process
ε'	Real part of permittivity (dielectric constant)
ε''	Imaginary part of permittivity (dielectric loss)
ε_0	Permittivity of vacuum
ε-PLL	ε-Polylysine
ν	Frequency
ξ_{helix}	Correlation length of helical parts
σ'	Real part of conductivity
σ_{dc}	DC conductivity
τ	Relaxation time

Broadband Dielectric and THz Spectroscopy on Bio-related Matter: Water, Amino Acids, Proteins, and Blood

Peter Lunkenheimer, Sebastian Emmert, Martin Wolf, and Alois Loidl

Abstract In the present work, we examine the relevance and proper interpretation of broadband dielectric and THz spectroscopy data for the investigation of various types of biological matter. We provide an overview of the rich variety of different dynamic processes that can be detected by these experimental methods. Several experimental examples are discussed in detail, helping to understand the information that can be drawn from such studies. This includes dielectric spectra, extending well into the GHz region, for pure water, which can be considered as a simple but highly important biological molecule. We also discuss results for a prototypical aqueous solution of a protein, belonging to one of the most important classes of biological macromolecules. Moreover, we examine broadband dielectric spectra on blood as an example of functional biological matter in organisms. To demonstrate the relevance of THz spectroscopy for the investigation of biological molecules, we finally treat such experiments applied to different amino acids.

Keywords Dielectric spectroscopy · THz spectroscopy · Water · Amino acids · Proteins · Blood · Lysozyme · L-asparagine · L-cysteine · L-serine · Dielectric relaxation · Maxwell–Wagner relaxation · Electrode polarization · Permittivity · Dc conductivity · Bound water · Protein dynamics · Boson peak · Vibrational excitations · Phonons

1 Introduction

Most biological systems reveal a rich variety of dynamic processes of their microscopic constituents, which often impact their vital functions in organisms, are relevant for electromagnetic dosimetry, and can be employed for diagnostic and therapeutic applications [1–7]. This includes the relatively slow inter- and intramolecular motions of large biomolecules like proteins, the translational hopping of dissolved

P. Lunkenheimer (✉) · S. Emmert · M. Wolf · A. Loidl
Experimental Physics V, Center for Electronic Correlations and Magnetism, University of Augsburg, Augsburg, Germany
e-mail: peter.lunkenheimer@physik.uni-augsburg.de

S. Cerveny and J. Swenson (eds.), *Dynamics of Biomolecules Studied by Dielectric Spectroscopy*, Advances in Dielectrics, https://doi.org/10.1007/978-3-032-30588-6_1

ions, interfacial effects, e.g., due to the presence of cell membranes, the motions of bound and free water molecules, phonon-like excitations, the stretching and bending of intramolecular bonds, and many more. These processes occur on partly very different time scales, extending from seconds or slower, down to picoseconds. When investigating them by spectroscopic methods, each process ideally should lead to a separate spectral feature in the detected susceptibilities, which can show up at frequencies ranging from sub-Hz to THz, depending on the particular type of motion. Thus, performing spectroscopy covering a broad frequency range is crucial for the investigation of the dynamics in biological systems. This requirement is most easily achieved by broadband dielectric measurements, complemented by THz experiments. Fortunately, most of the dynamic processes in biological systems involve dipolar fluctuations or charge transport, which is prerequisite for their investigation by these methods. Indeed, the application of dielectric and Terahertz (THz) spectroscopy to biological matter has been well establish for many decades [1–8]. The determination of the dielectric properties and the achievement of a better understanding of the dynamic processes in biological systems is highly relevant in many respects. Important examples are the fixing of limiting radiation values to control electromagnetic pollution, the development of new diagnostic and therapeutic techniques, and food monitoring [3, 6].

In the present work, we first provide a brief overview of the most important dynamic processes revealed in the dielectric spectra of biological matter. We then discuss the dielectric response of pure water, which governs the ubiquitous γ relaxation found in most biological materials, and treat the broadband dielectric spectra of aqueous protein solutions. We further consider results on blood as an example for the complex dielectric behavior of functional biological matter in organisms and, finally, discuss the vibrational modes in amino acids as detected by THz spectroscopy. Concerning experimental methods and dielectric quantities, we refer the reader to the numerous earlier publications treating these topics, e.g., Refs. [1–3, 6, 9–11].

2 Dielectric Properties of Biological Matter

2.1 Overview

Figure 1 shows a schematic dielectric-loss spectrum, $\varepsilon''(\nu)$, indicating several typical contributions arising in biological materials. Most of the latter include water with dissolved mobile ions, which can lead to considerable dc conductivity. As explained in Sect. 2.2, this generates a $1/\nu$ divergence in $\varepsilon''(\nu)$ at low frequencies (ν denotes the frequency of the applied ac field). As indicated by the dashed line in Fig. 1, this contribution can superimpose considerable parts of the relaxational contributions to the loss spectra and often is subtracted to more clearly reveal them (solid line). Ionic conductivity usually causes electrode polarization (EP), leading to huge low-frequency contributions in the real part of the permittivity ε', but it also affects ε''

(not shown in Fig. 1) as explained in Sect. 2.3. The denotations β, γ, and δ for the relaxational processes, as included in Fig. 1, follow the nomenclature used in large parts of the biophysical literature treating dielectric properties [2, 12, 13] (the α relaxation detected in some materials at very low frequencies is not shown as it usually is observed in ε' only; see Sect. 2.4). The underlying microscopic processes of these relaxations are discussed in Sects. 2.5–2.7. The abbreviation "res." in Fig. 1 stands for resonances due to vibrational excitations (Sect. 2.8) and "BP" denotes the boson peak (Sect. 2.9), which should be present in all disordered biological systems.

To avoid confusion, it should be mentioned here that the naming of loss peaks by Greek letters is also common practice within the field of supercooled liquids and other glass-forming materials [16, 17]. However, there their assignment is completely different and, for example, the biophysical β relaxation arising from protein reorientations would be denoted as α relaxation.

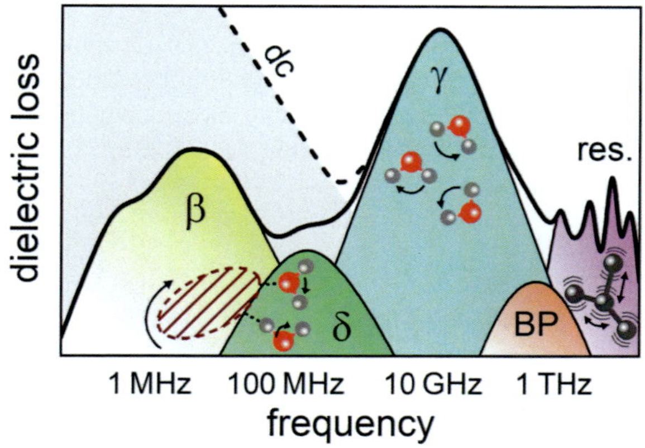

Fig. 1 Schematic representation of some typical contributions to dielectric-loss spectra of biological matter (double-logarithmic representation). The dashed line shows the spectrum including the dc conductivity and the solid line indicates the behavior after its subtraction. The denotations β, γ, δ follow the biophysics nomenclature [2, 12, 13]. β relaxations can be bimodal (visualized by the indicated double peak in the figure) and arise from reorientations of large dipolar entities (hatched ellipse in the figure) and/or Maxwell–Wagner relaxations due to internal heterogeneities (e.g., cell membranes). The δ relaxation also can be multimodal (not indicated) and mostly is ascribed to motions of bound water molecules and/or internal-protein dynamics. The γ relaxation reflects free water-molecule reorientations also occurring in bulk water. "BP" stands for the boson peak as suggested for unbound water molecules [14] and detected for other biological matter using scattering techniques [15]. Finally, "res." denotes high-frequency resonances, e.g., due to bond stretching and bending. The controversial α relaxation (Sect. 2.4) and possible sub-β processes due to conformational sampling [5] are not included in this figure

2.2 Dc Conductivity

Most dielectrically investigated biological matter such as tissues, cell suspensions (e.g., blood; see Sect. 3.3), or solutions of biomolecules (e.g., proteins; see Sect. 3.2) contain considerable amounts of water including dissolved ions. The high mobility of the latter in water [18] leads to considerable dc conductivity, σ_{dc}, i.e., biological samples usually are ionic conductors. There is a general relation between the dielectric loss ε'' and the real part of the conductivity σ', namely: $\varepsilon'' = \sigma'/(2\pi\,\nu\,\varepsilon_0)$ (here, ε_0 is the permittivity of vacuum). Hence, in $\varepsilon''(\nu)$, the dc conductivity leads to a contribution $\varepsilon''_{dc} = \sigma_{dc}/(2\pi\,\nu\,\varepsilon_0)$, implying a $1/\nu$ divergence in the loss which obscures the detection of possible relaxation processes at low frequencies (cf. dotted line in Fig. 1). A subtraction of the dc contribution can uncover these relaxations as indicated by the solid line in Fig. 1 at low frequencies. However, at very low frequencies, the insufficient precision of the experimental data can lead to considerable scatter in the subtracted results. Thus, very slow dynamic processes (e.g., the α relaxation) are difficult to detect in $\varepsilon''(\nu)$ in this way. This problem does not apply to the real part of the permittivity, ε', because the dc conductivity does not contribute to this quantity. However, in systems with high ionic conductivity, electrode polarization (Sect. 2.3) usually dominates the ε' spectra, also hampering the detection of intrinsic relaxation processes at low frequencies.

2.3 Electrode Polarization and Maxwell–Wagner Relaxations

In typical dielectric experiments, the ionic charge transport present in most biological materials leads to the inevitable occurrence of so-called electrode polarization as observed in all ionic conductors [19, 20]. It is caused by the accumulation of essentially immobile ions at the metallic sample electrodes, arising at low frequencies due to the simple fact that the ions cannot enter the metal. This leads to thin, poorly conducting layers close to the electrodes, an effect sometimes termed "blocking electrodes." They act as huge capacitors connected in series to the bulk sample. These effects lead to so-called Maxwell–Wagner relaxations causing a strong frequency dependence of the dielectric properties. Maxwell and Wagner [21, 22] showed that in inhomogeneous materials, consisting of two or more regions with different dielectric properties, non-intrinsic dispersion effects in the permittivity and conductivity can arise, closely resembling the behavior of intrinsic dipolar relaxation processes [i.e., a steplike decrease in $\varepsilon'(\nu)$ and a peak in $\varepsilon''(\nu)$]. The special situation of thin, poorly conducting layers in series to the bulk sample can be most easily modeled by an equivalent circuit consisting of two parallel RC circuits (one for the blocking-electrode layers and one for the bulk) that are connected in series to each other [20, 23, 24]) This leads to the signature of a relaxation process in the dielectric spectra, in accord with the experimental results. The resulting relaxation step in $\varepsilon'(\nu)$ reveals a very large value of the limiting low-frequency ε' (sometimes termed "colossal dielectric constant" [24]. The latter is dominated by the capacitance of the surface layers which is large due to their small thickness. At high frequencies, their capacitance becomes essentially shorted and the intrinsic bulk properties are detected. This effect is also responsible for a crossover of $\sigma'(\nu)$ from small values at low frequencies (dominated by the high layer resistance) to larger, intrinsic values at higher frequencies. Considering the relation $\varepsilon'' \propto \sigma'/\nu$, this causes a relaxation-like peak in the loss spectra. Its spectral shape follows the Debye formula, theoretically predicted for dipolar relaxations, although in fact for such non-intrinsic Maxwell–Wagner relaxations it is not related to dipolar dynamics. In some cases, perfect fits of the experimental spectra cannot be achieved by using a simple RC equivalent circuit to model the blocking electrodes. Introducing a distribution of RC elements or replacing the electrode-RC circuit by a so-called constant-phase element then leads to better fits of the data [19, 20, 25].

Notably, the above-described equivalent-circuit approach in principle should only provide an approximate description of EP. In fact, the capacitive layers only form at low frequencies, simply because most ions do not reach the sample boundaries at high frequencies. Nevertheless, the manifestations of EP in experimental spectra of ionic conductors can be excellently described using this somewhat simplified approach [20, 23, 26]. Overall, one should be aware that EP is a non-intrinsic phenomenon and its manifestations in the measured permittivity and conductivity spectra should not be misconceived to reflect intrinsic properties like dipole dynamics. To avoid such misinterpretations, e.g., additional measurements with different sample thicknesses can be helpful [20]. Finally, it should be noted that thin *internal* barriers in biological matter (e.g., cell walls) can cause similar effects, however, usually located at higher frequencies (see discussion of the β relaxation in Sect. 2.5).

2.4 *α Relaxation*

In various biological systems like tissues or cell suspensions, the occurrence of a low-frequency process (usually at $\nu < 1$ kHz), termed α relaxation, is reported [1, 2, 8, 12, 13, 27]. The associated absolute values of the limiting low-frequency dielectric constant are huge (»1000), clearly indicating that it is not due to the reorientation of dipolar entities within the material. A common explanation for the α relaxation ascribes it to the mobility of "counter-ions" along the surface of colloidal particles (e.g., cells) within electric double layers formed around them [1, 2, 12, 13, 28, 29]. In particular, it was pointed out that this relaxation is not due to a simple Maxwell–Wagner effect [28, 29]. Various alternative mechanisms for this process were also proposed [2, 30]. Notably, due to its occurrence at very low frequencies, usually the α relaxation cannot be detected in dielectric-loss spectra. This is due to the huge contribution of the dc conductivity to $\varepsilon''(\nu)$ at low frequencies ($\varepsilon''_{dc} \propto 1/\nu$, as noted above), preventing the generation of meaningful dc-corrected data by subtraction. Unfortunately, the ε' spectra in this low-frequency region are often dominated by EP, also hampering the unequivocal observation of any intrinsic low-frequency processes.

2.5 β Relaxation

Molecules of biological significance often carry a permanent dipole moment and, when dissolved in water, can reorient with a rate determined by temperature and viscosity of the fluid. As discussed in the following section, the water molecule also is characterized by a permanent dipole moment, and, hence, in biological fluids dielectric dispersion phenomena can be expected due to relaxations of the large biological and the considerably smaller water molecule. Indeed, since the early works by Oncley and co-workers [31–33], it is well documented that aqueous solutions containing biological molecules reveal at least two dispersion regions in the MHz and GHz frequency regimes, termed β and γ relaxation, respectively. The γ process, reflecting water dynamics, will be discussed in the following section. As schematically depicted in Fig. 1, specifically in protein solutions the β process can be attributed to reorientational motions of the whole protein molecule in the viscous environment of the aqueous solution. Since Oncley's pioneering works, many dielectric spectra were collected specifically for aqueous protein solutions (see, e.g., [34–37]), all of them pointing toward the relaxation of these macromolecules, characterized by a dipole moment of some 100 D and by a relaxational frequency of some MHz. The temperature dependence of the relaxation times, which can be derived from the loss-peak frequencies, in most cases (at least in the rather narrow temperature regimes being accessible in aqueous solutions) follows purely thermally activated Arrhenius-type of behavior, characterized by energy barriers of ~ 20 kJ/mol or ~ 0.2 eV.

Relaxations in a comparable frequency regime are also observed in all kinds of tissue, e.g., liver, fat, muscle, lung, and in blood [1, 2, 6, 12, 13, 26, 38]. In most cases, they are significantly stronger than those in protein solutions and must be explained by Maxwell–Wagner effects, analogous to the EP treated in Sect. 2.3. In this case, however, the relevant thin layers, giving rise to polarization effects, are thought to be the cellular membranes. They act like large capacitors and are charged through the electrolyte with the charging time varying inversely with the ionic conductivity. Again, the problem is that this type of dispersion, commonly observed in heterogeneous media, looks analogous to Debye-like relaxations of dipolar entities. Schwan pointed out that, in principle, it is possible to determine its microscopic nature from an analysis of the dependence of the dielectric parameters on temperature or cell concentration [39].

Nowadays, it seems quite generally accepted that most β relaxations in tissues and cell suspensions like blood must be ascribed to Maxwell–Wagner-like effects [1, 2, 26, 38]. Thus, dielectric spectra in this regime in principle should be analyzed in terms of equivalent circuits, just as commonly done for electrode-polarization effects [20, 23, 24, 26]. However, for biological matter such β relaxations are more commonly fitted by the functions developed for intrinsic dipolar relaxation processes. Then one should be aware that the relaxation times determined by such fits do not reflect the time scales of any intrinsic microscopic dynamics in the investigated materials. Fortunately, there are various models (see, e.g., [40–44]) enabling conclusions about the intrinsic dielectric properties of the different sample regions, based on the fitting parameters resulting from such an approach.

In principle, as there are at least two different mechanisms that may give rise to β relaxations, the corresponding features in the spectra of tissues or cell suspensions can be bimodal [1, 2]. However, in most cases the interfacial β loss peaks due to Maxwell–Wagner effects are significantly stronger than those caused by dipolar protein dynamics [2, 13]. Thus, in tissues and cell suspension, the protein-related β process is usually not detected in the spectra. Notably, even without assuming any dipolar reorientations of macromolecules, multiple Maxwell–Wagner-related β peaks can arise due to the non-spherical shape of cells such as erythrocytes [44]. For example, in blood (Sect. 3.3) some hints at a bimodal β relaxation were reported [26]. In Fig. 1, the possible bimodal nature of the β relaxation is taken into account by the schematic indication of a double peak at low frequencies.

Finally, we want to mention that in some nuclear magnetic resonance investigations of proteins, conformational selection dynamics was reported, which is slower than the reorientational β relaxation and seems to play an important role for molecular recognition [45–47]. In Ref. [5], indications for such dynamics were also found by dielectric measurements of a protein solution and termed "sub-β relaxation."

2.6 γ Relaxation

The γ relaxation occurs in water-containing biological matter and is observed at frequencies around 20 GHz [1, 2, 7, 8, 12, 13]. Pure water at room temperature reveals a prominent relaxational feature close to this frequency [14, 48, 49] (see Sect. 3.1 and Fig. 2). Therefore, it is obvious that the γ relaxation of biological matter and the main relaxation of pure water are caused by the same dynamical process. In the biophysics literature, this is commonly assumed to be the reorientational motion of essentially unbound water molecules, which indeed corresponds to the most common interpretation of the 20 GHz relaxation of pure water (see Sect. 3.1 for a brief discussion of alternative explanations).

In various studies of biological materials extending to sufficiently high frequencies, certain deviations of the spectral features of the γ relaxation, compared to the main relaxation of pure water, were revealed (see, e.g., [7, 26, 50–52]). In addition to a reduction of the peak amplitude, which may be explained by the trivial substitution effect of water by other constituents in the sample, this also includes the broadening of the loss peak and its occurrence at somewhat different frequencies. By investigating these changes, valuable information concerning the complex interactions of biological and water molecules can be gained, which are of high relevance for the functionality of biological systems [7].

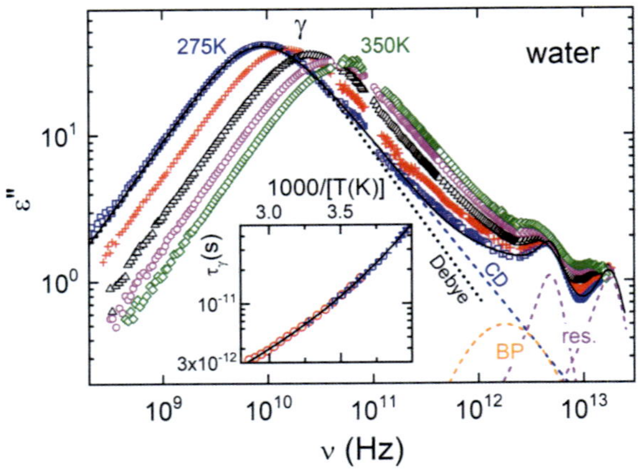

Fig. 2 Dielectric-loss spectra of pure liquid water as measured at various temperatures (from left to right: 275, 290, 310, 330, and 350 K) (data from Ref. [14]). The dotted line shows a fit of the main peak at 275 K by the Debye function, as often assumed for pure water, which, however, widely fails to describe the right peak flank. For the complete spectrum at 275 K, the solid line shows a fit with the sum of a CD function (for the main peak), a log-normal peak function (for the BP), and two Lorentz functions (for the resonances at ~ 5 and 19 THz) [14]. The dashed lines depict the different contributions to the overall fit curve. The inset presents the average relaxation times derived from the CD fit parameters (circles) (data from Ref. [68]) and additional $\tau(T)$ data from Ref. [105], extending into the supercooled region. The line represents a fit with the VFT function [68]. Please note that the main peak of water is named "γ" in the biophysical literature while it is denoted α or Debye peak in other fields

2.7 δ Relaxation

Between the β and γ dispersions, the dielectric spectra of some biological materials exhibit another process, which is roughly located in the 100 MHz regime [2, 12, 13, 19]. Its existence was unequivocally revealed only after the discovery of the β and γ relaxations [53–55], and, thus, it was termed δ relaxation. Often, its signature in the real part, $\varepsilon'(\nu)$, is rather weak and evidenced by a shallow continuous decrease with increasing frequency, while in $\varepsilon''(\nu)$ a small, separate peak or at least a shoulder shows up, as indicated in Fig. 1. In the loss spectra, in most cases the δ-relaxation peak is significantly weaker than the neighboring β and γ peaks and, in addition, it is often hidden by the conductivity contribution. The δ process is mostly detected in protein solutions (e.g., Refs. [37, 53–59]; see Sect. 3.2 for an example) but was also reported, e.g., for blood [12, 38] (however, the comprehensive broadband spectra of blood reported in Ref. [26] could be explained without invoking a δ relaxation; see Sect. 3.3).

The most common mechanism considered for the explanation of the δ relaxation, especially in the earlier literature, is the dynamics of bound water, existing in various types of biological matter [37, 53, 55, 57, 60]. For example, proteins are known to exhibit a hydration shell of water molecules bound to the surface of the protein molecule via hydrogen bonds. In the spectra, the δ relaxation shows up at frequencies below that of the γ relaxation, which is reasonable because bound water molecules should move more slowly than free molecules. Over the years, the interpretation of the δ relaxation in terms of bound water gained increasing complexity, e.g., assuming that bound water relaxations are supplemented by fluctuating polar side groups [60]. There are also reports of multiple δ relaxations which were ascribed to shells with different degrees of bonding to the protein [13, 35, 36, 57]. Interestingly, the dynamics of bound water was also discussed (e.g., in Refs. [61–65]) in the context of the glass transition and the controversial fragile-to-strong crossover of supercooled water [66, 67], which cannot be directly investigated in bulk water due to rapid crystallization occurring within the so-called no-man's land at about 150–235 K [68].

Especially in recent years, there are hints that δ relaxations can also arise due to other mechanisms such as internal-protein or protein-side-chain dynamics, which can exist in addition to bound water relaxations [19, 69, 70]. Moreover, collective protein-water motions were considered [37, 64, 70]. It was even suggested that, in aqueous solutions, the dynamics of hydration water cannot be directly detected using dielectric spectroscopy [69].

Overall, it seems clear that multiple contributions can lead to the δ relaxation which, thus, often is multimodal. However, due to their mutual overlap in the spectra and the superposition with the adjacent β and γ relaxations, it is impossible to unequivocally resolve all these contributions using dielectric spectroscopy. Thus, additional methods like nuclear magnetic resonance or molecular-dynamics simulations are helpful for definite conclusions on the dynamics of biological matter in the intermediate frequency range between β and δ relaxation [15, 69].

2.8 Resonances

Phonon-like or vibrational excitations of intra- and intermolecular origin are characteristic of the short-time dynamics in all types of condensed matter and, hence, also occur in bio-related materials. As schematically indicated in Fig. 1, these Lorentzian-type excitations, with well-defined eigenfrequencies and in most cases low damping, usually occur at frequencies beyond 1 THz, corresponding to ~ 33 cm^{-1} or ~ 4 meV. These excitations can be subdivided into inter- and intramolecular excitations: The former can be characterized as phonon-like translational or librational modes with wave-vector dependent dispersion. Intramolecular excitations are local excitations of stretching (ν), scissoring (δ), rocking (ρ), wagging (ω), or torsional (τ) type, where the former two can have symmetric and antisymmetric character. These local intramolecular excitations, which usually depend mainly on the bonding to next-nearest atomic neighbors, reveal almost no dispersion effects, appear in a frequency regime from 10 to 100 THz, and very often are well-separated in energy from the phonon-like intermolecular excitations. This enormous zoo of excitations mainly is studied using optical (THz, FIR, IR, and Raman) and inelastic neutron-scattering

(INS) techniques. The crystal symmetry usually determines whether the modes are infrared (IR) or Raman active.

A three-dimensional molecular crystal with M molecules per unit cell and N atoms per molecule will exhibit $3MN$ degrees of freedom. Clearly, even the spectra of simple molecular crystals reveal an astonishing complexity, and often the microscopic origin of specific excitations can hardly be determined. Even a single molecule with N atoms will show $3N$ - 6 internal modes. This will be discussed in some detail in the analysis of the THz spectra of amino acids (see Sect. 3.4). Amino acids are the building blocks of proteins which consist of hundreds of them, and it seems impossible to realistically model the excitation spectra of proteins on the basis of microscopic details [71, 72].

In the light of the following section, here, we would like to add a comment on low-lying optical phonons in molecular crystals. Very often intermolecular optical excitations involve displacements of complete molecules with relatively heavy masses and, hence, will be characterized by very low eigenfrequencies. These excitations will significantly contribute to the phonon density of states yielding characteristic and sometimes strong deviations from the expected Debye-type T^3 behavior of the specific heat, which should not be confused with the BP expected in amorphous solids. These large excess-heat capacities were clearly documented for some amino acids [11]. These excess contributions at low temperatures, resulting from well-defined optical phonons, raise some doubts whether the BP reported to occur in some proteins [73] results from the glassiness of the samples or just stems from low-lying phonons, which seem to be generic excitations in bio-related matter.

2.9 Boson Peak

It is well-known that dielectric spectra of disordered matter extending up to THz frequencies reveal a universal additional loss-peak around 1 THz. In the literature on supercooled liquids, it is often termed boson peak [16, 74–77], analogous to the corresponding feature observed in the scattering function and susceptibility measured by neutron and light scattering [78, 79]. Various competing explanations of its microscopic origin were proposed (e.g., Refs. [80–85]). In the frequency range of the BP, *crystalline* matter usually reveals vibrational excitations like phonons, showing up by rather narrow Lorentz peaks in the loss spectra. While the BP is much broader than such typical resonance excitations, most of the proposed explanations for its occurrence assume a phonon- or vibration-related origin.

The BP is mostly investigated for supercooled liquids, i.e., at temperatures below the melting point. There, the structural relaxation process (termed α relaxation in glass physics and usually closely coupled with the reorientational motions of the molecules in dipolar liquids) is located at low frequencies, not obscuring the BP. Recently, the latter was proposed to even show up in all kinds of liquids, including water around room temperature [14]. However, in the dielectric spectra, water's BP is partly superimposed by its very strong main relaxation peak (usually ascribed to its dipolar relaxation, see Sect. 3.1) which, at room temperature, is located at very high frequencies (~ 20 GHz).

As discussed in Sect. 2.6, biological systems containing water exhibit essentially the same process around 20 GHz as pure water, which is termed γ relaxation in biophysics. Thus, it is reasonable that their dielectric response also resembles that of water concerning the BP (Sect. 3.1), which should give rise to an additional contribution at the right flank of the γ relaxation. Indeed, data extending to sufficiently high frequencies seem to be consistent with such a scenario [51, 86], and the corresponding loss peak is schematically indicated in Fig. 1 (denoted "BP"). The occurrence of the BP in biological matter is also corroborated by its recent unobscured detection in dry protein powder using THz spectroscopy [87]. A BP was also found by scattering experiments for proteins and other biomolecules [15] (however, see the critical remarks concerning the BP in biological matter in the preceding section).

3 Experimental Examples

3.1 Pure Water

More than half of the body mass of most animals is made up of water [88]. Thus, to understand the properties of biological matter in organisms, one has to understand its interactions with water. As mentioned in Sect. 2.6 and recently pointed out in Ref. [7], information about such interactions can be obtained, e.g., by a comparison of the main relaxation process, revealed in the dielectric spectra of pure water, with the corresponding dynamics in biosystems, the γ relaxation. Therefore, knowledge of water's dielectric spectra is essential. This is also valid for other fields, e.g., the everyday application of microwave cooking or the absorption of radar waves by clouds [89–91].

Figure 2 shows dielectric-loss spectra of pure water covering frequencies from 100 MHz to 20 THz and measured at various temperatures [14]. Data obtained using different experimental setups were combined to arrive at these spectra. In contrast to most other broadband data in literature, results for different temperatures are provided, which were collected by a single workgroup using the same sample preparation. The main feature of the spectra of Fig. 2 is the well-known prominent loss peak of water, which at room temperature is located at about 20 GHz. As mentioned in Sect. 2.6, it is observed within the same frequency range as the ubiquitous γ relaxation of water-containing biological matter, clearly pointing to the same principle microscopic origin. Nevertheless, amplitude, peak frequency, and width of the γ relaxation can be somewhat modified compared to pure water due to interactions with ions, macromolecules, or cellular structures [7]. The temperature-induced shift of the peak revealed in Fig. 2 is typical for relaxational processes and essentially mirrors the slowing down of the molecular dynamics due to reduced thermal activation at low temperatures [16, 17].

As mentioned in Sect. 2.6, in the biophysical literature, the γ dispersion is commonly ascribed to the dipole dynamics of unbound water molecules [1, 2, 13, 57] as also existing in pure water. Nevertheless, one should be aware that part of the current literature on pure water also considers other interpretations of its main relaxation process (e.g., Refs. [92–99]). The most prominent one is analogous to the interpretation of the dominant Debye relaxation detected in many monohydroxy alcohols [100–102]: It is believed to mirror the motions of dipolar clusters formed by hydrogen-bonded molecules [103], while the structural relaxation from essentially single-molecule motions (termed "α relaxation" outside of biophysics) shows up as a minor relaxation peak at higher frequencies only. However, for pure water such an interpretation is disputed and arguments against this scenario were provided, e.g., in Refs. [14, 104].

Traditionally, the 20 GHz relaxation of water is fitted by a Debye function derived within the Debye theory of dipolar relaxation [106]. However, newer measurements with sufficient precision at frequencies up to the THz range revealed deviations of the experimental data from the Debye function at the high-frequency flank of the 20 GHz loss peak [92, 93, 95, 107]. They cannot be explained by contributions of the known vibration-related resonances observed beyond THz, usually described by Lorentz functions. This becomes especially clear for spectra close to the freezing point of water as provided in Ref. [14] (see 275 K curve in Fig. 2) where the main peak has shifted to lower frequencies while the resonances remain largely unaffected by temperature. As revealed by the dotted line in Fig. 2, at the right flank of the main loss peak, already for frequencies that are about four times larger than the peak frequency, deviations from the Debye fit start to show up. It should be noted that there is no fundamental reason to assume the validity of the Debye model for pure water. This becomes obvious when considering that for nearly all other dipolar liquids, deviations from Debye behavior, especially pronounced at the high-frequency flanks of their main relaxation peaks, are very common [16, 17, 108]. It is nowadays quite widely accepted that the corresponding broadening of the relaxation peaks, exceeding the half width of 1.14 decades predicted by the Debye model, is due to a distribution of relaxation times caused by the disordered structure of liquids [109, 110]. This should apply to water just as to other dipolar liquids. Indeed, the water spectra can be significantly better fitted by the empirical Cole-Davidson (CD) function [111], often used for dipolar liquids [16, 17, 112] (see blue dashed line denoted "CD" in Fig. 2). It matches, e.g., the 275 K spectrum for frequencies up to about 1.5 decades beyond the peak frequency.

Anyway, Fig. 2 demonstrates that, even when using a CD function for the main peak (blue dashed line), deviations from the fits at high frequencies still show up, due to significant excess intensity in $\varepsilon''(\nu)$ at ~ 1 THz. In literature, the main loss feature of water, including this excess contribution, was mostly fitted by the sum of two Debye functions [92, 113, 114], probably sticking to the Debye shape of the main peak due to historical reasons. In general, broadband spectra of water were described by many different combinations of various relaxation and resonance contributions, and good fits of the experimental data were achieved with partly very different approaches [93, 95–99, 107, 114]. Of course, one should be aware that, by combining a sufficiently large number of peak functions such as Debye or Lorentz, almost any experimental curve can be fitted.

In Ref. [14], we have pointed out the similarity of the water dielectric spectra to those of many glass-forming dipolar liquids [16, 17, 74, 112, 115, 116] (e.g., glycerol at 413 K [112]) if the latter are heated to temperatures where their main loss peaks are located in the GHz region (cf. Figure 3a). These glass formers can be easily supercooled below their melting point T_m. As schematically indicated in Fig. 3b, this supercooling shifts their main relaxation peak to lower frequencies, which reveals the origin of the excess intensity around 1 THz: It is due to a separate, essentially temperature-independent loss peak, clearly observed in their low-temperature spectra just in this frequency region. It was termed the boson peak [74] in analogy to the corresponding feature found by scattering methods [78, 79] (see Sect. 2.9). Its microscopic origin is still controversial. Unfortunately, pure water cannot be easily supercooled to relieve its suggested BP from the superposition of the main relaxation at 20 GHz [68]. However, supercooling becomes possible when adding certain amounts of salt to water [14, 68, 117]. At room temperature, the dielectric-loss spectra of such solutions exhibit just the same excess contribution as pure water at the right flank of their main peak at ~ 20 GHz [14]. When they are supercooled, indeed this shoulder develops into a peak at about 1 THz, which can be interpreted as the BP of water [14]. As mentioned in Sect. 2.9, it seems reasonable that this contribution is also present in biological matter containing water (cf. Fig. 1).

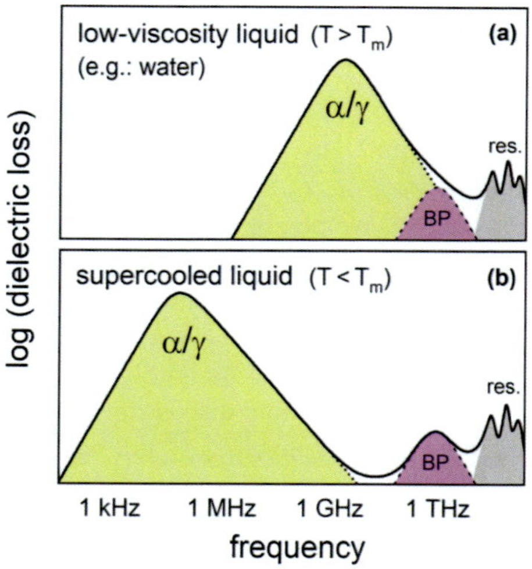

Fig. 3 Schematic dielectric-loss spectra of a low-viscosity dipolar liquid (**a**) (e.g., water around room temperature [14] or glycerol at T ~ 410 K [112]) and of a supercooled liquid (**b**) (e.g., water with LiCl at ~ 170 K [14] or glycerol at ~ 240 K [112]). The main peak is named α relaxation in the field of supercooled liquids and γ relaxation in biophysics for materials containing unbound water. "BP" stands for the boson peak and "res." denotes high-frequency resonances. The fast process of supercooled liquids, discussed in glass physics to appear in the minimum region between α and BP [16, 17], is not indicated in the figure

To illustrate the suggested approach for pure water, the solid line in Fig. 2 is a fit of $\varepsilon''(\nu)$ at 275 K with the sum of four contributions [14]: A CD function to cover the main relaxation, two Lorentz functions for the vibrational IR resonances, and a log-normal peak function for the BP. The latter is entirely phenomenological and is able to account for the fact that the BP in dipolar liquids is narrower than a Debye peak but broader than a resonance peak [16, 17, 74, 112, 115, 116].

From fits of the loss peaks in Fig. 2, the temperature-dependent relaxation times of the main dynamic process of pure water can be deduced. The inset of Fig. 2 shows an Arrhenius plot of the resulting $\tau(T)$, also based on spectra collected at further temperatures [14, 68] (circles). The plusses represent additional data reported in Ref. [105], extending the temperature range into the supercooled region, down to 252 K, below which the mentioned no-man's land starts [68]. The curved shape of this combined data set provides clear evidence for significant deviations from Arrhenius temperature dependence, $\tau(T) \sim \exp(E/T)$. Here, E is an energy barrier (in K) that has to be overcome by thermal activation for the corresponding motion to happen. Assuming that the main relaxation of water is due to reorientations of the water molecules, E should be largely determined by the energy needed to break the intermolecular hydrogen bonds. However, the found deviations from Arrhenius behavior imply that the situation in water is more complex. Interestingly, non-Arrhenius behavior as found for $\tau(T)$ of pure water is also a very common property of other dipolar liquids where it was mainly investigated in the supercooled state but also exists above T_m [16, 17]. It is often ascribed to an increase of the cooperativity of molecular motions with decreasing temperature [118, 119], which also seems to apply to water.

The line in the inset of Fig. 2 is a fit with the empirical Vogel-Fulcher-Tammann (VFT) function, $\tau(T) \sim \exp[DT_{VF}/(T - T_{VF})]$, often applied to dipolar liquids [16, 17, 112, 115, 118, 120, 121]. Here, D is termed strength parameter and T_{VF} is the Vogel-Fulcher temperature. From D, the so-called fragility index of a liquid can be calculated [120], which is a measure of the degree of deviations from Arrhenius temperature dependence of $\tau(T)$. We obtain a high value of $m = 175$ [68], making liquid water a very "fragile" liquid within the often-employed classification scheme introduced by Angell [122], in accord with earlier reports [66]. However, one should be aware that the fragility of pure water and the closely related fragile-to-strong crossover suggested in its supercooled state are highly controversial topics [66, 67, 123–127] (see Ref. [68] for a more detailed discussion). Notably, one approach to clarify the open questions about the behavior of pure water in the no-man's land and its controversial glass transition is the investigation of the dynamics of water within biological matter which often can be easily supercooled (e.g., [61–64, 128]). However, a detailed discussion of such studies is out of scope of the present work.

3.2 Broadband Spectra of a Protein Solution

Proteins are major components of all cells and play crucial roles in almost every biological process. Via dielectric spectroscopy, they are studied not only in the pure, solid state (mostly as powder), but also dissolved in water, which more closely resembles in vivo conditions. This is essential because water is known to strongly affect the biological functions of proteins, although the details of these interactions are only partly understood [7, 36, 65, 129, 130]. Therefore, investigations of dissolved and hydrated proteins and of protein-water interactions are very active research fields.

Figure 4 shows an example for permittivity and conductivity spectra of a typical aqueous protein solution (5 mmol/l lysozyme) at body temperature, covering a very broad frequency range [86]. Such broadband spectra are essential for the investigation of the many different aspects of protein dynamics, e.g., dipolar relaxation processes of the protein molecules themselves and of bound and unbound water molecules (cf. Fig. 1). At low frequencies, the dielectric constant $\varepsilon'(\nu)$ approaches unrealistically large values beyond 10^8 (Fig. 4a), clearly indicating a non-intrinsic origin. In fact, EP, caused by the blocking of translational ionic motions at the metallic electrodes, generates this huge increase (see Sect. 2.3). A detailed discussion of this effect, as detected in a similar lysozyme solution and various other materials, in terms of a equivalent-circuit analysis was provided in Ref. [20]. Moreover, the decrease of the conductivity below about 1 kHz, revealed in Fig. 4c, is also due to EP and simply caused by the concomitant reduction of the time-averaged ionic mobility at low frequencies. Above 1 kHz, $\sigma'(\nu)$ remains approximately constant up to about 1 GHz, signifying the ionic dc conductivity of this solution. Due to the general relation $\varepsilon'' = \sigma'/(2\pi\, \nu\, \varepsilon_0)$, mentioned in Sect. 2.2, the dc charge transport leads to $1/\nu$ behavior in the loss spectra (evidenced by the linear decrease between about 1 kHz and 1 GHz in Fig. 4b) while EP causes the crossover to weaker $\varepsilon''(\nu)$ variation seen at lower frequencies, $\nu < 1$ kHz.

The steplike decrease in $\varepsilon'(\nu)$ and the peak in $\varepsilon''(\nu)$, revealed at the highest frequencies in Fig. 4a, b, are typical signatures of a relaxation process. Due to the mentioned direct relation of ε'' and σ', the loss peak corresponds to a shoulder in the conductivity spectra (Fig. 4c). This process can be identified with the γ relaxation of this protein solution [57, 58], commonly assumed to arise from reorientational motions of unbound water molecules (but see Sect. 2.6 for a possible alternative scenario discussed for pure water). This assignment becomes immediately obvious from the close agreement of the loss-peak frequencies and amplitudes in Fig. 4b with those of pure water (Fig. 2). This is illustrated in the inset of Fig. 4 showing $\varepsilon''(\nu)$ of the 5 mmol lysozyme solution (open circles) and of pure water (line) in the γ-peak region which agree very well. As discussed in Ref. [58], in fact the peak amplitude of the protein solution is a few percent smaller than that of pure water (not visible in the inset of Fig. 4, covering more than one decade in ε''), which is due to its slightly reduced water concentration.

At first glance, based on Fig. 4, only three dynamic processes seem to exist in this protein solution, none of them providing direct information on the protein itself: EP which is non-intrinsic, the γ relaxation, mirroring motions of free water molecules just as in pure water, and the dc conductivity which is due to translational motions of dissolved mobile ions stemming from residual salt contents. However, as noted in Sects. 2.1 and 2.2, subtracting the dc contribution in the loss spectra can uncover additional processes (see Refs. [5, 58] for possible caveats concerning the dc

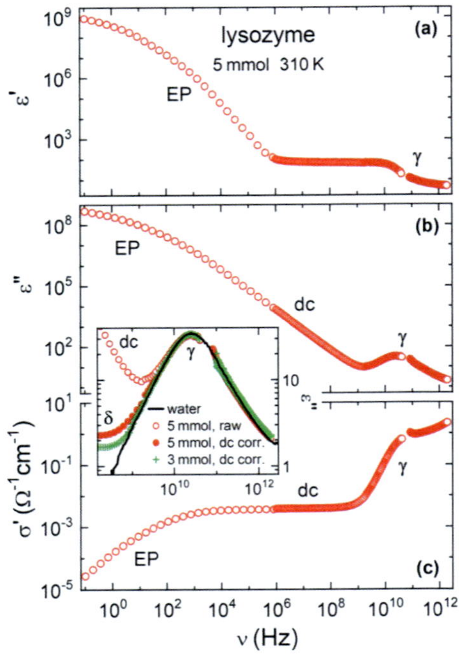

Fig. 4 Broadband dielectric spectra of a 5 mmol/l lysozyme aqueous solution, showing **a** the dielectric constant (real part of the permittivity), **b** the dielectric loss, and **c** the real part of the conductivity as measured at 310 K (data from Ref. [86]). The contributions from EP, γ relaxation, and dc conductivity (the latter only existing in ε'' and σ') are denoted in the figure. The inset provides a magnified view of the γ-relaxation region (open circles). The closed circles represent the same data set, corrected for the dc conductivity, and the plusses indicate dc-corrected data for a 3 mmol solution at the same temperature. Both corrected curves reveal the onset of the δ relaxation (cf. Fig. 5) at low frequencies. In addition, the line in the inset shows $\varepsilon''(\nu)$ of pure water at 310 K [14]

subtraction). The inset of Fig. 4 shows such corrected data for the 5 mmol solution (closed circles) and for a 3 mmol solution (plusses). The dc subtraction exposes additional data points lying at the left flank of the γ peak. Deviations from this flank revealed below about 1 GHz indeed point to at least one additional process at lower frequencies. In fact, these deviations are due to the δ relaxation as will be discussed below.

Figure 5a shows $\varepsilon'(\nu)$ of a 3 mmol lysozyme solution above 1 MHz, where EP can be widely neglected [58, 86]. Data for three temperatures are provided, revealing the typical shift of the γ-relaxation step, observed around 10 GHz, to lower frequencies upon cooling, as expected for thermal activation. In Fig. 5b, we show the corresponding loss spectra, corrected for the dc conductivity contribution by subtracting $\varepsilon''_{dc} \propto \sigma_{dc}/\nu$ (see Sect. 2.2). In addition to the γ peak, which also exhibits the temperature-induced shift seen in ε', two additional relaxations at lower frequencies are revealed in these dc-corrected data. Most obvious is a peak occurring around 10 MHz. It can be ascribed to the β process of this protein solution [58]. As mentioned in Sect. 2.5, β relaxations typically show up just in this frequency region and are mostly considered to arise from two possible mechanisms: Maxwell–Wagner relaxations due to internal heterogeneities, e.g., cell membranes, and/or reorientational motions of large dipolar molecules [1, 2, 6, 13, 37]. Obviously, the latter applies to the present solutions containing dipolar protein macromolecules. The shift of the β-relaxation peaks with temperature, seen in Fig. 5b, again indicates that the protein dynamics is essentially thermally activated.

Between the two well-pronounced β and γ peaks, the loss spectra in Fig. 5b reveal significant additional intensity that is due to the δ relaxation [58]. As discussed in Sect. 2.7, the origin of the δ process is controversial, and different processes may contribute in this frequency region. Thus, in principle it could be multimodal, e.g., composed by a sum of several Debye contributions as assumed in some publications [35, 37, 57, 59, 70]. In fact, an investigation of a frozen lysozyme solution, where the ionic conductivity is strongly reduced, revealed unequivocal evidence for its bimodal nature [64]. A comparison of those results with data for liquid lysozyme solutions [58] and hydrated powders [86] provided evidence that the faster component of the δ relaxation arises from the dynamics of loosely bound hydration water [64]. Moreover, in Ref. [64] collective protein-water motions [37, 70, 131] were suggested to be the origin of the slower component. In the spectra for liquid lysozyme solutions shown in

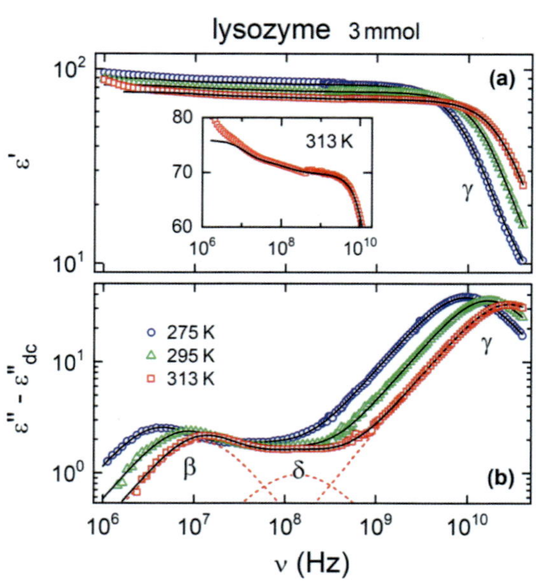

Fig. 5 Dielectric spectra of a 3 mmol/l lysozyme aqueous solution, showing **a** the dielectric constant and **b** the dc-corrected dielectric loss, as measured at three temperatures (data from Refs. [58, 86]). The solid lines are fits with the sum of a Debye and two CC functions for the β, δ, and γ relaxations, respectively. They were simultaneously performed for the real and imaginary part. For 313 K, the dashed lines in (**b**) show the contributions of the three processes to the overall fit. The inset presents a magnified view of $\varepsilon'(\nu)$ at 313 K in the region of the low-frequency plateau

Fig. 5, the superposition of the δ relaxation by the β and γ processes obviously is too strong to enable conclusions about its possible bimodal nature (see also discussion of the performed fits given below).

The loss-peak amplitudes of the β and δ processes are more than one decade smaller than for the γ relaxation. Therefore, corresponding relaxation steps are not immediately evident in $\varepsilon'(\nu)$ (Fig. 5a), which, between about 1 MHz and 3 GHz, is dominated by the large static dielectric constant $\varepsilon_s \approx$ 70–80 of the γ process, mirroring the large ε_s of pure water [14]. Nevertheless, the inset of Fig. 5a, showing a magnified view of this region for 313 K, reveals faint indications for a sigmoidal decay of $\varepsilon'(\nu)$ around 10 MHz, which is due to the β process. At lower frequencies, it is partly superimposed by the onset of EP and at higher frequencies, the weak δ dispersion shows up, followed by the onset of the much stronger γ-relaxation step. (The anomaly close to 500 MHz is due to a slight mismatch of about 1% between the results from the employed low and high-frequency devices; see Refs. [58, 86] for experimental details.) Thus, from the ε' spectra alone, it would be impossible to draw unequivocal conclusions about the existence of the β and δ relaxations in this protein solution. This does not hold for protein samples where the ionic conductivity is less dominant, e.g., for solutions measured below the freezing point of water as shown in Ref. [64].

The solid lines in Fig. 5 are fits with the sum of one Debye and two Cole–Cole (CC) [132] functions, accounting for the β, δ, and γ relaxations, respectively. The empirical CC function gives rise to symmetrically broadened loss peaks and is often employed for disordered matter [17, 132, 133]. As mentioned in Sect. 3.1, such non-Debye behavior is assumed to be caused by different environments sensed by each molecule, giving rise to a distribution of relaxation times [109, 110]. Concerning the β process, which is ascribed to protein reorientations, concentrations of few mmol/l imply such small amounts of dissolved protein molecules that direct intermolecular protein interactions appear unlikely. Thus, each macromolecule basically "sees" the same time-averaged environment of very rapidly (related to the β-relaxation time) fluctuating water molecules [58]. This rationalizes the use of the Debye function for the description of the β peaks, in agreement with previous publications [35, 37, 51, 57, 59]. As noted in Ref. [58], the γ relaxation in Fig. 5 can be best accounted for by a CC function, in accord with later reports on different protein solutions [52, 69]. Notably, just as for pure water (Sect. 3.1), in some publications multiple Debye peaks were used to fit dielectric spectra of protein solutions in the γ-relaxation region (e.g., [51, 59]). However, when considering Occam's razor, we think a single CC function, assuming only one instead of two dynamic processes and involving fewer fit parameters, is preferable. For the δ process, a CC function was employed in the fits, too. As mentioned above, in fact it may be multimodal, e.g., composed by a sum of several Debye contributions as assumed in some publications [37, 57, 59]. Nevertheless, the δ relaxation in the spectra of Fig. 5 can be well modeled by a single CC function, and using multiple peaks would lead to parameters with limited significance in the present case.

The temperature dependences of the relaxation times obtained from the fits of the dielectric spectra are shown in Fig. 6 (open symbols), using an Arrhenius representation, $\log_{10}(\tau)$ versus $1000/T$. The found linear behavior indicates thermally activated behavior. However, one should be aware that the temperature range accessible in these aqueous liquids is quite limited (compared, e.g., to glass-forming liquids which can be supercooled [16, 17]), which is why non-Arrhenius behavior in a broader temperature range cannot be excluded. The stars in Fig. 6 are data for pure water [68]. They agree well with $\tau(T)$ of the γ process, again evidencing its origin due to unbound water-molecule dynamics. Interestingly, the similar slopes in Fig. 6 point to similar hindering barriers for the three relaxational processes in this protein solution (between 0.19 and 0.22 eV). As discussed in detail in Ref. [58], the β and γ relaxations have almost identical barriers and, thus, are most closely coupled, as expected for rotations of molecules in a solution. In contrast, the δ dynamics seems to be less governed by that of the solvent molecules and appears to be influenced by interactions with the protein molecules. A more detailed discussion of the relaxation times and various other parameters deduced from the dielectric spectra of different lysozyme solutions is provided in Ref. [58]. For example, the results enable conclusions about the hydrodynamic radius and dipole moment of the protein molecules. Comparing the obtained dc conductivity and the γ relaxation time reveals that the ionic conductivity is decoupled from the solute-molecule dynamics, implying a breakdown of the Debye Stokes–Einstein relation [134, 135].

Overall, the results exemplarily discussed in the present section demonstrate that broadband dielectric spectroscopy, extending well into the GHz range, can provide valuable information about the complex dynamics in aqueous protein solutions, helping to uncover their microscopic origins.

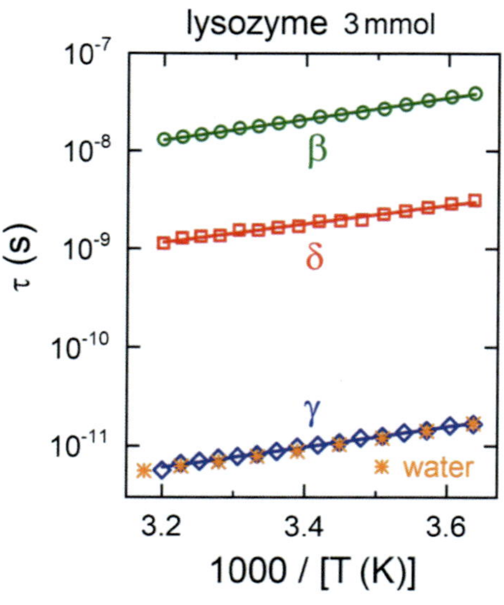

Fig. 6 Temperature-dependent relaxation times of the three processes in the 3 mmol lysozyme solution as deduced from the dielectric spectra (open symbols; data from Ref. [58]). The lines are Arrhenius fits with energy barriers 0.22, 0.19, and 0.21 eV for the β, δ, and γ processes, respectively. The stars show data for pure water [68]

3.3 Broadband Spectra of Blood

Needless to say, blood is one of the most important liquids ("Blut ist ein ganz besonderer Saft", J. W. von Goethe, Faust I), and in living organisms it serves as a highly functional body fluid. It delivers oxygen to the vital parts of the body, transports nutrients, vitamins, and metabolites, and it also acts as a fundamental part of the immune system. Therefore, the precise knowledge of its constituents and their biological, physical, and chemical properties including the involved dynamic processes are of great importance. On average, approximately 8% of the body weight is blood, consisting mainly of blood plasma (55%) and red blood cells (RBCs) (45%), with an additional small amount of white blood cells and platelets. RBCs or erythrocytes have a biconcave shape with a cell membrane composed of proteins and lipids, filled with erythrocyte cytoplasm with high hemoglobin concentration and without a nucleus. Naively, one could say that RBCs are bags filled with hemoglobin. Blood plasma consists of 90% water with solutes, like proteins, lipids or salts. In modeling the dielectric response of cell suspensions including blood, usually heterogeneous media are assumed: The RBCs are approximated by ellipsoids, with well-defined inner radius, filled with medium, encapsuled by a shell of given thickness and embedded in a homogeneous solution. Cell interior, shell, and solution are assumed to have well-defined dielectric parameters [41].

The dielectric parameters of blood, which have been investigated since more than 100 years, are of relevance for various biotechnological and medical applications, e.g., for the separation of infected from normal blood cells [136], cancer diagnosis [137], for glucose monitoring of diabetes patients [138], for the examination of possible deterioration of preserved blood [139], to determine the sedimentation rate of erythrocytes [140], and to estimate the precise dielectric properties of blood as prerequisite for fixing limiting values for electromagnetic pollution [141], to name just a few. The detailed investigation of the analysis of dielectric properties of blood started with the works of Höber [142] and Fricke [143] measuring capacitance and conductance of cell suspensions at kHz to MHz frequencies. Shortly after World War II, with microwave technology being developed to a high standard, experiments on blood were extended well into the GHz regime [144–147], allowing to study the dynamics of cell suspensions, including blood at high frequencies. Since then, detailed dielectric work was performed, partly with the emphasis on determining dielectric material parameters of the blood constituents, as the capacitance of the cell membranes and the conductivities of plasma and cytoplasm, but partly also to establish number and microscopic character of the various relaxation processes detected by broadband dielectric spectroscopy. Informative surveys and detailed references to published dielectric work on proteins, tissues, and specifically on blood can be found in [1, 2, 12, 19, 26, 30, 138, 148–152].

Broadband dielectric spectra on blood, important for a critical analysis of the involved relaxation processes, were published by Gabriel et al. [149] and Wolf et al. [26]. The former, however, represent a collection of results from various groups, using different samples, partly with relatively large scatter, and at a single temperature only.

In contrast, Wolf et al. [26] have investigated blood using identical samples for all frequency ranges, measured as function of temperature and hematocrit value. Results on whole blood collected within the framework of that study are shown in Fig. 7, covering more than 10 decades in frequency and including all important relaxation regimes. The three frames display the frequency dependence of dielectric constant (a), dielectric loss (b), and real part of the conductivity (c) of whole blood on double-logarithmic scales at two temperatures (280 and 330 K), as well as the spectra of blood plasma at 280 K. Whole blood clearly exhibits three major dispersion regions as often observed in raw dielectric spectra of all kinds of tissues and cell suspensions [1, 2, 12, 13, 30, 153, 154]. Wolf and co-workers [26] assigned these relaxations with increasing frequency to EP, β, and γ relaxation. Qualitatively similar results were also obtained for lower RBC concentrations, i.e., smaller hematocrit values [26]. Figure 7 reveals that blood plasma only exhibits two relaxation steps, namely EP and γ process, with the β process being completely absent [26].

The dipolar strength and temperature dependence of the γ dispersion in blood, as shown in Fig. 7, resembles the relaxational dynamics of water molecules as observed in pure water with a maximum of the loss peak close to 20 GHz (Sect. 3.1) [14]. As discussed in Sect. 2.6, this experimental result is similar to most of the high-frequency observations in biological matter in solution. According to Wolf et al. [26], the γ relaxation of blood can be well described by the symmetrically broadened CC function close to pure mono-dispersive Debye behavior, with a slightly increasing

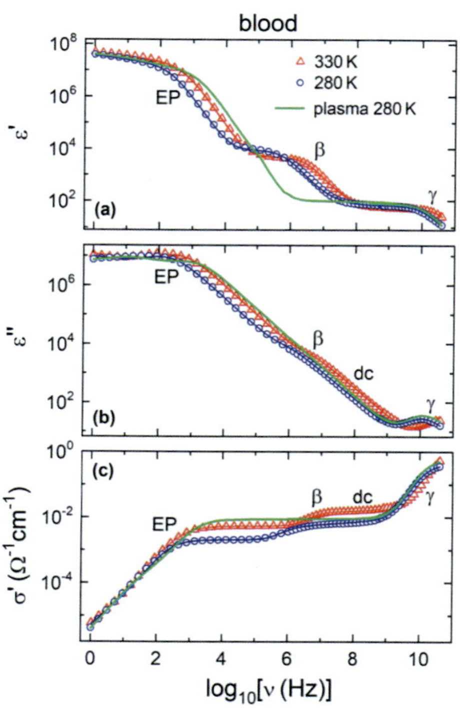

Fig. 7 Frequency dependence of dielectric constant **a**, dielectric loss **b**, and real part of the conductivity **c** of whole blood at two different temperatures and of blood plasma at 280 K (the latter from Ref. [26]). The three well-separated dispersion regimes are indicated as EP, and as β and γ relaxations. The lines are fits (simultaneously performed for ε' and ε'') assuming a distributed RC equivalent circuit to account for the electrodes [20] and two Cole–Cole functions for the β- and γ-relaxations. For the plasma data, only a single Cole–Cole function was used

width upon cooling. The temperature dependence of the relaxation time, inversely proportional to the loss-peak frequency, can be well fitted by thermally activated Arrhenius behavior with a hindering barrier of about 0.2 eV. Figure 7 documents that blood plasma, lacking any cells, at 280 K reveals a γ relaxation that is of similar strength and peak frequency as found for blood at the same temperature. This further corroborates the assignment of the γ relaxation to the dynamics of unbound water molecules (Sect. 3.1).

On decreasing frequency, in the loss spectra (Fig. 7b) the γ relaxation exhibits a smooth transition into $1/\nu$ behavior due to dc conductivity (see Sect. 2.2), signified by a dc plateau in $\sigma'(\nu)$ around 100 MHz (Fig. 7c). In blood, this dc plateau depends significantly on temperature as routinely observed in ionic conductors. In blood plasma, this dc plateau extends until the onset of the electrode-polarization contribution at low frequencies, while in whole blood it is limited by the onset of the β relaxation (Sect. 2.5). Interestingly, as seen for 280 K in Fig. 7c, σ_{dc} of the plasma is of similar magnitude as for blood.

This additional dispersion in whole blood, ascribed to a Maxwell–Wagner effect governed by the cell membranes [12, 26, 38], shows up at MHz frequencies (Fig. 7). In $\sigma'(\nu)$ (Fig. 7c), it is limited by two regions of frequency-independent conductivity. The intrinsic dc conductivity of the sample seen at high frequencies is governed by ionic charge transport within the plasma and cytoplasm As discussed in Ref. [26], the lower conductivity plateau found at frequencies below the β dispersion (at about 1 kHz–1 MHz, depending on temperature) is affected by the presence of RBCs in blood, in particular their membranes, locally blocking the charge transport. Typical for many Maxwell–Wagner effects (Sects. 2.3 and 2.5) [20, 23, 24], at high frequencies, the capacitive interfaces (here: the cell membranes) become effectively short-circuited and the higher, intrinsic dc conductivity value is detected. The latter is in fact a mixture of the conductivity of the plasma outside of the cells and of the cytoplasm within the cells, the latter having somewhat lower conductivity [26]. This scenario is well supported by the mentioned approximate agreement of this high-frequency plateau value with the dc conductivity of the pure plasma [26] as demonstrated in Fig. 7c for 280 K (the remaining small deviations were suggested to arise from the volume fraction of blood's cytoplasm [26]).

The complete absence of the β-dispersion effects in the plasma sample (Fig. 7) is also well consistent with Maxwell–Wagner-like effects. However, this could also be said for possible intrinsic relaxations due to dipolar reorientation dynamics of macromolecules (e.g., proteins within the RBCs), which also can lead to β relaxations in biological matter (see Sects. 2.5 and 3.2). Dispersion effects due to these two very different phenomena in principle look similar. However, in the present case, the very high static dielectric constant of blood's β relaxation of order 10^4 revealed in Fig. 7a again strongly speaks in favor of the discussed Maxwell–Wagner-related origin due to cell membranes. One should further note that, for a reorientational relaxation, the conductivity plateau above the corresponding dispersion in $\sigma'(\nu)$ is not at all related to the dc conductivity of the sample, simply because it arises from the right flank of a dipolar relaxation peak in the loss. Thus, within such a scenario it could not be explained why this plateau value in blood approximately agrees with that in the plasma sample (Fig. 7c).

In Ref. [26], the frequency regime of the β relaxation of blood was phenomenologically analyzed by a CC function, with parameters that come very close to the mono-dispersive Debye case. It was found that the width parameter and the dipolar strength are almost temperature independent. The relaxation time was revealed to follow Arrhenius-type temperature dependence with energy barriers between 0.11 and 0.15 eV of similar order as for the determined dc conductivity. This agreement is expected for Maxwell–Wagner relaxations, for which the intrinsic conductivity should govern the temperature dependence of τ [20]. Wolf et al. [26] analyzed the detected dispersion effects utilizing the Pauly-Schwan model [41] and obtained reasonable values for the conductivity inside and outside of the cells and for the capacitance of the cell membranes. Interestingly, in Ref. [26] some hints were found indicating that the β relaxation of blood is composed of two different, closely overlapping relaxation processes. Such a bimodal β relaxation may be explained by Maxwell–Wagner-related models accounting for the non-spherical geometry of the RBCs [44].

Finally, one should note that there are contradicting reports about the presence of an α and a δ relaxation in blood and blood-related solutions [2, 12, 38, 149, 151, 155–157]. Concerning the δ relaxation, its presence in blood seems possible because it contains proteins, which should be surrounded by bound water and which may also reveal intramolecular degrees of freedom, both suggested as possible origins of this process (see Sect. 2.7). However, Fig. 7 documents that excellent fits of the broadband spectra of blood can be obtained assuming three distinct dispersion regimes only, namely EP, β, and γ dispersion [26]. It is not necessary to assume the existence of a further δ relaxation, located between the β and γ processes to describe the smooth transition between these two processes which is simply caused by a bare superposition of the wings of the two neighboring processes. Thus, it was concluded that the δ dispersion observed in many protein solutions is absent in blood or at least strongly superimposed by the other contributions [26].

As discussed in Sect. 2.4, the α process, representing the lowest-frequency quasi-intrinsic dispersion regime sometimes reported in tissues, is often assumed to arise from counter-ion diffusion along cell membranes [1, 2, 12, 13, 29] (but also other origins were discussed, see, e.g., [2, 30, 153]). Thus, it may also occur in blood, and, indeed, strong low-frequency dispersion is revealed in its spectra (Fig. 7). However, Wolf et al. [26] attributed this dispersion regime solely to EP (Sect. 2.3). As this dispersion is also detected in blood plasma where RBCs are completely absent, it is quite evident that this low-frequency relaxation of whole blood is not related to the presence of any dynamics at the cell membranes.

To support the interpretation in terms of EP, in Ref. [26] a careful analysis of the low-frequency dispersion of whole blood was provided. As demonstrated in Fig. 8 by the dash-dotted magenta line, the low-frequency dispersion clearly cannot be fitted by a Debye relaxation. Even the assumption of additional dc conductivity (blue dotted line) cannot explain the shallow minimum appearing toward the lowest frequencies. Identical frequency dependence as the dotted line also results when

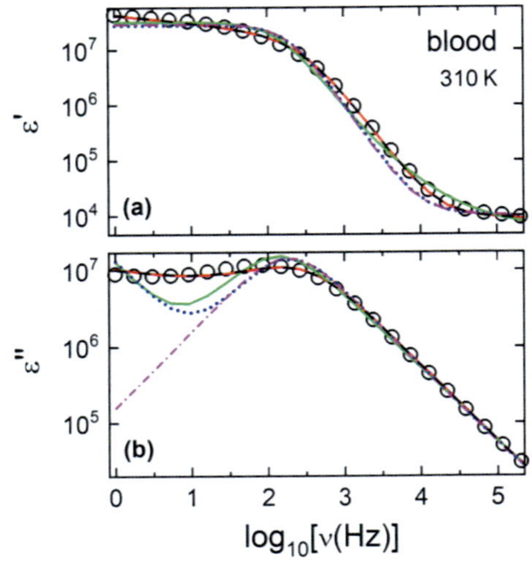

Fig. 8 Frequency dependence of dielectric constant **a** and dielectric loss **b** of whole blood at low frequencies and body temperature [26]. The lines represent fits using a Debye function (dash-dotted magenta line), a Debye (dotted blue line) and CC function (green line), both with additional dc conductivity, a distributed RC equivalent circuit (red line), as well as a constant-phase element (black dashed line). This figure is reproduced from Ref. [26] Copyright (2011) with permission from Elsevier

assuming a simple RC equivalent circuit in series to the bulk sample, as is often done to describe canonical Maxwell–Wagner effects of inhomogeneous matter [20, 158]. Thus, obviously such an approach also cannot account for the measured data. Substituting the Debye model by a CC function (green solid line) also does not lead to a reasonable fit of the spectrum. Good fits, however, were achieved, assuming a distributed RC circuit [20] (red dashed line) or a constant-phase element (black dashed line), both models pointing toward the existence of EP, which describes blocking electrodes in ion conductors (see Sect. 2.3). Wolf et al. [26] concluded that in broadband dielectric spectra on blood there is no evidence for α dispersion due to counter-ion diffusion as sometimes reported for tissues. However, they also stated that a weak α relaxation cannot be fully excluded to be hidden by the other dominating contributions in the spectra.

3.4 THz Spectra of Amino Acids

Amino acids are the fundamental building blocks of proteins and, hence, are essential for the evolution of life on Earth. They are linked together to polypeptide chains, which finally are folded into proteins. Amino acids are defined as molecules forming a chain of at least two carbon atoms, one belonging to an acidic carboxy, and the second to a basic amino group. In α amino acids, these two carbon atoms are directly linked together. In aqueous solutions, amino acids exist in their zwitterionic form with a net proton transfer from the carboxy to the amino group. α amino acids have a common structure, consisting of a central α carbon atom, connected to a charged amino group (NH^{3+}), a carboxyl group (COO^-), a hydrogen atom, and a variable organic side chain *R*. The side group determines the different chemical properties and the specific biological role of each amino acid. 22 α amino acids are genetically encoded (proteinogenic) with 20 of them belonging to the standard genetic code. Interestingly, despite their importance for the evolution of life, it still is a matter of controversy whether they were synthesized on primitive Earth or were transported via meteorites to our home planet [159, 160].

Terahertz radiation, covering a frequency regime from 0.1 to 10 THz, corresponding to ~ 3–300 cm^{-1}, defines the long wavelength part of the IR spectral range partly overlapping with the far-IR (FIR) regime. For a long time, this frequency range was only accessible by elaborate backward-wave oscillator techniques [161] in the laboratory and, hence, was termed "terahertz gap." Approximately during the past twenty years, based on the time-resolved detection of ultrashort radiation bursts of electric fields by photoconductive switches, this technique has undergone an outstanding development, nowadays known as terahertz time-domain spectroscopy (THz-TDS) [162], and plays a dominant role in optical spectroscopy in a variety of different fields in condensed-matter physics as well as in material and biological sciences. In optical spectroscopy on molecular crystals, it is especially sensitive to intermolecular excitations (see Sect. 2.8).

Vibrational spectroscopy on crystalline amino acids routinely was performed utilizing Raman and IR spectroscopy, and INS techniques. Infrared spectroscopy on amino acids started with the work of Klotz and Gruen [163] and Thompson et al. [164]. Raman spectroscopy started even earlier and was initiated by a series of publications by Edsall and co-workers [165, 166]. Terahertz spectroscopy on biological samples started in the early twenty-first century and largely helped to explore the hardly accessible frequency range between the FIR and the microwave regime [167, 168]. In addition, THz-TDS with its ability to concomitantly monitor real and imaginary parts of the optical constants and its high transmission through the material, combined with little damage, is of utmost importance for the investigation of biological matter.

Recent THz work on the amino acids L-serine (Ser) and L-cysteine (Cys) in the frequency regime from 40 to 120 cm^{-1} was published by Emmert et al. [11], with a detailed analysis of the temperature dependence of eigenfrequencies, damping, and dipolar strength, and including a complete reference list of Raman, FIR, and THz results specifically with reference to Cys and Ser. In addition, in Ref. [11] by referring to existing model calculations and computer simulations, it was attempted to assign the observed eigenfrequencies to the microscopic nature of specific vibrational modes. Here, we refer to this work and add unpublished THz results of L-asparagine (ASN; $C_4H_8N_2O_3$). Experimental details can be found in Ref. [11].

L-asparagine has a side chain R=CH_2CONH_2, and, hence, this molecule contains four carbon atoms, with the terminating carbon in the side chain bound to oxygen and a NH_2 group. L-asparagine crystallizes in the monocline space group $P2_1$ with two molecules per unit cell ($M = 2$) [169]. The single molecule exhibits $3N$ - 6 modes (see Sect. 2.8). In crystalline form one expects a total of $3MN$ - 3 external plus internal modes. That is, in the case of ASN with 17 atoms per molecule and two molecules per unit cell, 99 modes are expected. The lattice constants are $a = 5.0622$ Å, $b = 6.7001$ Å, and $c = 8.0543$ Å, with an angle between (a,c) $\beta = 91.706$. The volume of the unit cell at room temperature $V = 273.06$ Å^3 [169]. Vibrational spectroscopy utilizing FIR and THz, Raman, as well as INS results on ASN were reported in Refs. [170–177]. The measurements were mostly performed at room temperature only, sometimes complemented by model calculations, and, in some cases, attempts were made to assign specific modes to characteristic excitation patterns. We think that an interpretation of the microscopic displacement patterns of observed eigenfrequencies can best be made following the temperature dependence of the eigenmodes and, hence, we recorded the detailed temperature evolution of the real and imaginary parts of the complex dielectric constant.

In the forthcoming figures, we discuss the temperature dependence of the dielectric loss ε'', which seems to be most informative with respect to eigenfrequencies and widths of the involved excitations. Figure 9 shows the temperature dependence of the dielectric loss of ASN in a wavenumber regime from 45 to 115 cm^{-1} (Fig. 9c), compared to the results obtained in Ser (a) and Cys (b), the latter being published in Ref. [11]. In good approximation, the eigenfrequencies of the observed vibrations correspond to the center of the Lorentzian peaks. In all spectra, mainly at low temperatures, experimental data points in the center of the most intensive peaks are missing, due to high absorption in the investigated samples. In these experiments, we had to make a compromise and tried to choose the sample thickness in a way that allows measurements of strong and weak intensities concomitantly at low as well as at high temperatures. With respect to the crystallographic structure, ASN crystallizes in a monoclinic space group, while Ser and Cys are orthorhombic and, hence, one might expect similarities in the spectra of Ser and Cys, different to those observed in ASN. However, experimentally the frequency dependencies of Cys and ASN reveal astonishing similarities, with the sequence of modes II, III, IV, V, and VI of Cys, recaptured in modes I, II, III, IV, and V in ASN, with the latter three modes shifted to higher frequencies.

The spectra of all amino acids, investigated in the course of this work and documented in Fig. 9, reveal well-defined Lorentzian peaks, which undergo a red shift and concomitant broadening on increasing temperatures, canonical anharmonic effects as observed in many condensed-matter systems. Ser and Cys were discussed in some detail by Emmert et al. [11]. Specifically, it remained unclear if the very weak absorption features detected in Ser close to ~ 62, 86, and 100 cm^{-1} at the lowest temperatures correspond to weak intrinsic excitations, result from impurities, or are due to background scatter. Focusing on ASN, in the frequency regime investigated, we observe five well-defined excitations at low temperatures, all of them showing considerable red shift and broadening on heating. However, most of the peaks are still well-defined at room temperature, only with mode III being slightly overdamped, appearing as hump in the low-frequency wing of mode IV. Mode IV possibly represents a double

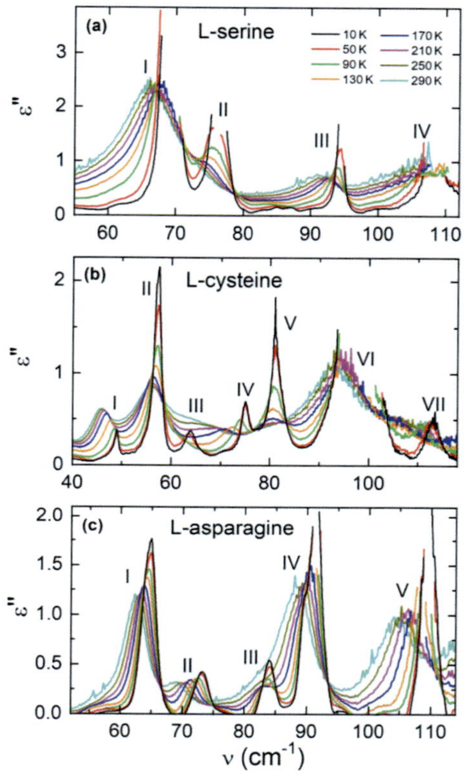

Fig. 9 THz data of the wavenumber/frequency dependence of the dielectric loss of Ser (**a**), Cys (**b**), and ASN (**c**) as function of temperature between 10 and 290 K. The sequence of eigenmodes on increasing frequency is labeled by Roman numerals. For the strongest peaks, close to resonance, data points are missing due to too high absorption for the given sample thickness (see text). The results on Ser and Cys were taken from Ref. [11]

peak, best visible in the 10 K spectra, which, due to experimental uncertainties, could not be finally clarified.

For a more systematic analysis, we performed Lorentzian fits to the observed spectra, with eigenfrequency ν, damping γ (half width at half maximum), and optical weight Δ, which characterizes the dipolar strength (for more details on the Lorentzian model used and on the fitting procedure, please see Ref. [11]). We always concomitantly fitted real (not shown) and imaginary part of the dielectric constant. Results of these fits are indicated in Fig. 10 by solid lines: As representative examples, we provide the frequency dependences of the dielectric loss at three temperatures. As documented in Fig. 10, the fits work reasonably well, with deviations mainly in the low-loss regions of the spectra, close to the background contributions characterized by low absorption. The observed red shifts and increasing dampings are well described, and at room temperature mode III is only visible as shoulder in the low-frequency wing of mode IV. In the temperature and frequency regime investigated, all the modes seem to be well behaved and their red shifts and increasing line widths on increasing temperature signal canonical anharmonic behavior of these modes as routinely observed in crystalline materials.

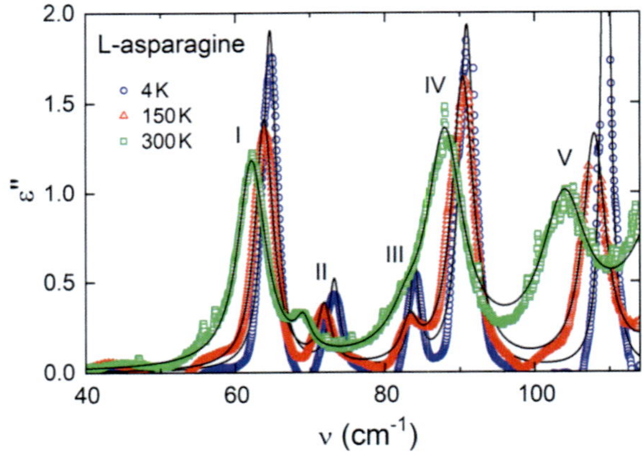

Fig. 10 Temperature dependence of the dielectric loss ε" of ASN as function of wavenumber ν at three selected temperatures 4, 150, and 300 K. The lines represent the concomitant fits of the real (not shown) and imaginary parts of the dielectric permittivity using Lorentzian line shapes. Deviations in the low-loss regions result from background scatter of the polycrystalline samples

In Fig. 11, we document the results of these fits, namely temperature dependencies of the normalized eigenfrequencies [$\nu(T)/\nu(4\text{K})$], of the optical weight or dipolar strength $\Delta(T)$, and of the damping of the eigenmodes $\gamma(T)$. In condensed-matter systems, we expect a continuous increase of the eigenfrequencies with decreasing T due to anharmonic effects, saturating toward low temperatures and exhibiting a linear decrease toward temperatures that are higher than the Debye temperature. Concomitantly, we expect a similar increase in the temperature-dependent damping of the eigenmodes. The dipolar strength or optical weight mainly depends on the static or dynamic dipole moments of the involved vibrations and, in first approximation, is expected to be temperature independent. Figure 11 reveals that this expected canonical anharmonic behavior describes the experimentally observed results reasonably well. The observed scatter in the data, specifically observed for modes III and V, results from experimental uncertainties. Figure 11a shows normalized frequencies only, and the characteristic eigenfrequencies at 4 K as indicated in this frame are: 65, 73, 84, 91, and 110 cm^{-1}.

Finally, by comparing these results to existing model calculations as well as to published results on vibrational excitations in ASN, we would like to gain some insight into the microscopic nature of the observed excitations. First, the continuous temperature evolution of the relevant parameters of the eigenmodes signals that in ASN there is no phase transition at low temperatures (< 300 K), as was observed e.g., in Cys [11]. In literature, there are several published data sets of excitations in ASN utilizing FIR, Raman, and INS techniques, in some cases complemented by model calculations [170, 172, 173, 175–177]. Most of the published experiments focus on excitations > 100 cm^{-1} and only a few experimental results of excitations with wavenumbers < 110 cm^{-1} are available. Matei et al. [172] reported FIR spectra of ASN at room temperature and found weak absorption lines at 62, 87, and 103 cm^{-1} coming close to our results of modes I, IV, and V. As documented in Fig. 9, these are the strongest peaks in our frequency regime and Matei et al. [172] obviously missed the weak lines denominated as mode II and III in our work. These authors ascribed their observed excitations to hydrogen-bond modes. In this frequency regime and

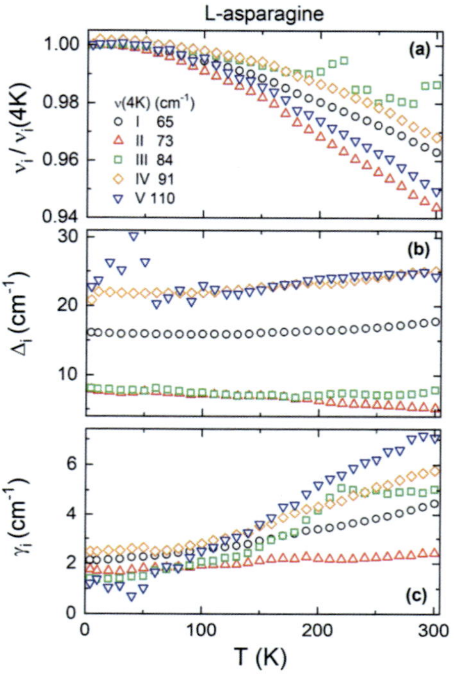

Fig. 11 Temperature dependence of **a** the normalized eigenfrequencies $\nu_i/\nu_i(4\text{ K})$, **b** the oscillator strengths Δ_i, and **c** the damping constants γ_i derived from the Lorentzian fits of the five modes observed in ASN

at room temperature, Sylvestre et al. [175] reported the observation of two modes by FIR techniques at wavenumbers of 68 and 84 cm^{-1}, probably corresponding to modes I and IV, however, with relatively large uncertainties. These authors describe the 68 cm^{-1} excitation to a backbone bending of the carbon atoms coupled to a torsional motion of some end groups. By performing detailed INS and FIR experiments, Pawlukoj et al. [176] reported eigenfrequencies detected by IR techniques at room temperature at 62 and 88 cm^{-1} and by INS at 30 K at 75 and 87 cm^{-1}. The high-frequency excitation probably corresponds to our mode IV, which they ascribe to a torsional motion of the CO_2^- group, and the low-frequency mode at 62 cm^{-1} reported from their FIR experiments, which comes close to our mode I, they assign to lattice vibrations, without further specification. The mode they observe by INS at 30 K, cannot be assigned in our THz experiments. Navarette et al. [178] studied the excitation spectra of ASN in solution. They calculated all the eigenfrequencies of the molecule, and found the three lowest excitations at 48, 84, and 91 cm^{-1}, which is only partly compatible with our experimental results. They ascribe these eigenfrequencies mainly to torsional and bending excitations of the carbon backbone and the carboxy group. This discussion and comparison to other work makes clear how difficult it is to assign modes, specifically at low wave numbers, to identify the nature of a specific molecular excitation.

The similarity of the spectra of ASN and Cys mentioned before makes it reasonable to identify most of these modes resulting from coupled intermolecular translational and librational excitations of the rigid amino acid molecules. This certainly coincides with a detailed INS study of single-crystalline L-alanine by Micu et al. [179]. These authors document that the vibrational spectra below 100 cm^{-1} are dominated by intermolecular translational optical and librational modes of the center of mass of the molecules (see Sect. 2.8). However, these excitations certainly are strongly coupled with rotational excitations of the backbone, of the side chain or of molecular side groups.

4 Concluding Remarks

As discussed in the introduction, performing dielectric spectroscopy on biological matter can reveal information that not only is of interest from an academic point of view but also highly valuable for applications in the fields of dosimetry, diagnostics, and therapeutic medicine. Arriving at a proper microscopic understanding of such spectra, thus certainly is an important task. However, the overview of the many dynamic processes that can contribute to the dielectric response of biological matter, provided in Sect. 2 of the present work, and the experimental examples discussed in Sect. 3 demonstrate that this task can be challenging. Even for such simple molecules as water (Sect. 3.1), several different dynamics must be considered to understand the experimental results, and, despite many decades of research efforts, no consensus is reached about their correct interpretation. For larger biological molecules, the situation usually becomes worse. For example, the amino-acid investigations treated in the present work (Sect. 3.4) show that, even in their solid form and in the restricted THz frequency range dominated by resonances only (Sect. 2.8), the behavior can be very complex and the interpretation quite demanding. Dissolving biological macromolecules like proteins in water, which comes closer to in vivo conditions, generates various additional dispersion effects, especially in the classical dielectric-spectroscopy frequency range (Sect. 3.2): Aside from the tumbling dynamics of the dipolar bio- and water molecules (Sects. 2.5 and 2.6), further processes can be expected, arising, e.g., from bound water and from intramolecular degrees of freedom of macromolecules, both considered to be important for the functionality of proteins (Sect. 2.7). As the aqueous solvent usually contains considerable amounts of ions, significant spectral contributions from EP can arise at low frequencies (Sect. 2.3), which should not be mistaken for intrinsic dynamics. In the dielectric constant, they strongly superimpose the other dynamics of interest, analogous to the influence of the dc conductivity on the low-frequency dielectric loss (Sect. 2.2), which also can obscure intrinsic spectral features. Thus, only careful analysis can unveil the true dielectric behavior at low frequencies. For actually functional biological matter in organisms, like tissues or blood (Sect. 3.3), the presence of cells introduces mesoscopic inhomogeneities in the samples that can give rise to various additional relaxational features in the dielectric spectra. They partly have to be considered as essentially non-intrinsic, arising from heterogeneity-triggered Maxwell–Wagner mechanisms (Sect. 2.5). As then the frequencies and other parameters of the associated loss peaks do not necessarily reflect the time scales of any microscopic motions within the material, only a proper evaluation in terms of model calculations can reveal the intrinsic properties.

Listing up some of the most important possible mechanisms that were considered to manifest in dielectric spectra of biological matter, we have: EP, ionic dc conductivity in the solvent and within the cellular cytoplasm, counter-ion diffusion along cell membranes causing the elusive α relaxation, conformational sampling motions (sub-β), reorientations of dipolar macromolecules like proteins (β relaxation), Maxwell–Wagner relaxation of cell membranes (β), bound water dynamics (maybe bimodal due to loosely and tightly bound layers; δ relaxation), collective protein-water motions (δ), intramolecular motions of macromolecules (δ), unbound water reorientations (γ), and phonon-like or vibrational resonances of intra- and intermolecular origin. Here, we omitted the possible additional secondary dynamics like the excess wing [16, 17, 112] or the Johari-Goldstein β relaxation [180], and the BP [16, 74–77], all mainly known from the field of disordered materials such as glass-forming liquids. However, generally there is no reason to dismiss their existence in biological matter, which usually also is disordered.

The above list is quite overwhelming, especially when considering that, in principle, all of these processes can exist in tissues and blood, many of them in solutions of macromolecules, and at least several of them in simpler molecules like water or solid amino acids. Nevertheless, even for dissolved proteins, tissues, or blood, the experimentally determined dielectric spectra are astonishingly simple. Obviously, the spectral features caused by many of the listed processes are absent or totally obscured by those from other, stronger processes and, thus, undetectable in the spectra. The remaining, detectable ones often are superimposed to each other. Therefore, to perform meaningful dielectric investigations of biological materials, three requirements should be fulfilled: (i) A sufficiently broad frequency range should be covered by the experiments. To obtain correct parameters for the dynamics of interest, then one is able to correctly account for overlapping contributions from the neighboring processes in the fits of the data. (ii) A proper modeling of the experimental results employing: suitable equivalent circuits, the broadened relaxation functions known from glass physics (instead of multiple Debye functions, which often is misleading), and the models developed for the analysis of apparent relaxations arising from the presence of suspended cells. (iii) The collection of dielectric spectra in dependence of various parameters, such as temperature, sample geometry, cell concentration. This can help to discriminate between alternative origins of a detected relaxation feature.

We think the present work demonstrates that complying with these requirements can indeed help in drawing significant conclusions from broadband dielectric and THz measurements on bio-related matter, and we hope it will further advance the art of performing proper dielectric investigations of this class of materials.

References

1. Pethig R, Kell DB (1987) Phys Med Biol 32:933
2. Foster KR, Schwan HP (1989) Crit Rev Biomed Eng 17:25
3. Kaatze U, Feldman Y (2006) Meas Sci Technol 17:R17
4. Frauenfelder H, Chen G, Berendzen J, Fenimore PW, Jansson H, Mcmahon BH, Stroe IR, Swenson J, Young RD (2009) Proc Natl Acad Sci U S A 106:5129
5. Ban D, Funk M, Gulich R, Egger D, Sabo TM, Walter KFA, Fenwick RB, Giller K, Pichierri F, de Groot BL, Lange OF, Grubmüller H, Salvatella X, Wolf M, Loidl A, Kree R, Becker S, Lakomek N-A, Lee D, Lunkenheimer P, Griesinger C (2011) Angew Chem Int Ed 50:11437
6. Sasaki K, Porter E, Rashed EA, Farrugia L, Schmid G (2022) Phys Med Biol 67:14TR01
7. Zuev YF, Galindo CJ, Zueva OS, Feldman YD (2025). Biophys Rev. https://doi.org/10.1007/s12551-025-01355-1
8. Schwan HP (1957) Adv Biol Med Phys 5:147
9. Böhmer R, Maglione M, Lunkenheimer P, Loidl A (1989) J Appl Phys 65:901
10. Schneider U, Lunkenheimer P, Pimenov A, Brand R, Loidl A (2001) Ferroelectrics 249:89
11. Emmert S, Lunkenheimer P, Loidl A (2024) J Chem Phys 160:145103
12. Schwan HP (1983) Blut 46:185
13. Bone S, Bogumil Z (1992) Bioelectronics. Wiley, New York
14. Lunkenheimer P, Emmert S, Gulich R, Köhler M, Wolf M, Schwab M, Loidl A (2017) Phys Rev E 96:062607
15. Khodadadi S, Sokolov AP (2015) Soft Matter 11:4984
16. Lunkenheimer P, Schneider U, Brand R, Loidl A (2000) Contemp Phys 41:15
17. Lunkenheimer P, Loidl A (2018). In: Kremer F, Loidl A (eds) The scaling of relaxation processes. Springer, Cham, p 23
18. Gulich R, Köhler M, Lunkenheimer P, Loidl A (2009) Radiat Environ Biophys 48:107
19. Feldman Y, Ermolina I, Hayashi Y (2003) IEEE Trans Dielectr Electr Insul 10:728
20. Emmert S, Wolf M, Gulich R, Krohns S, Kastner S, Lunkenheimer P, Loidl A (2011) Eur Phys J B 83:157
21. Maxwell JC (1873) A treatise on electricity and magnetism. Clarendon Press, Oxford
22. Wagner KW (1914) Arch Electrotech 2:371
23. Macdonald JR (1987) Impedance spectroscopy—emphasizing solid materials and systems. Wiley, New York
24. Lunkenheimer P, Bobnar V, Pronin AV, Ritus AI, Volkov AA, Loidl A (2002) Phys Rev B 66:052105
25. Ben Ishai P, Sobol Z, Nickels JD, Agapov AL, Sokolov AP (2012) Rev Sci Instrum 83:083118
26. Wolf M, Gulich R, Lunkenheimer P, Loidl A (2011) Biochim Biophys Acta 1810:727
27. Grimnes S, Martinsen ØG (2010) J Phys Conf Ser 224:012073
28. Schwarz G (1962) J Phys Chem 66:2636
29. Schwan HP, Schwarz G, Maczuk J, Pauly H (1962) J Phys Chem 66:2626
30. Schwan HP (1994) Proceedings of 16th annual international conference of the IEEE engineering in medicine and biology society (Baltimore. MD, USA 1:A70
31. Oncley JL (1938) JACS 60:1115
32. Ferry JD, Oncley JL (1938) JACS 60:1123
33. Oncley JL, Ferry D, Shack J (1940) Ann N Y Acad Sci 100:371
34. Grant EH, South GP (1972) Advan Mol Relax Process 3:355
35. Grant EH, Mcclean VER, Nightingale NRV, Sheppard RJ, Chapman MJ (1986) Bioelectromagnetics 7:151
36. Pethig R (1992) Annu Rev Phys Chem 43:177
37. Oleinikova A, Sasisanker P, Weingärtner H (2004) J Phys Chem B 108:8467
38. Zhbanov A, Yang S (2018) Anal Methods 10:168
39. Schwan H (1941) Ann Physik 40:509
40. Bruggemann DAG (1935) Ann Phys 24:636

41. Pauly H, Schwan HP (1959) Z Naturforsch 14B:125
42. Hanai T (1960) Kolloid Z 171:23
43. Hanai T (1961) Kolloid Z 175:61
44. Asami K (2002) Prog Polym Sci 27:1617
45. Lange OF, Lakomek NA, Farès C, Schröder GF, Walter KFA, Becker S, Meiler J, Grubmüller H, Griesinger C, de Groot BL (2008) Science 320:1471
46. Lakomek N-A, Lange OF, Walter KFA, Farès C, Egger D, Lunkenheimer P, Meiler J, Grubmüller H, Becker S, de Groot BL, Griesinger C (2008) Biochem Soc Trans 36:1433
47. Pratihar S, Sabo TM, Ban D, Fenwick RB, Becker S, Salvatella X, Griesinger C, Lee D (2016) Angew Chem Int Ed 55:9567
48. Kaatze U (1989) J Chem Eng Data 34:371
49. Ellison WJ, Lamkaouchi K, Moreau JM (1996) J Mol Liq 68:171
50. Rodríguez-Arteche I, Cerveny S, Alegría A, Colmenero J (2012) Phys Chem Chem Phys 14:11352
51. Charkhesht A, Regmi CK, Mitchell-Koch KR, Cheng S, Vinh NQ (2018) J Phys Chem B 122:6341
52. Latypova L, Puzenko A, Levy E, Feldman Y (2020) J Chem Phys 153:045102
53. Schwan HP (1965) Ann N Y Acad Sci 125:344
54. Grant EH (1962) Nature 196:1194
55. Grant EH (1966) J Mol Biol 19:133
56. Knocks A, Weingärtner H (2001) J Phys Chem B 105:3635
57. Cametti C, Marchetti S, Gambi CMC, Onori G (2011) J Phys Chem B 115:7144
58. Wolf M, Gulich R, Lunkenheimer P, Loidl A (2012) Biochim Biophys Acta 1824:723
59. Hartl J, Friesen S, Johannsmann D, Buchner R, Hinderberger D, Blech M, Garidel P (2022) Mol Pharm 19:494
60. Miura N, Asaka N, Shinyashiki N, Mashimo S (1994) Biopolymers 34:357
61. Chen S-H, Liu L, Fratini E, Baglioni P, Faraone A, Mamontov E (2006) Proc Natl Acad Sci USA 103:9012
62. Doster W, Busch S, Gaspar AM, Appavou MS, Wuttke J, Scheer H (2010) Phys Rev Lett 104:098101
63. Swenson J, Cerveny S (2015) J Phys: Condens Matter 27:033102
64. Wolf M, Emmert S, Gulich R, Lunkenheimer P, Loidl A (2015) Phys Rev E 92:032727
65. Cerveny S, Swenson J (2019) J Chem Phys 150:234904
66. Ito K, Moynihan CT, Angell CA (1999) Nature 398:492
67. Palmer JC, Martelli F, Liu Y, Car R, Panagiotopoulos AZ, Debenedetti PG (2014) Nature 510:385
68. Lunkenheimer P, Reuter D, Schulz A, Wolf M, Loidl A (2025) Phys Rev E 111:065408
69. Latypova L, Puzenko A, Poluektov Y, Anashkina A, Petrushanko I, Bogdanova A, Feldman Y (2021) J Chem Phys 155:015101
70. Braun D, Schmollngruber M, Steinhauser O (2017) Phys Chem Chem Phys 19:26980
71. Soprani L, Giunchi A, Bardini M, Meier QN, D'Avino G (2025) J Chem Theory Comput 21:8073
72. Shrestha UR, Mamontov E, O'Neill HM, Zhang Q, Kolesnikov AI, Chu X (2022) Innovation 3:100199
73. Perticaroli S, Nickels JD, Ehlers G, Sokolov AP (2014) Biophys J 106:2667
74. Schneider U, Lunkenheimer P, Brand R, Loidl A, Non-Cryst J (1998) Solids 235–237:173
75. Kojima S, Wada Takeda M, Nishizawa S (2003) J Mol Struct 651:285
76. Yomogida Y, Sato Y, Nozaki R, Mishina T, Nakahara J (2010) Phys B 405:2208
77. Gabriel J, Pabst F, Helbling A, Böhmer T, Blochowicz T (2018). In: Kremer F, Loidl A (eds) The scaling of relaxation processes. Springer, Cham, p 203
78. Malinovsky VK, Sokolov AP (1986) Solid State Commun 57:757
79. Wuttke J, Hernandez J, Li G, Coddens G, Cummins HZ, Fujara F, Petry W, Sillescu H (1994) Phys Rev Lett 72:3052
80. Elliott SR (1992) Europhys Lett 19:201

81. Schirmacher W, Diezemann G, Ganter C (1998) Phys Rev Lett 81:136
82. Lunkenheimer P, Loidl A, Non-Cryst J (2006) Solids 352:4556
83. Shintani H, Tanaka H (2008) Nat Mat 7:870
84. Baggioli M, Zaccone A (2019) Phys Rev Lett 122:145501
85. Schirmacher W, Paoluzzi M, Mocanu FC, Khomenko D, Szamel G, Zamponi F, Ruocco G (2024) Nat Commun 15:3107
86. Wolf M (2014) The dynamics of proteins, water and their interactions—a dielectric study. Verlag Dr. Hut, Munich
87. Mori T, Jiang Y, Fujii Y, Kitani S, Mizuno H, Koreeda A, Motoji L, Tokoro H, Shiraki K, Yamamoto Y, Kojima S (2020) Phys Rev E 102:022502
88. Takei Y (2000) Jpn J Physiol 50:171
89. Huang W, Richert R (2009) J Chem Phys 130:194509
90. Liebe HJ, Manabe T, Hufford GA (1989) IEEE Trans Antennas Propag 37:1617
91. Zhao Z, Wu Z (2000) Int J Infrared Millim Waves 21:1607
92. Kaatze U, Behrends R, Pottel R, Non-Cryst J (2002) Solids 305:19
93. Ben Ishai P, Tripathi SR, Kawase K, Puzenko A, Feldman Y (2015) Phys Chem Chem Phys 17:15428
94. Hansen JS, Kisliuk A, Sokolov AP, Gainaru C (2016) Phys Rev Lett 116:237601
95. Popov I, Ben Ishai P, Khamzin A, Feldman Y (2016) Phys Chem Chem Phys 18:13941
96. Elton DC (2017) Phys Chem Chem Phys 19:18739
97. Shiraga K, Tanaka K, Arikawa T, Saito S, Ogawa Y (2018) Phys Chem Chem Phys 20:26200
98. Volkov AA, Chuchupal SV (2022) J Mol Liq 365:120044
99. Alvarez F, Arbe A, Colmenero J (2023) J Chem Phys 159:134505
100. Hansen C, Stickel F, Berger T, Richert R, Fischer EW (1997) J Chem Phys 107:1086
101. Gainaru C, Kastner S, Mayr F, Lunkenheimer P, Schildmann S, Weber HJ, Hiller W, Loidl A, Böhmer R (2011) Phys Rev Lett 107:118304
102. Sillrén P, Matic A, Karlsson M, Koza M, Maccarini M, Fouquet P, Götz M, Bauer T, Gulich R, Lunkenheimer P, Loidl A, Mattsson J, Gainaru C, Vynokur E, Schildmann S, Bauer S, Böhmer R (2014) J Chem Phys 140:124501
103. Gainaru C, Meier R, Schildmann S, Lederle C, Hiller W, Rössler EA, Böhmer R (2010) Phys Rev Lett 105:258303
104. Ngai KL (2018) J Mol Liq 253:113
105. Bertolini D, Cassettari M, Salvetti G (1982) J Chem Phys 76:3285
106. Debye P (1929) Polare Molekeln. S. Hirzel, Leipzig
107. Vij JK, Simpson DRJ, Panarina OE (2004) J Mol Liq 112:125
108. Kudlik A, Benkhof S, Blochowicz T, Tschirwitz C, Rössler E (1999) J Mol Struct 479:201
109. Sillescu HJ, Non-Cryst J (1999) Solids 243:81
110. Ediger MD (2000) Annu Rev Phys Chem 51:99
111. Davidson DW, Cole RH (1951) J Chem Phys 19:1484
112. Lunkenheimer P, Loidl A (2002) Chem Phys 284:205
113. Buchner R, Bathel J, Stauber J (1999) Chem Phys Lett 306:57
114. Yada H, Nagai M, Tanaka K (2008) Chem Phys Lett 464:166
115. Kastner S, Köhler M, Goncharov Y, Lunkenheimer P, Loidl A, Non-Cryst J (2011) Solids 357:510
116. Lunkenheimer P, Wehn R, Köhler M, Loidl A, Non-Cryst J (2018) Solids 492:63
117. Angell CA (2002) Chem Rev 102:2627
118. Debenedetti PG, Stillinger FH (2001) Nature 410:259
119. Bauer T, Lunkenheimer P, Loidl A (2013) Phys Rev Lett 111:225702
120. Böhmer R, Ngai KL, Angell CA, Plazek DJ (1993) J Chem Phys 99:4201
121. Angell CA (1995) Science 267:1924
122. Angell CA (1985). In: Ngai KL, Wright GB (eds) Relaxation in complex systems. Office of Naval Research, Washington, DC, p 3
123. Cerveny S, Schwartz GA, Bergman R, Swenson J (2004) Phys Rev Lett 93:245702
124. Johari GP, Tombari E, Salvetti G, Mallamace F (2009) J Chem Phys 130:126102

125. Shi R, Russo J, Tanaka H (2018) Proc Natl Acad Sci USA 115:9444
126. Capaccioli S, Ngai KL (2011) J Chem Phys 135:104504
127. Kringle L, Thornley WA, Kay BD, Kimmel GA (2020) Science 369:1490
128. Jansson H, Swenson J (2010) Bioch. Biophy. Acta 1804:20
129. Ball P (2008) Chem Rev 108:74
130. Jungwirth P (2015) J Phys Chem Lett 6:2449
131. Löffler G, Schreiber H, Steinhauser O (1997) J Mol Biol 270:520
132. Cole KS, Cole RH (1941) J Chem Phys 9:341
133. Puzenko A, Ishai PB, Feldman Y (2010) Phys Rev Lett 105:037601
134. Stickel F, Fischer EW, Richert R (1996) J Chem Phys 104:2043
135. Johari GP, Andersson O (2006) J Chem Phys 125:124501
136. Becker FF, Wang XB, Huang Y, Pethig R, Vykoukal J, Gascoyne PRC (1995) Proc Natl Acad Sci USA 92:860
137. Fahmy HM, Hamad AM, Sayed FA-Z, Abdelaziz YS, Serea ESA, El-D. Mustafa AB, Mohammed MA, Saadeldin AM (2020) Microw Opt Technol Lett 62:3739
138. Caduff A, Dewarrat F, Talary M, Stalder G, Heinemann L, Feldman Y (2006) Biosens Bioelectron 22:598
139. David M, Levy E, Barshtein G, Livshits L, Arbell D, Ishaid PB, Feldman Y (2020) Biochim Biophys Acta Biomembr 1862:183410
140. Schwan H (1948) Colloid Polym Sci 111:53
141. Bassen H (2002) IEEE std c95.3–2002 (The Institute of Electrical and Electronics Engineers, New York). https://doi.org/10.1109/IEEESTD.2002.94226
142. Höber R (1913) Arch Gesamte Physiol 150:15
143. Fricke H (1923) Phys Rev 21:708
144. Rajewsky B, Schwan HP (1948) Naturwissenschaften 5:315
145. Cook HF (1951) Nature 168:247
146. Schwan HP (1952) Fed Proc Fed Am Soc Exp Biol 11:142
147. Schwan HP (1953) Am J Phys Med 32:144
148. Trautmann ED, Newbower RS (1983) IEEE Trans Biomed Eng 30:141
149. Gabriel S, Lau RW, Gabriel C (1996) Phys Med Biol 41:2271
150. Nandi N, Bhattacharyya K, Bagchi B (2000) Chem Rev 100:2013
151. Chelidze T, Non-Cryst J (2002) Solids 305:285
152. Zhbanov A, Yang S (2017) Anal Methods 9:3302
153. Schwan HP (1988) Ann Biomed Engineering 16:245
154. Gabriel S, Lau RW, Gabriel C (1996) Phys Med Biol 41:2251
155. Bothwell TP, Schwan HP (1956) Nature 178:265
156. Schwan HP, Carstensen EL (1957) Science 125:985
157. Abdalla S, Al-Marzouki F, Al-Ghamdi AA, Al-Malki AL (2012) J Mol Liq 173:144
158. Lunkenheimer P, Krohns S, Riegg S, Ebbinghaus SG, Reller A, Loidl A (2010) Eur Phys J Special Topics 180:61
159. Ehrenfreund P, Charnley SB (2000) Annu Rev Astron Astrophys 38:427
160. Krasnokutski SA, Chuang K-J, Jäger C, Ueberschaar N, Henning T (2022) Nat Astron 6:381
161. Gorshunov B, Volkov A, Spektor I, Prokhorov A, Mukhin A, Dressel M, Uchida S, Loidl A (2005) Int J Infrared Millimeter Waves 26:1217
162. Jepsen PU, Cooke DG, Koch M (2011) Laser Photonics Rev 5:124
163. Klotz IM, Gruen DM (1948) J Phys Chem 52:961
164. Thompson HW, Nicholson L, Short LN (1950) Discuss Faraday Soc 9:222
165. Edsall JT, Scheinberg H (1940) J Chem Phys 8:520
166. Edsall JT, Otvos JW, Rich A (1950) J Am Chem Soc 72:474
167. Plusquellic DF, Siegrist K, Heilweil EJ, Esenturk O (2007) ChemPhysChem 8:2412
168. Ueno Y, Ajito K (2008) Anal Sci 24:185
169. Yamada K, Hashizume D, Shimizu T, Yokoyama S (2007) Acta Cryst. E Struct Rep Online 63:3802
170. Husain SK, Hasted JB, Rosen D, Nicol E, Birch JR (1984) Infrared Phys 24:201

171. Casado J, Ramirez FJ, Lopez Navarrete JT (1995) J Mol Struct 349:57
172. Matei A, Drichko N, Gompf B, Dressel M (2005) Chem Phys 316:61
173. Nishizawa J, Sasaki T, Suto K, Tanabe T, Yoshida T, Kimura T, Saito K (2006) Int J Infrared Millimeter Waves 27:923
174. Culka A, Jehlicka J, Edwards HGM (2010) Spectrochim Acta Part A 77:978
175. Sylvestre S, Sebastian S, Edwin S, Amalanathan M, Ayyapan S, Jayavarthanan T, Oudayakumar K, Solomon S (2014) Spectrochim Acta Part A Mol Biomol Spectrosc 133:190
176. Pawlukoj A, Hołderna-Natkaniecc K, Batord G, Natkaniec I (2014) Vib Spectrosc 72:1
177. Shkir M, Muhammad S, AlFaify S, Ahmad Irfan I, Yahia S (2015) Spectrochimica acta part A: molecular and biomolecular spectroscopy 137:432
178. Navarrete JTL, Casado J, Hernandez V, Ramirez FJ (1997) J RAMAN Spectroscopy 28:501
179. Micu AM, Durand JD, Quilichini M, Field MJ, Smith JC (1995) J Phys Chem 99:5645
180. Johari GP, Goldstein M (1970) J Chem Phys 53:2372

The Glass “Transition” Temperature and the Dynamics of Polypeptides

Marianna Spyridakou and George Floudas

Abstract The liquid-to-glass temperature (broader—but incorrectly—known as “glass transition”) in polypeptides and proteins has traditionally been interpreted through the lens of solvent-mediated dynamics. There is growing experimental evidence indicating that glass formation is an intrinsic feature of polypeptides/proteins. This chapter first reviews key experimental work on hydrated and nearly dry proteins, illustrating how hydration modulates protein mobility. Interestingly, nearly solvent-free systems still display a liquid-to-glass temperature, T_g. Subsequently, this chapter focuses on the dynamics of poly(γ-benzyl-L-glutamate) (PBLG) polypeptides with a broad range of molar masses and different end groups. PBLG is a model synthetic polypeptide, regarded as the *drosophila* in this area. Distinct *local* and *global* dynamics associated with α-helical and β-sheet assemblies are resolved and two glass temperatures (T_gs) are identified in oligopeptides exhibiting both secondary structures. The effect of the type of the secondary structure on the fragility, the pressure sensitivity, and the viscoelastic response reveal that α-helical and β-sheet polypeptides possess distinctly different dynamics. Finally, the impact of nanoconfinement on PBLG polypeptides is discussed. The results show that the effect is the acceleration of the segmental dynamics and the partial destabilization of the secondary structure. Overall, the chapter establishes a direct link between secondary structure organization and glass-forming behavior in polypeptides.

Keywords Glass transition · Polypeptides · Secondary structure · α-helices · β-sheets · Dynamics · Viscoelasticity

M. Spyridakou · G. Floudas (✉)
Department of Physics, University of Ioannina, Ioannina, Greece
e-mail: gfloudas@uoi.gr

G. Floudas
Max Planck Institute for Polymer Research, Mainz, Germany

S. Cerveny and J. Swenson (eds.), *Dynamics of Biomolecules Studied by Dielectric Spectroscopy*, Advances in Dielectrics, https://doi.org/10.1007/978-3-032-30588-6_2

1 Introduction

1.1 Early Dynamic Transitions in Proteins and the Emergence of the "Slaving" Concept

Protein function is intimately linked to molecular dynamics. Early investigations into the temperature-dependent mobility of proteins demonstrate that these systems undergo a characteristic liquid-to-glass transformation. The concept of T_g and the associated freezing of the dynamics has since become central to understanding biomolecular flexibility, stability, and function. Myoglobin, a primarily alpha helical protein and one of the first proteins structurally resolved by X-ray crystallography [2], provided the initial platform for probing internal motions. Inelastic neutron scattering experiments on hydrated myoglobin by Doster and co-workers [3, 4] revealed an increase in the mean-square displacement of hydrogen atoms near 180 K, indicating a crossover from harmonic to anharmonic vibrations and collective motions (Fig. 1a). This behavior was interpreted in terms of two dynamic regimes: a low-temperature β-process and a higher-temperature α-process, the latter resembling the glass transition observed in amorphous polymers. The results were further supported by molecular dynamics simulations [5]. The research also indicated that this state was hydration-dependent.

Comparable behavior was later identified in predominantly β-sheet heme proteins, like Nitrophorin 4 (NP4), demonstrating that the dynamic transition is not restricted to α-helical secondary structures [9]. Subsequently, Mössbauer spectroscopy and X-ray crystallography showed that NP4 exhibits a similar increase in atomic fluctuations around 200 K, reinforcing the idea that the onset of large-scale motions is strongly modulated by hydration and solvent mobility rather than by secondary-structure type.

In a follow-up study, Iben et al. used infrared (IR) spectroscopy to monitor carbon-monoxymyoglobin (MbCO) in a solvent environment as a function of temperature and pressure [6]. They observed the same "transition," now termed the "slaved glass transition" (T_{sg}) (Fig. 1b). Through DSC results, they claimed to detect two distinct glass temperatures, one for the solvent and one for the protein. The proximity of the two T_gs (i.e., $T_g^{solvent} = 175$ K and $T_{sg}^{protein} = 178$ K), and the same non-Arrhenius temperature dependence of the main relaxations, suggested that the protein dynamics was "slaved" to that of the solvent. This study was one of the first to introduce the term "slaving" to describe the coupling of the protein motions to the dynamic properties of the surrounding solvent. Subsequent studies of various proteins, such as ribonuclease A, showed that, below this "transition", the protein loses its ability to bind to substrates, rendering its functionally inactive [10].

Calorimetric studies soon establish that the apparent T_g of proteins depends sensitively on water content [7, 8, 11, 12]. Nearly hydrated proteins exhibited transitions in the 300–400 K range, whereas fully hydrated ones showed T_g values near 180–220 K (Fig. 1c). These results indicated that the "transition" is dependent on water content,

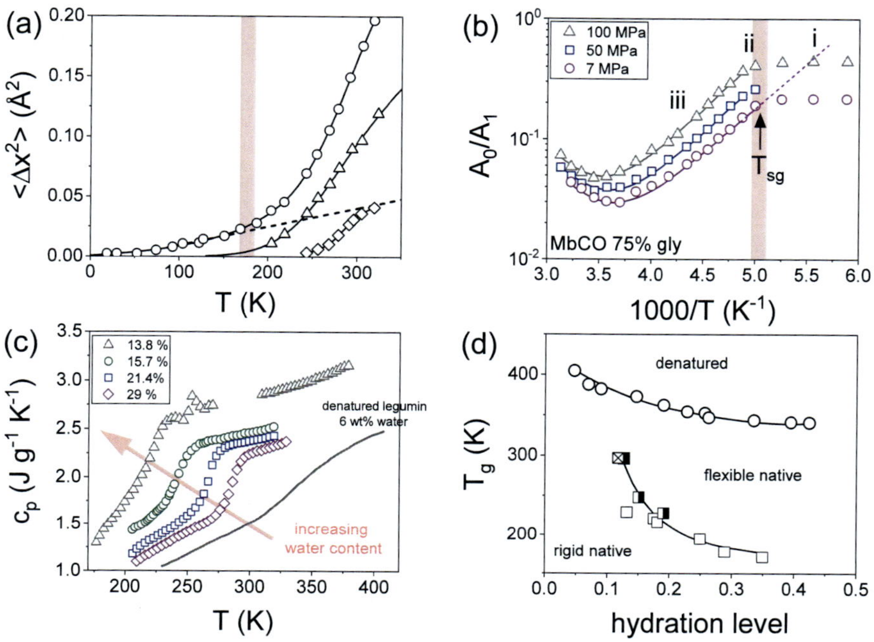

Fig. 1 **a** Average mean-square hydrogen displacements of hydrated myoglobin as a function of temperature. Two contributions were identified. At low and intermediate T, vibrational contributions dominate, indicated by the dashed straight line (Debye–Waller factor). With increasing T (> 200 K), a deviation from the Debye–Waller behavior was observed, signaling a possible dynamic transition in the system. [Adapted with permission from Ref. [3]. Copyright (1989) (Springer Nature)]. **b** Ratio of the relative area of the infrared CO stretch band A_0 (A_0/A_1) in hydrated carbonmonoxymyoglobin (MbCO) as a function of temperature for three different pressures. (A_0 and A_1 correspond to different conformational substates of the CO bound heme in MbCO.) The transition temperature, T_{sg}, indicates the onset of metastable behavior. [Adapted with permission from Ref. [6]. Copyright (1989) (American Physical Society)]. **c** Temperature dependence of the heat capacity of poly-L-asparagine/water solutions for four different water contents and denatured legumin in water. [Adapted with permission from Ref. [7]. Copyright (1994) (American Chemical Society)]. **d** Hydration dependence of the "glass transition" temperature of lysozyme. The data show a decrease in T_g with increasing hydration, marking transitions from the rigid glassy native state to the flexible native state (with flexible matrices and glassy knots), and finally to the denatured state in which the knot regions are disrupted and the native fold is lost. [Adapted with permission from Ref. [8]. Copyright (1995) (Taylor and Francis Group LLC (Books) US)]

a behavior expected if water acted as a plasticizer enhancing protein flexibility and conformational transitions to lower temperatures (Fig. 1d).

Molecular dynamics simulations by Karplus and collaborators provided a mechanistic interpretation of this hydration dependence [13]. Their work demonstrated that global protein fluctuations are governed primarily by the temperature and mobility of the solvent, not by the intrinsic thermal energy of the protein itself. At lower temperatures, water-protein hydrogen bonds restricted motion; at higher temperatures, solvent mobility drives the increase in protein fluctuations (Fig. 2). These

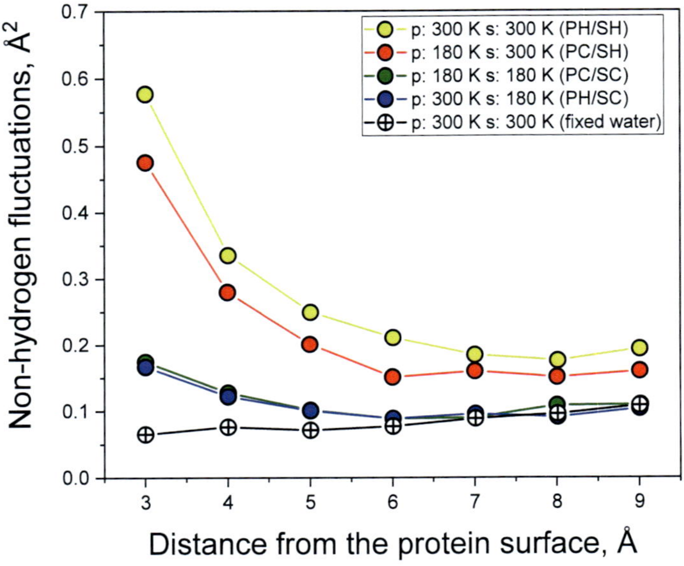

Fig. 2 Mean-square fluctuations of the non-hydrogen myoglobin atoms as a function of the distance from the protein surface for different protein and solvent temperatures. An increase in the fluctuations was only observed when the solvent temperature was increased. [Adapted with permission from Ref. [13]. Copyright (1969) (Springer Nature)]

results established the conceptual framework that proteins behave as glass-like solids only at low temperatures, while at ambient conditions their dynamics is dominated by solvent-coupled motions.

Building on these observations, Frauenfelder and Fenimore focused on explaining the concept of slaving and the role of the hydration shell (the thin layer of water surrounding the protein) in proteins. They suggested that protein motions are dynamically coupled to the solvent relaxation [14, 15]. According to their findings [16], large-scale protein motions were slaved to the α-fluctuations in the bulk solvent, i.e., were dependent on the presence of bulk-like water, while internal protein motions were slaved to the β-fluctuations of the hydration shell, i.e., were coupled to surface-bound water. Although similarities in temperature dependence between protein and solvent relaxations supported a degree of coupling, the processes often differ by several orders of magnitude in timescale [17]. As a result, slaving was viewed as a modulation of peptide dynamics by the solvent, rather than a strict one-to-one "synchronization" [18]. Owing to its complexity, the concept of slaving will be discussed in more detail in the next chapter.

1.2 Insights in the Dynamics of *β*-Sheet-Rich Proteins

While most early dynamic studies focused on α-helical proteins, β-sheet systems provide equally important evidence for intrinsic glass "transitions" in polypeptides. Thermal and structural investigations suggested the presence of a high-temperature T_g (near 175 °C) in nearly dehydrated fibroin, attributed to random-coil configurations within the protein [19–22]. At even higher temperatures, a cold-crystallization was observed. It was interpreted as a transition from random-coil to the β-sheet secondary structure, indicating that thermal activation can promote ordering within previously disordered regions. Similar behavior reported in other protein systems suggested that glass formation and subsequent ordering may represent general features of both fibrous and globular proteins [23]. Subsequent work demonstrated that hydration profoundly affected this behavior: water acted as a plasticizer, lowering T_g by up to ~ 140 °C at high relative humidity, with the effect being fully reversible upon drying [24]. These findings established silk fibroin as an early model system illustrating how water content modulates the thermal response, structural transitions, and glass-forming behavior of β-sheet-rich proteins.

Systematic studies by the Cebe group clarified the relationship between β-sheet crystallization, influence of water, and molecular mobility in silk fibroin [25]. Using TM-DSC, they demonstrated that increasing β-sheet "crystallinity" progressively suppressed the heat-capacity increment at T_g, reflecting a reduction in the fraction of mobile amorphous domains (Fig. 3a). On this basis, they proposed a direct calorimetric approach for quantifying β-sheet crystallinity (ϕ_C), describing the system within a simplified two-phase model, consisting of crystalline and amorphous regions. The linear dependence of the heat-capacity step, ΔC_p, on crystallinity ($\Delta C_p = 0.475 - 0.494\phi_C$), linked the magnitude of the heat-capacity step directly to β-sheet content, providing a thermodynamic method for quantifying crystallinity (Fig. 3b).

Follow-up studies on hydrated silk fibroin [26] revealed a complex, hydration-dependent thermal behavior characterized by two distinct glass "transitions": a low-temperature, water-mediated T_g attributed to the transient silk-water network, and a higher-temperature T_g attributed to the dry protein. Complementary FTIR measurements showed progressive conversion of random-coil and α-helical structures into β-sheets during thermal treatment, highlighting the central role of bound water in precursor formation and structural reorganization. Dielectric spectroscopy further demonstrated water-dependent relaxation processes and reduced amorphous mobility following crystallization [27]. Subsequent work on recombinant spider silk-like block copolymers confirmed the generality of dual glass temperatures in protein-water systems, with the first T_g reflecting water-mediated conformational rearrangements and the second T_g corresponding to transitions in the dry protein [28, 29].

Another computational study [30] examined the macrodipole moments of polypeptides in β-sheet conformations, focusing on systems composed of glycine and alanine residues. Recognizing the importance of dipole–dipole interactions in protein

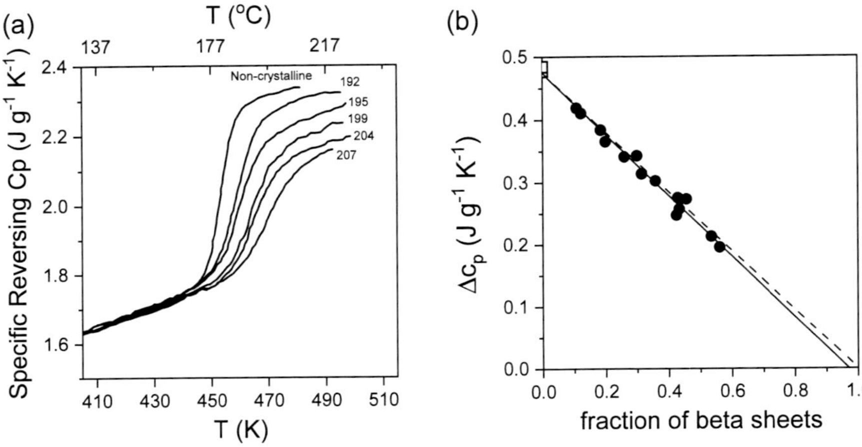

Fig. 3 **a** Temperature dependence of the specific reversing heat capacity of silk fibroin (rate of 2 °C min^{-1}). The non-crystalline sample was scanned before the isothermal crystallization protocol, whereas the other samples were scanned after they had been isothermally crystallized at the respective temperatures. **b** Specific reversing heat-capacity increment as a function of the β-sheet crystallinity. Filled and empty symbols represent the crystalline and non-crystalline samples, respectively. The solid line is a linear fit to the data. The dashed line represents the extrapolation of ΔC_p from a non-crystalline sample to a fully crystalline one (two-phase model). [Adapted with permission from Ref. [25]. Copyright (2006) (American Chemical Society)]

structure and function, the authors developed a predictive model (VSHBβ, Vectorial Summation of Hydrogen-Bonded β-strands). The model estimated the macrodipole moment of a β-sheet by vectorially summing the dipole moments of individual amino acid residues. The later were treated as building blocks, while accounting for their local hydrogen-bonding environment. The study systematically evaluated the dipole moments of both antiparallel and parallel β-sheets with varying numbers of amino acid residues and strands. Contrary to the prevailing assumption that the β-sheet macrodipole moments are negligible, due to the alternating residue orientations, the results showed that the macrodipole moment depends strongly on the number of strands, their length, and their arrangement (Fig. 4). In antiparallel β-sheets, the dipole moments largely cancel out when an even number of strands are present. In contrast, parallel β-sheets exhibit significant macrodipole moments that scale with strand number and length. The predictions from the VSHBβ model showed agreement with those obtained from ab initio molecular orbital (MO) calculations. Additionally, the model accurately predicted the direction of the macrodipole moment. The study also revealed that the dipole moment of an amino acid residue in a β-sheet is smaller than in an α-helix, suggesting weaker polarization in β-sheets due to interstrand hydrogen bonding compared to the more cooperative hydrogen bonding in α-helices.

Overall, these findings underscore that β-sheet assemblies can carry non-negligible macrodipoles whose magnitude and direction depend sensitively on strand

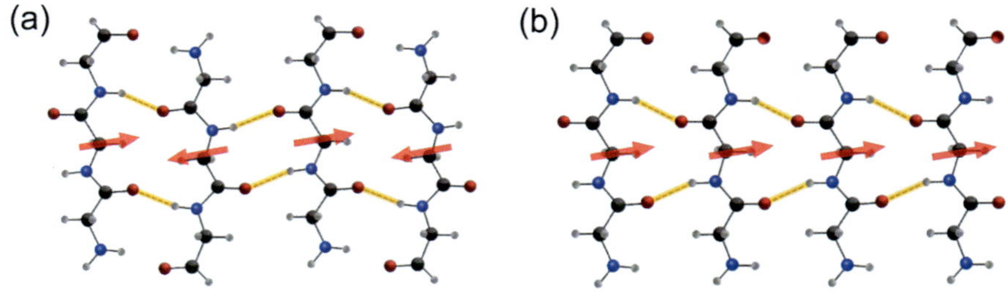

Fig. 4 Geometries and dipole moment vectors of the central residues in **a** antiparallel and **b** parallel β-sheets of glycine. Each strand consists of three amino acids, while the β-sheet structures consist of four β-strands. Different colors indicate the different atoms: carbon (black), nitrogen (blue), oxygen (red), hydrogen (gray). The hydrogen bonds are indicated with yellow dashed lines. Red arrows indicate the axes of the central residue dipole moments

architecture. Therefore, differences between α-helical and β-sheet polypeptides are expected to arise both in local and global dynamics.

2 Glass Transition in Dry Model Polypeptides

Studies of model polypeptides in their bulk state confirmed the existence of a liquid-to-glass "transition" in the absence of solvent, e.g., a property intrinsic to the polypeptides [31–33]. In α-helical polypeptides, the liquid-to-glass "transition" was assigned to defected hydrogen bonds interrupting the α-helical regions. Furthermore, the α-helices themselves were found to be dielectrically active, because of their axial dipole moment, which gives rise to a characteristic relaxation process commonly referred toas the "slow helix process." Among these systems, poly(γ-benzyl-L-glutamate) (PBLG) has emerged as a model polymer. Its ability to adopt either α-helical or β-sheet conformations—stabilized by intra- and intermolecular hydrogen bonds, respectively—makes it uniquely suited for probing how structural motifs influence the thermodynamics and the concomitant dynamics in the solid state [33]. The balance between α-helices and β-sheets depends sensitively on molecular parameters such as the degree of polymerization and chain-end chemistry, making PBLG an exceptionally tunable platform for exploring the fundamental principles that govern polypeptide behavior in the absence of solvent.

2.1 Thermodynamics and Self-assembly

Two series of PBLG samples were investigated through precise oligomer synthesis having different chains ends (Fig. 5) [1]. The first set of PBLG samples terminated with a dimethylamino group (–N(CH_3)$_2$) at one end and a hydrogen atom at the other.

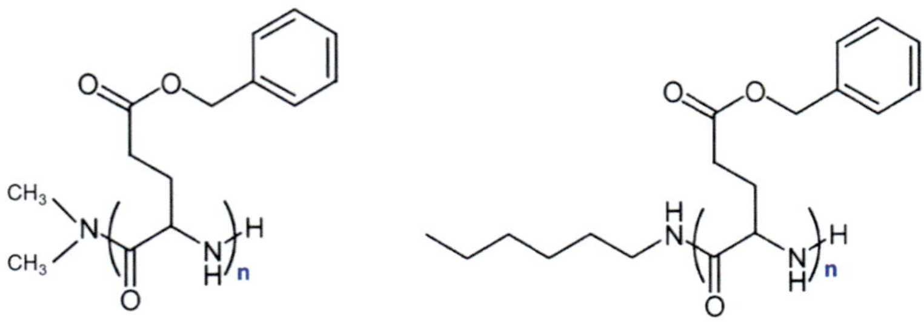

Fig. 5 Chemical structures of two poly(γ-benzyl-L-glutamate) (PBLG) series synthesized via conventional ring-opening polymerization. (Left) PBLG, terminated with a dimethylamino group (–N(CH_3)$_2$) at one end and a hydrogen atom at the other. (Right) PBLG, terminated with a secondary amine substituted with an *n*-hexyl chain (–NH–(CH_2)$_6$$CH_3$) at one end and a hydrogen atom at the other. The end groups result from the different initiators used during synthesis. [Reuse from Ref. [1]. Copyright 2026, American Chemical Society (https://creativecommons.org/licenses/by/4.0/)]

The second set carried a bulkier end group consisting of a secondary amine substituted with an *n*-hexyl chain (–NH–(CH_2)$_6$$CH_3$) at one end and a hydrogen atom at the other. Because of the low molar masses, chain-end chemistry may influence local mobility, packing efficiency, and the stabilization of specific secondary structures. The chemical structures of the two PBLG series are shown in Fig. 5.

Thermodynamic measurements have demonstrated that PBLG exhibits a rich dynamic behavior that depends strongly on chain length (Fig. 6). At low degrees of polymerization, two distinct thermal signatures were observed. A lower T_g appeared across all samples and was associated with the relaxation of amorphous peptide segments that interrupt the α-helical segments as well as to segments at the chain ends [32]. This transition followed a Fox–Flory dependence, consistent with its origin in a cooperative segmental dynamics. A second, higher T_g emerged only in β-sheet-containing oligomers. This feature was attributed to the dynamics of less-ordered or completely disordered regions within β-sheet assemblies, marking one of the first clear demonstrations of a β-sheet-associated T_g in a fully solvent-free peptide. The coexistence of two thermodynamically active relaxation processes—referred to as α and α^* segmental dynamics—highlighted the heterogeneous nature of polypeptide dynamics and underscored the distinct configurational entropies associated with different structural motifs.

Further insight into the two T_gs could be obtained from quantitative analysis of the thermodynamic traces (Inset of Fig. 6). Deconvolution of the heat-capacity derivative into two Lorentzian contributions allowed the relative fractions of α-helical and β-sheet segments to be estimated directly from calorimetry. Consistent with the structural interpretation (below), the lower T_g corresponds to amorphous regions that interrupt the α-helical segments, whereas the higher T_g reflects the dynamics of less-ordered or disordered parts of β-sheets. The relative areas of the two components ($A_{\alpha\text{ - helix}}$ and $A_{\beta\text{ - sheet}}$, respectively) provided a measure of the α-helical and β-sheet fractions, according to $f_{\alpha\text{ - helix}}^{DSC} = \frac{A_{\alpha\text{ - helix}}}{A_{\alpha\text{ - helix}} + A_{\beta\text{ - sheet}}}$ and $f_{\beta\text{ - sheet}}^{DSC} = \frac{A_{\beta\text{ - sheet}}}{A_{\beta\text{ - sheet}} + A_{\alpha\text{ - helix}}}$.

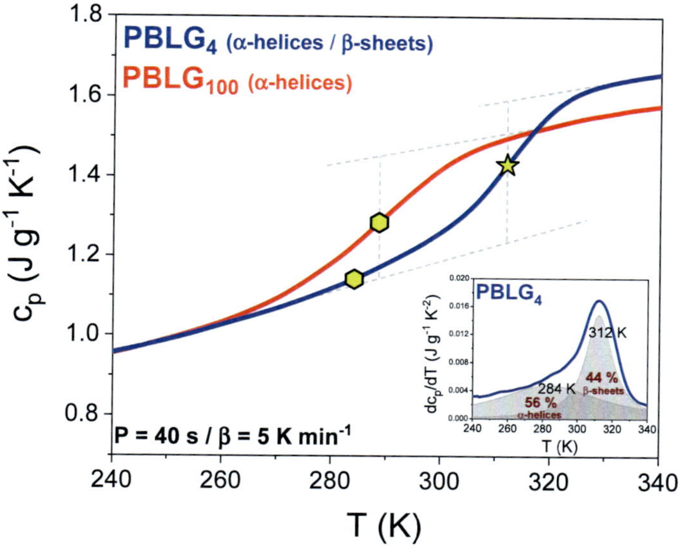

Fig. 6 Temperature dependence of the specific (reversing) heat of $PBLG_4$ (blue) and $PBLG_{100}$ (red), obtained from TM-DSC at a period of modulation of $P = 40$ s ($\beta = 5$ K•s^{-1}). Dashed lines correspond to linear extrapolations of the low- and high-temperature baselines used to determine the onset of the thermal transitions. Polygons indicate the main liquid-to-glass temperature of PBLG. Star symbol corresponds to the additional glass temperature, e.g., the high T_g, evident only in lower molar mass PBLG. Inset: Derivative of the specific heat capacity with respect to temperature plotted as a function of temperature for $PBLG_4$. Gray integrated areas under each Lorentzian peak were used to estimate the relative $f_{\alpha-\mathrm{helix}}^{\mathrm{DSC}}$ and $f_{\beta-\mathrm{sheet}}^{\mathrm{DSC}}$ of the oligopeptide. [Adapted from Ref. [1]. Copyright 2026, American Chemical Society (https://creativecommons.org/licenses/by/4.0/)]

These values showed good agreement with the WAXS and NMR results. This thermodynamic analysis also highlighted the influence of chain-end chemistry at low degrees of polymerization. Interestingly, short oligomers bearing bulky *n*-hexyl termini (i.e., $PBLG_{7(n\text{-hexyl})}$) displayed substantially higher β-sheet content than their dimethylamino-terminated counterparts. Quantitatively, DSC indicated that the *n*-hexyl terminated sample consisted of approximately 64% β-sheets and 36% α-helices. In comparison, the dimethylamino-terminated $PBLG_7$ had approximately 62% α-helices and ~ 38% β-sheets. At intermediate chain lengths, the distinction between end groups became less apparent, and the calorimetric signatures converged toward a single glass transition associated with α-helical segments. Overall, the results indicated that β-sheets were formed only in short oligomers, whereas increasing n promoted the stabilization of α-helices, eventually dominating the conformational landscape of PBLG [1]. In the dimethylamino-terminated samples, however, α-helices were the predominant secondary structure, as $f_{\alpha\text{ - helix}}^{\mathrm{DSC}} > f_{\beta\text{ - sheet}}^{\mathrm{DSC}}$, for all molar masses investigated.

A more detailed picture of this structural evolution comes from wide-angle X-ray scattering (WAXS), as shown in Fig. 7. Two distinct groups were identified: (i) short to intermediate oligopeptides ($n = 4, 6, 7, 8, 10, 12, 14, 16, 18, 20$) that consisted of both secondary structures and (ii) polypeptides ($n = 40, 60, 80, 100$) where only the α-helical secondary structure was present. In the WAXS curves, all dimethylamino-terminated PBLG samples displayed, at low scattering vectors, a set of strong Bragg reflections with characteristic ratios $1:3^{1/2}:4^{1/2}$, with the primary peak at $q \sim 4.7$ nm^{-1} (intercylinder distance of ~ 1.54 nm). These features were consistent with the hexagonally packed α-helices, known as the nematic-like paracrystalline form C of PBLG, in which α-helices are packed periodically in the direction lateral to the chain axis with a nematic-like paracrystalline order [33, 34]. In the polypeptides, an additional sharp reflection appeared at higher q (= 12.6 nm^{-1}) corresponding to the helical pitch of 0.5 nm along the chain axis. Oligopeptides showed an additional Bragg reflection at $q = 3.7$ nm^{-1} ($d = 1.7$ nm), corresponding to the lamellar spacing or the inter-sheet spacing of β-sheets. The $PBLG_{7(n\text{-hexyl})}$ showed reflections exclusively associated with the β-sheet structure, consistent with its high β-sheet fraction (~ 80% according to NMR results). Moreover, n-hexyl terminated oligopeptides exhibited an additional peak at $q = 13.4$ nm^{-1}, corresponding to an inter-strand spacing of 0.47 nm between adjacent peptide chains within the β-sheets. Across all chain lengths, a broad amorphous halo is present, reflecting correlations between the side-group atoms and contributions from the amorphous PBLG regions. These disordered segments along the PBLG chains coexist with the ordered configurations identified as α-helices or β-sheets, and, as shown later in the DS results (see Fig. 7), they have a distinct dynamic signature [1].

Complementary insight into the secondary structure in PBLG can be provided by ^{13}C solid-state NMR, which offers site-specific resolution of the ordered motifs. The characteristic chemical shifts associated with α-helices and β-sheets are well established (Fig. 8) [35]: helical segments give rise to resonances near 176 ppm (C=O) and 58 ppm (C_α), whereas β-sheet structures appear at the more upfield positions of ~ 172 ppm (C=O) and ~ 53 ppm (C_α). In addition, the C_α carbon adjacent to the terminal amine can be distinguished from the other β-sheet C_α sites, appearing as a separate resonance around 49 ppm. These spectral markers allow the absolute fractions of α-helices and β-sheets to be quantified directly from the C_α region. Across the series, the spectra revealed a progressive increase in α-helical signal intensity with chain length, accompanied by a corresponding decrease of β-sheet resonances. For dimethylamino-terminated samples, this trend followed a logarithmic dependence, with β-sheet resonances decreasing substantially and eventually disappearing for chains with $n \geq 40$, indicating that β-sheets become unstable in this regime. As expected, intermediate oligomers display coexisting α-helical and β-sheet signals, consistent with the structural motifs identified by WAXS and with the two T_gs observed thermodynamically.

Taken together, the calorimetric, scattering, and NMR results converged on a coherent structural picture: from β-sheet-rich domains at low n to α-helical structures at higher n. This naturally raises the question of which secondary structure is thermodynamically preferred in PBLG. Between the different secondary structures, the α-helix represents the thermodynamically preferred state, as it is the only structure detected at higher molar masses. β-sheets are formed only in short oligomers, reflecting the fact that the criteria for stable helices (e.g., the 18/5 or 7/2) are not

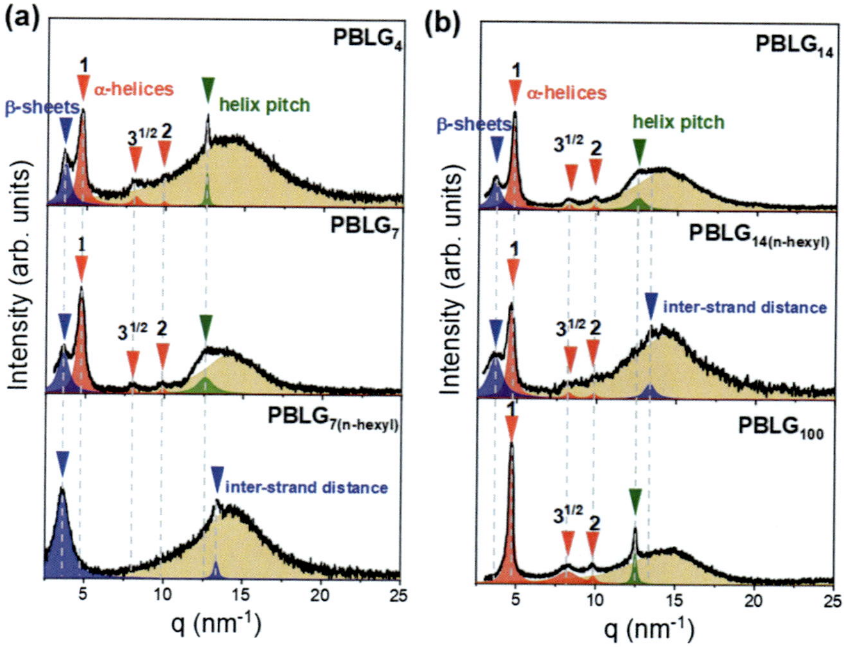

Fig. 7 WAXS patterns of the investigated samples **a** top: $PBLG_4$, middle: $PBLG_7$, bottom: $PBLG7_{(n\text{-hexyl})}$ and **b** top: $PBLG_{14}$, middle: $PBLG_{14(n\text{-hexyl})}$, bottom: $PBLG_{100}$. Blue arrows correspond to the lamellar spacing of β-sheet secondary structure and the inter-stand distance of β-sheets, respectively. Red arrows indicate Bragg reflections of the hexagonally packed α-helices, while green arrows give the reflection corresponding to the pitch of the α-helix. The amorphous halo is indicated in yellow. [Reuse from Ref. [1]. Copyright 2026, American Chemical Society (https://creativecommons.org/licenses/by/4.0/)]

yet satisfied at low chain lengths. This behavior reflects the different cooperativity requirements of the two motifs: intramolecular hydrogen bonds, which stabilize α-helices, can form even in relatively short chains, whereas β-sheets rely on extensive intermolecular hydrogen bonding, requiring higher cooperativity and therefore becoming unfavorable as chain length increases [1].

Beyond chain length alone, the NMR data revealed that the α-helix and β-sheet fractions are also sensitive to the end-group chemistry, particularly in the oligomeric regime. Comparisons between dimethylamino-terminated and n-hexyl-terminated samples showed that the nature of the terminal group can shift the balance between secondary structures. In $PBLG_{14}$, for example, the dimethylamino-terminated chain adopts predominantly α-helical conformations, whereas its n-hexyl-terminated analogue exhibits a majority of β-sheets. The upfield resonances associated with the terminal carbons confirm the presence and distinct environments of the respective end groups. These observations suggested that bulky termini promote local segregation that favors β-sheet formation, while smaller end groups stabilize

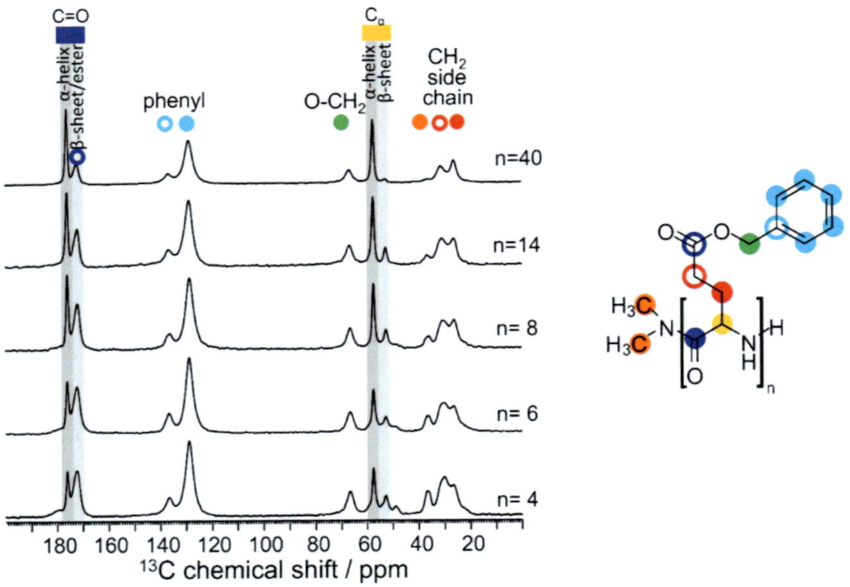

Fig. 8 ^{13}C solid-state NMR spectra of the dimethylamino-terminated samples: $PBLG_4$, $PBLG_6$, $PBLG_8$, $PBLG_{14}$ and $PBLG_{40}$. The highlighted areas refer to the chemical shifts used for the calculation of the absolute fractions of peptide secondary structures. The molecular structure of the repeat unit is also shown, including the color scheme employed for assignment purposes. [Reuse from Ref. [1]. Copyright 2026, American Chemical Society (https://creativecommons.org/licenses/by/4.0/)]

the α-helical secondary structure [1]. Thus, end-group chemistry acts as an additional molecular parameter that modulates the self-assembly of PBLG.

2.2 *Molecular Dynamics*

The structural picture established above naturally leads to the question of how these different secondary motifs behave dynamically. Temperature-dependent ^{13}C solid-state NMR provided direct insight into this aspect, revealing a clear contrast between the mobility of α-helical and β-sheet segments (Fig. 9a). The C_α resonances showed that the α-helical signal at ~ 58 ppm undergoes pronounced broadening and a reduction in intensity with increasing temperature. This behavior reflected a progressive temperature-induced softening of the helical segments, especially at the ends of the helices where they are connected to more disordered or amorphous segments. As the system approached the segmental (α) process—whose relative timescales are shown in Fig. 10—the segmental motions enter the dynamic window of NMR. These thermally activated motions reduced the local rigidity of helices, leading to a broadening

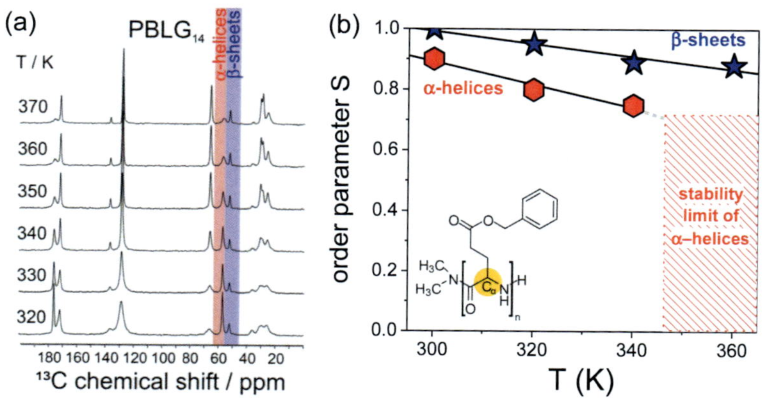

Fig. 9 **a** ^{13}C solid-state NMR spectra of PBLG$_{14}$ as a function of temperature. The highlighted areas refer to the chemical shifts used for the calculation of the absolute fractions of α-helices and β-sheets. **b** Temperature dependence of the dynamic order parameter S for the C_α-H bonds in the two secondary structures, α-helices and β-sheets. $S = 0$ indicates isotropic motion, while $S = 0.7$ indicates strongly anisotropic motion with a gaussian orientation distribution of 73 deg width at half height. (inset) The given order parameter S have been determined for the CH bond of the C_α position highlighted in molecular structure for clarity. [Adapted from Ref. [1]. Copyright 2026, American Chemical Society (https://creativecommons.org/licenses/by/4.0/)]

of resonances and to a reduction in signal intensity. In contrast, the β-sheet resonances remained sharp—and in some cases even increased in intensity—over the entire temperature range studied (up to 370 K). The absence of broadening or weakening indicated that the motions associated with the β-sheet-related α^* process were not thermally activated in the studied temperature range up to the frequency ($f \sim 10^5$ Hz) required to average local NMR interaction. Thus, the results indicated that β-sheets remained practically frozen on the NMR timescale.

Additional insight into the local mobility was obtained from site specific ^{1}H–^{13}C REPT-HDOR spinning sideband pattern measurements as a function of temperature [36, 37]. This experiment provided access to the motional averaged dipolar coupling constants, $D_{\mathrm{CH}}(t)_t$, from which the dynamic order parameter, S, can be extracted. This parameter [38], defined in terms of the time-averaged second order Legendre polynomial, $S_{\mathrm{CH}} = \frac{1}{2}\left(3\cos^2\theta_{\mathrm{CH}}(t) - 1\right)_t = \frac{D_{\mathrm{CH}}(t)_t}{D_{\mathrm{CH,static}}}$, and quantifies the residual motional anisotropy of a given molecular segment. The parameter.

The resulting $S(T)$ profiles highlighted a clear difference between the two secondary structures, as shown in Fig. 9b. For β-sheets, S remained nearly constant across the entire temperature range, decreasing only slightly from 1.0 (22.7 kHz) at 300 K to 0.88 (19.7 kHz) at 360 K. This small change indicated only minimal local fluctuations and confirmed that β-sheet structures remained highly rigid and dynamically arrested up to 360 K. The α-helical segments, however, displayed a markedly different behavior. Starting from an S value of 0.90 (20.3 kHz) at 300 K, the order parameter decreased steadily to 0.75 at 340 K, reflecting increasing local

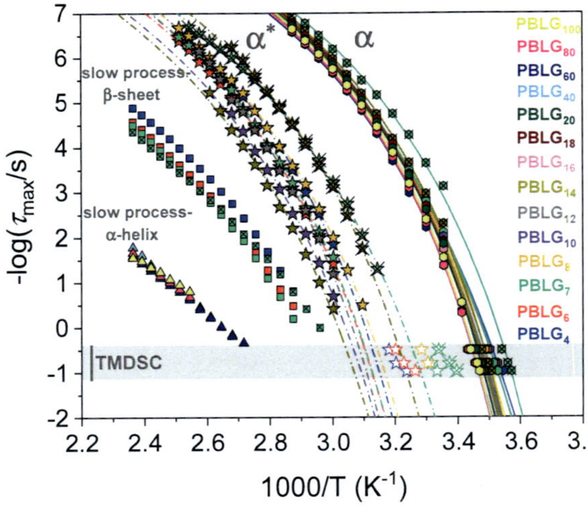

Fig. 10 Characteristic relaxation times as a function of the inverse temperature of PBLGs with different degrees of polymerization: $PBLG_4$ (blue), $PBLG_6$ (red), $PBLG_7$ (green), $PBLG_8$ (orange), $PBLG_{10}$ (purple), $PBLG_{12}$ (gray), $PBLG_{14}$ (dark yellow), $PBLG_{16}$ (pink), $PBLG_{18}$ (wine), $PBLG_{20}$ (dark green), $PBLG_{40}$ (light blue), $PBLG_{60}$ (navy), $PBLG_{80}$ (magenta), $PBLG_{100}$ (yellow). Crossed symbols correspond to $PBLG_{7(n\text{-hexyl})}$ (green) and $PBLG_{14(n\text{-hexyl})}$ (dark yellow). Hexagons represent the segmental α process, stars the segmental α^* process, squares the "slow" β-sheet process and triangles the "slow" α-helical process (see text). TM-DSC data are also shown. The lines represent fits to the VFT equation. [Reuse from Ref. [1]. Copyright 2026, American Chemical Society (https://creativecommons.org/licenses/by/4.0/)]

mobility and partial dynamic softening of the helical structure. At 360 K, the REPT-HDOR signal for α-helices disappeared entirely due to a substantial reduction in residual dipolar coupling and shortened ^{13}C T_2 times, consistent with the onset of large-amplitude segmental motion (see below). This loss of signal marks the point at which α-helical segments become too dynamic to be probed on the timescale of the experiment. As a result, this divergence in mobility is quantitatively demonstrated [1]: while β-sheets remained largely immobilized, α-helices underwent a clear temperature-driven relaxation into a dynamically softened state.

NMR clearly distinguishes mobile from rigid segments, but it cannot directly resolve the timescales of these motions. To access the characteristic relaxation times and to probe molecular dynamics over a broader frequency window, dielectric spectroscopy (DS) can be employed. First, the molecular dynamics across different molar masses was examined and presented in Fig. 10. At low temperatures, all samples—independent of chain length or end-group chemistry—displayed the α process. This process originated from amorphous segments interrupting the α-helices as well as from segments at the chain-end giving rise to the lower glass temperature identified by TM-DSC [31–34]. For oligomers and short polypeptides ($n \leq 20$) that stabilize both α-helical and β-sheet structures, two additional relaxations appeared. The first was the α^* process, attributed to the dynamics of segments located in less-ordered or completely disordered parts of β-sheets. Notably, the α^* process appeared at lower frequencies, i.e., longer timescales, than the α process at the same temperature (Fig. 10), indicating that β-sheet-related motions were significantly more cooperative. The second process, termed the "slow" process, was associated with a global relaxation of β-sheets; its dielectric strength and implications will be discussed below.

For polypeptides with $n \geq 40$, which adopt exclusively α-helical structures, the dielectric response resembled earlier reports on PBLG [31–34]. At higher T, a slower and intense relaxation was observed, attributed to the relaxation of macrodipoles in the helical parts, the so-called slow α-helix process (Fig. 12). This behavior was interpreted with the "defected helix" model [32], in which the α-helix can be viewed as consisting of distinct "ideal" helical parts of uniform correlation length, denoted as ξ_{helix}, that are able to rotate on the surface of a cone at a defined angle θ. Because of the hexagonal packing of helices, each part is assumed to rotate independently, while the axes of the rotational cones remain parallel. The dipole moment associated with each helical part is $\mu = 3.4D(\xi/0.15\,\mathrm{nm})$, where 0.15 nm corresponds to the helix length per repeat unit. Within this framework, the ξ_{helix} can be estimated from the dielectric strength of the slow helical process as, $\Delta\varepsilon = \frac{N_A\rho}{3\varepsilon_0 kTM_0}(3.4Debye)^2(\xi/0.15\,\mathrm{nm})\sin^2\theta$. X-ray scattering from oriented fibers provide an upper value for θ, which is subsequently used to estimate the helical correlation length. The resulting value is approximately 2 nm, which is considerably smaller than the theoretical length expected for an "ideal" helix. This suggested that α-helices consist of broken helices rather than behaving as a single rigid unit [32].

The characteristic relaxation times for the different processes are shown in the Arrhenius plot of Fig. 10. The α and α^* processes conform to the usual Vogel-Fulcher-Tammann (VFT) equation, $\tau = \tau_o^{\#}\exp\left(\frac{B}{T-T_o}\right)$, where $\tau_o^{\#}$ is the relaxation time in the limit of very high temperatures, B is the activation parameter, and T_o is the "ideal" glass temperature.

The polypeptide dynamics bear additional similarities with the segmental dynamics of amorphous polymers. The dependence of the glass temperatures on molar mass is shown in Fig. 11a. The T_gs associated with the α and α^* segmental processes, as determined by both TM-DSC and DS, are shown as a function of the degree of polymerization n. The T_g of the α process increased with increasing chain length, following a Fox–Flory dependence, indicating the influence of chain ends on the segmental dynamics. Despite a more limited dataset, the T_g of the α^* process exhibited a similar trend. Both processes were attributed to the dynamics of amorphous segments, especially those located at the chain ends or in disordered regions in-between the more ordered domains. However, the key difference arises from their structural environment: the α^* process occurs at higher temperatures, implying that these amorphous segments are subject to additional constraints. The reduced mobility of segments associated with the α^* process in β-sheets associate with h.b. defects between different chains in their β-sheet secondary structure.

In addition, the *n*-hexyl terminated samples displayed lower T_gs than the corresponding dimethylamino-terminated samples, both for the α-helical (α) and β-sheet (α^*) secondary structures.

Attention is now turned to the analysis of the "slow" process, observed at elevated temperatures in PBLG with low and intermediate molar masses. In α-helical polypeptides, this process is assigned to the relaxation of macrodipoles in the α-helical parts (Fig. 12). For low n ($4 \leq n \leq 10$), the ratio $\Delta\varepsilon_{\beta\text{ - sheet}}/\Delta\varepsilon_{\alpha^*} \sim 2$, suggesting that the relaxation exceeds a repeat unit in the β-sheet configuration. Such process can result from the parallel β-sheet structures, where the vectorial sum of individual

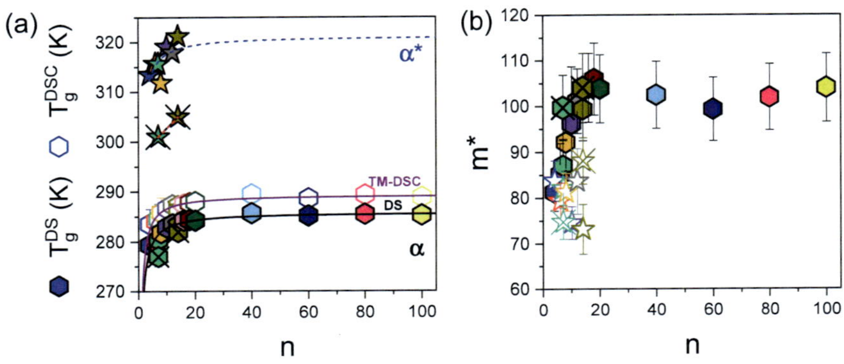

Fig. 11 **a** Liquid-to-glass temperatures of the segmental α process (hexagons) and the segmental α^* process (stars) as a function of molar mass. Crossed symbols correspond to the *n*-hexyl terminated oligopeptides. Filled symbols refer to DS (T_g defined at $\tau = 100$ s) and open symbols to TM-DSC. Lines represent fits to the Fox–Flory equation. **b** Fragility or steepness index, m^*, of the two segmental relaxations as a function of molar mass. *N*-hexyl terminated samples are indicated with crossed symbols. [Adapted from Ref. [1]. Copyright 2026, American Chemical Society (https://creativecommons.org/licenses/by/4.0/)]

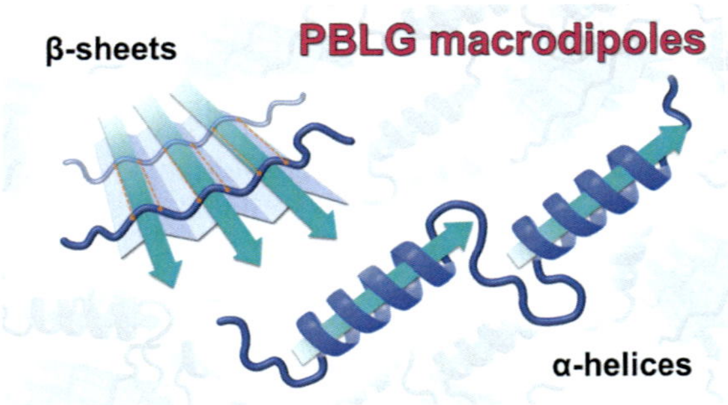

Fig. 12 Schematic representation of two PBLG chains forming (i) α-helices and (ii) a parallel β-sheet. The green arrows indicate respective macrodipoles. Amorphous segments interrupting the α-helices/β-sheets, as well as amorphous segments at the chain ends, are also shown. [Reuse from Ref. [1]. Copyright 2026, American Chemical Society (https://creativecommons.org/licenses/by/4.0/)]

amino acid residues in a β-sheet creates a macrodipole (Fig. 12). With respect to macrodipoles, β-sheets can be considered as type-B polymers according to the Stockmayer classification (α-helical peptides are of type-A) [39]. Type-B polymers have rigid dipoles perpendicular to the chain and give rise to normal modes. Similar behavior has been reported for functionalized poly-p-phenylenes bearing ultrastrong dipoles perpendicular to the backbone [40].

Another parameter that provides insight into the dynamic behavior of the β-sheets versus the α-helices is the fragility of the corresponding segmental process, shown in Fig. 11b. The fragility or steepness index [41], m, is defined as $\vartheta \log\tau/\vartheta\left(T_g/T\right)|_{T=T_g}$, and quantifies the deviation of the temperature dependence of the relaxation times from the Arrhenius behavior at $T > T_g$. The extracted values are presented in Fig. 11b. Interestingly, the steepness index increases systematically with increasing molar mass, a trend most pronounced for the α process. In terms of the effect of secondary structure on the fragility, it can observed that the slower process (α^*), associated with the local relaxation of β-sheets, shows consistently lower fragility than the faster process (α), associated with the local relaxation in α-helices, indicating that β-sheets display a stronger glass behavior. Although both processes originate from disordered segments, it is the structural environment, and, particularly, the extent and type of hydrogen bonding (inter vs intra), that determines their fragility [1].

2.3 *Pressure Effects*

Pressure-dependent measurements were performed on two representative peptides to further elucidate the nature of the α and α^* processes, as well as of the slower processes associated with the α-helical and β-sheet secondary structures. The systems examined were $PBLG_{7(n\text{-hexyl})}$, an oligopeptide consisting a majority of β-sheets, and $PBLG_{40}$, which forms exclusively α-helices.

The pressure dependence of the characteristic frequencies, under isothermal conditions, is shown in Fig. 13a, b for the two samples. The P-dependence of the segmental α and α^* processes could be described by the pressure equivalent to the VFT equation [42], $f_{max} = f_\infty exp\left(-\frac{D_P P}{P_0 - P}\right)$, where f_∞ is the segmental relaxation frequency at atmospheric pressure at a given temperature, D_P is a dimensionless parameter, P_0 is the pressure corresponding to the "ideal" glass. In $PBLG_{7(n\text{-hexyl})}$, the segmental (α) process exhibited a stronger pressure dependence than the α^* process. By contrast, the slow processes associated with macrodipole relaxation in the β-sheet and α-helical secondary structures (indicated as β-sheet and as α-helix, respectively) exhibited considerably weaker pressure dependences.

Pressure-dependent measurements allow access to the apparent activation volume, $\Delta V^{\#}$ [43, 44]. This quantity was extracted from the pressure dependence of the relaxation times/frequencies (Fig. 13a, b), as the slope at each pressure, according to $\Delta V^{\#} = 2.303RT\left(\frac{\partial \log\tau}{\partial P}\right)_T$. In systems without hydrogen bonding, $\Delta V^{\#}$ has been interpreted as reflecting the molecular volume of the underlying dynamic processes.

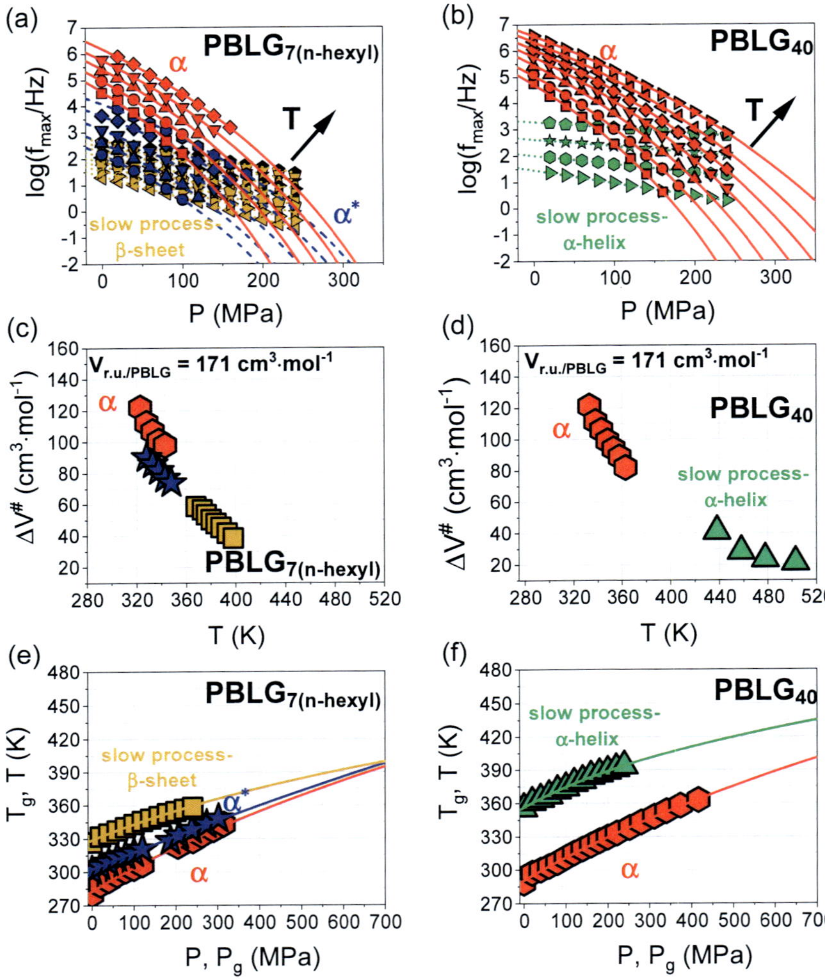

Fig. 13 **a** Frequency at a maximum loss as a function of pressure for $PBLG_{7(n\text{-hexyl})}$, corresponding to the segmental (α) process (red); from 323 to 343 K in 5 K steps, α^* process (blue); from 328 to 348 K in 5 K steps, and β-sheet process (orange); for temperatures 368 K, 373 K, 376 K, 380 K, 384 K, 388 K, 393 K, and 398 K. The solid and dashed lines are fits to Eq. 3, while the dotted lines are fits to an Arrhenius dependence. **b** Pressure dependence of the characteristic frequencies at maximum loss for the $PBLG_{40}$, indicating the segmental (α) process (red); from 333 to 363 K in 5 K steps, and the α-helix process (green); for temperatures 438 K, 458 K, 478 K, and 503 K. The solid and dotted lines are fits to the pressure equivalent of the VFT equation and to an Arrhenius dependence, respectively. **c, d** Temperature dependence of the apparent activation volumes ($\Delta V^{\#}$) for the α (red), α^* (blue), β-sheet (orange) and α-helix (green) processes. **e, f** Pressure dependence of the glass temperatures (T_g) as obtained from the isothermal and isobaric representations, corresponding to the freezing of the respective processes, at a characteristic time of $\tau = 10$ s. Solid and dashed lines represent fits to the empirical equation reported in the text. [Adapted from Ref. [1]. Copyright 2026, American Chemical Society (https://creativecommons.org/licenses/by/4.0/)]

When hydrogen bonds are present, it may reflect a characteristic volume of defects. The apparent activation volume at ambient pressure was calculated as a function of temperature for the investigated processes, with the results plotted in Fig. 13c, d. For all processes studied, the calculated $\Delta V^{\#}$ values were found to be smaller than the molar volume of PBLG repeat unit ($\Delta V^{\#}_{\text{r.u.}} = 171\ \text{cm}^3\ \text{mol}^{-1}$). However, at T_g, $\Delta V^{\#}\ (\alpha) \approx \Delta V^{\#}_{\text{r.u.}}$ and $\Delta V^{\#}\ (\alpha^*) < \Delta V^{\#}_{\text{r.u.}}$. The α process showed similar activation volumes in both $\text{PBLG}_{7(n\text{-hexyl})}$ and PBLG_{40}, consistent with its assignment to the relaxation of amorphous segments in the α-helical peptides. In $\text{PBLG}_{7(n\text{-hexyl})}$, the α^* process had a lower activation volume than the α process at the same temperature but followed a similar temperature dependence, suggesting a smaller defect volume in the case of β-sheets.

Lastly, the pressure sensitivity of the glass temperature was examined, extracted for each process at $\tau = 10$ s, thereby avoiding long extrapolations. The P and T dependence of T_g and P_g are presented in Fig. 13e, f, respectively. The data were fitted using the empirical equation [34, 43] $T_g(P) = T_g(0)\left(1 + \frac{\nu}{\mu}P\right)^{1/\nu}$, with $T_g(0)$ being the glass temperature at atmospheric pressure and μ and ν are fitting parameters. The pressure sensitivity of T_g, can be quantified through the $\left(dT_g/dP\right)_{P\to 0}$ coefficient. The segmental α process of both $\text{PBLG}_{7(\text{m})}$ and PBLG_{40} showed similar $\left(dT_g/dP\right)_{P\to 0}$ parameters (~ 200 K•GPa^{-1}), in agreement with previously reported values for α-helical polypeptides such as PBLG and PZLL [31]. In contrast, the segmental α^* process of $\text{PBLG}_{7(n\text{-hexyl})}$ exhibited a lower pressure coefficient of 158 K GPa^{-1}, comparable to the behavior observed for the β-sheet-forming PGly [31]. These findings further emphasized that the secondary structure governs pressure sensitivity. The network-like structure of β-sheets imposes stronger constraints on the amorphous segments, thereby limiting the pressure response of the segmental α^* process.

In summary, the local dynamics in polypeptides that exhibit both secondary structures revealed two distinct glass temperatures, each associated to the relaxation of amorphous segments interrupting α-helices and β-sheets [1]. ^{13}C NMR and DS demonstrated that β-sheets relax at longer timescales, reflecting a more cooperative relaxation. The global dynamics of the α-helical and β-sheet macrodipoles was also evident at even longer timescales. Further analysis revealed consistently lower fragility values for β-sheets, which can be attributed to the constraints imposed by their inherent network-like structure stabilized by intermolecular hydrogen bonds. In contrast, α-helices (stabilized by intramolecular h.b.) allow for greater mobility in nearby amorphous segments, leading to a more thermally sensitive dynamics. These structural differences were also reflected in the lower pressure coefficient of β-sheets compared to α-helices.

2.4 Viscoelastic Response

The viscoelastic properties of PBLG can be discussed as a function of molar mass and chain end-group chemistry. Master curves of the storage (G') and the loss (G'') moduli as a function of frequency are shown in Fig. 14a. For both oligopeptides and polypeptides, the storage modulus consistently exceeds the loss modulus ($G' > G''$), indicating a solid-like (elastic) behavior. The solid-like response reflects the underlying structural organization, i.e., the α-helical and β-sheet secondary structures and their supramolecular self-assembly into ordered domains. The respective van Gurp-Palmen (vGP) plot, plotting the phase angle (δ) again the complex modulus ($|G^*|$), shown in Fig. 14b, provides deeper insight into the applicability of time–Temperature superposition (tTs) and of the elastic plateau. For homopolymers following the tTs, all isothermal data in a vGP plot overlap in a single master curve. Such systems typically exhibit two minima in δ: one at the plateau modulus G_N^0,, which reflects the elastic properties of the entanglement "network" (typically $G_N^0 \sim 10^5$ Pa), and another in the vicinity of the liquid-to-glass temperature (T_g) where $|G^*| \sim 10^9$ Pa. Regarding the former, it was observed that, well-ordered α-helices (in $PBLG_{60}$ and $PBLG_{100}$) and β-sheets (in $PBLG_{7(n\text{-hexyl})}$) display similar plateau values (~ 10^7 Pa), reflecting increasing topological constraints through percolation/aggregation. Between the two minima, δ increases and goes through a maximum.

Overall, the results demonstrated that chain length and end-group chemistry can be strategically employed to favor the stabilization α-helices and/or β-sheets, providing precise control over the structural, dynamical, and viscoelastic properties of polypeptides [1]. This insight could offer a basis for designing polypeptides tailored to specific biomedical applications; from rigid, load-bearing scaffolds, implantable supports,

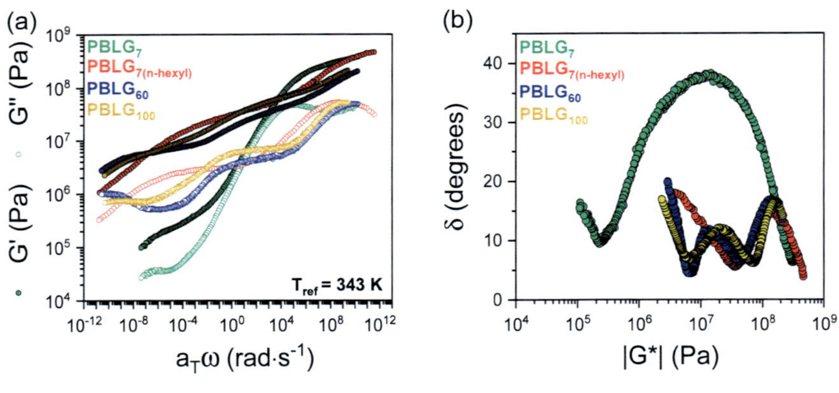

Fig. 14 **a** Master curves of the storage (filled circles) and loss (open circles) moduli of $PBLG_7$ (green), $PBLG_{7(n\text{-hexyl})}$ (red), $PBLG_{60}$ (blue) and $PBLG_{100}$ (yellow). There is a solid-like behavior ($G' > G''$) across all measured frequencies. **b** The van Gurp-Palmen plot for the corresponding samples: $PBLG_7$ (green), $PBLG_{7(n\text{-hexyl})}$ (red), $PBLG_{60}$ (blue), and $PBLG_{100}$ (yellow). [Adapted from Ref. [1]. Copyright 2026, American Chemical Society (https://creativecommons.org/licenses/by/4.0/)]

and drug delivery matrices (harnessing β-sheet rigidity) to soft tissue engineering platforms, such as cell-support matrices and biointerfaces (exploiting the tunability of α-helical networks) [45, 46].

3 Polypeptides Under Confinement

The behavior of polypeptides under nanometer confinement represents a distinct regime compared to the bulk state. Structural organization and dynamics can be influenced by geometric constraints and surface interactions. Indeed, studies of flexible biopolymers or DNA macromolecules translocating through nanoscopic pores have shown that the interaction between the macromolecule and the pore walls can affect their functionality [47–53]. In this context, controlling the peptide secondary structure and dynamics under confinement is a challenging task for biotechnological applications.

To probe these confinement effects, solution-infiltrated PBLG homopolymers were investigated under confinement in alumina nanopores with three pore diameters; 400 nm, 100 nm and 40 nm [54]. X-ray diffraction analysis revealed that increasing confinement lead to a progressive weakening and broadening of the characteristic low-q Bragg reflection associated with the hexagonally packed α-helices. Correspondingly, the lateral coherence length of the ordered helical domains decreased with decreasing pore size, indicating a reduction in long-range order. Changes were also observed in the higher-q region: the peak corresponding to van der Waals correlations between the atoms of the side-groups incorporated into helices becomes weaker, while the peak associated with correlations between the atoms of the amorphous segments increased in relative intensity. Overall, confinement strongly affected the PBLG self-assembly as α-helices were destabilized giving rise to increasingly amorphous configurations.

DS measurements performed on the same PBLG under confinement revealed that the structural changes were accompanied by changes in the dynamics. Broader curves and lower dielectric strengths were observed with increasing confinement. The segmental dynamics of confined PBLG (Fig. 15b) speed-up with decreasing pore diameter, with this effect being more evident in the smaller pores ($\Delta T_g = 10$ K) (Fig. 15c). A faster segmental dynamics is not uncommon in confined polymers [43, 55]. It associates with non-equilibrium dynamics [56]. Earlier work on confined PBLG also reported a reduction in T_g, although in that case the polymer was synthesized directly inside the nanopores [55]. This led to α-helices formed perpendicular to the pore walls and to different interfacial interactions than those present in the solution-infiltrated samples. The stronger decrease in T_g observed in that system likely reflects these distinct packing arrangements and peptide-surface interactions.

Additional information about the self-assembly of PBLG under confinement could be obtained from the normalized dielectric strength of the slow helix process. The results showed a decrease with increasing confinement, indicating that smaller α-helices (helical segments with a smaller correlation length) were formed within the

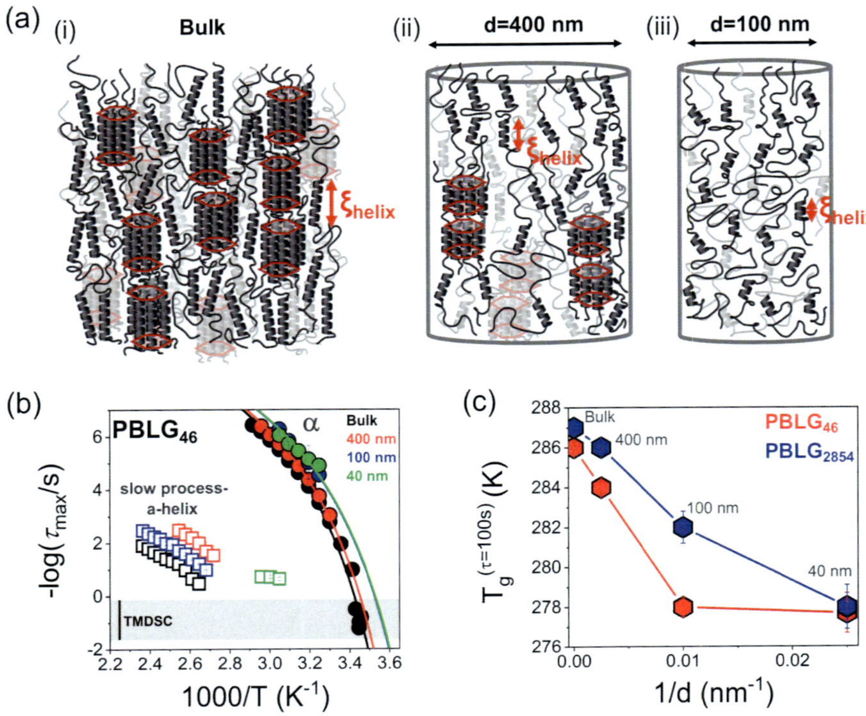

Fig. 15 **a** Schematic representation of PBLG self-assembly (i) in the bulk state and under confinement in AAO nanopores with diameters of (ii) 400 nm and (iii) 100 nm. **b** Activation plot of the dynamic processes in bulk and confined $PBLG_{46}$. Filled circles correspond to the segmental process (bulk: black spheres, 400 nm: red spheres, 100 nm: blue spheres, 40 nm: green spheres) and squares to the slow process that reflects the motion of α-helical segments. Lines represent fits to the VFT equation. **c** Glass temperature (defined at $\tau \sim 100$ s) as a function of the inverse pore diameter for confined $PBLG_{46}$ and $PBLG_{2854}$. [Adapted with permission from Ref. [54]. Copyright (2022) (American Chemical Society)]

smaller pores (Fig. 15a). In the bulk, longer α-helical segments were packed in a hexagonal lattice, interrupted by amorphous segments (Fig. 15(i)). Confinement reduced the number and the correlation length of α-helices (Fig. 15(ii)). Within 100 nm pores, only a small number of α-helices survived (Fig. 15(iii)).

Overall, the results showed that the polypeptide secondary structure is strongly altered upon imbibition into nanopores, with α-helical segments reduced in favor of more amorphous configurations. Such effects are important to consider when biomacromolecules with defined secondary structures enter narrow pores. The resulting "defected" helices make PBLG resemble intrinsically disordered proteins (IDPs), that are characterized by conformational flexibility [53, 57, 58]. Unlike ordered proteins, IDPs lack a fixed 3D structure which enables them to engage in unique biological activities. Because of their natural abundance and functional significance, IDPs, as well as protein hybrids that consist of both intrinsically disordered protein regions (IDPRs) as well as ordered regions, have gained immense importance and have changed protein science. In this respect, PBLG can be considered as a synthetic analogue to IDPs and IDPRs. Understanding the configurations and dynamics of simple synthetic polypeptides confined at the nanoscale may allow for a better understanding of the properties of much more complex protein structures in confinement.

References

1. Spyridakou M, Stavrakaki IG, Tsagkaraki E, Varfi C, Graf R, Iatrou E, Floudas G (2026) Biomacromol 27:855–872
2. Kendrew JC, Bodo G, Dintzis H, Parrish RG, Wyckoff H, Phillips DC (1958) Nature 181:662–666
3. Doster W, Cusack S, Petry W (1989) Nature 337:754–756
4. Doster W, Cusack S, Petry W, Non-Cryst J (1991) Solids 131:357–361
5. Smith J, Kuczera K, Karplus M (1990) Proc Natl Acad Sci 87:1601–1605
6. Iben IET, Braunstein D, Doster W, Frauenfelder H, Hong MK, Johnson JB, Luck S, Ormos P, Schulte A, Steinback PJ, Xie AH, Young RD (1989) Phys Rev Lett 62:1916–1919
7. Green JL, Fan J, Angell CA (1994) J Phys Chem 98:13780–13790
8. Gregory RB (1995) In: Gregory RB (ed) Protein-solvent interactions. Marcel Dekker, New York
9. Schmidt M, Achterhold K, Prusakov V, Parak FG (2009) Eur Biophys J 38:687–700
10. Rasmussen BF, Stock AM, Ringe D, Petsko GA (1992) Nature 357:423–424
11. Doster W, Bachleitner A, Dunau R, Hiebl M, Lüscher E (1986) Biophys J 50:213–219
12. Rupley JA, Careri G (1991) Adv Protein Chem 41:37–172
13. Vitkup D, Ringe D, Petsko GA, Karplus M (2000) Nat Struct Biol 7:34–38
14. Fenimore PW, Frauenfelder H, McMahon BH, Parak FG (2002) Proc Natl Acad Sci 99:16047–16051
15. Fenimore PW, Frauenfelder H, McMahon BH, Young RD (2004) Proc Natl Acad Sci 101:14408–14413
16. Frauenfelder H, Chen G, Berendzen J, Fenimore PW, Jansson H, McMahon BH, Stroe IR, Swenson J, Young RD (2009) Proc Natl Acad Sci 106:5129–5134
17. Swenson J, Jansson H, Bergman R (2006) Phys Rev Lett 96:247802
18. Cerveny S, Combarro-Palacios I, Swenson J (2016) J Phys Chem Lett 7:4093–4098
19. Magoshi J, Nakamura S (1975) J Appl Polym Sci 19:1013–1015
20. Magoshi J, Magoshi Y (1975) J Polym Sci Polym Phys Ed 13:1347–1351
21. Magoshi J, Magoshi Y, Nakamura S (1977) J Appl Polym Sci 21:2405–2407
22. Magoshi J, Magoshi Y, Nakamura S, Kasai N, Kakudo M (1977) J Polym Sci Polym Phys Ed 15:1675–1683
23. Magoshi J, Nakamura S, Murakami KI (1992) J Appl Polym Sci 45:2043–2048
24. Agarwal N, Hoagland DA, Farris RJ (1997) J Appl Polym Sci 63:401–410
25. Hu X, Kaplan D, Cebe P (2006) Macromolecules 39:6161–6170
26. Hu X, Kaplan D, Cebe P (2008) Macromolecules 41:3939–3948
27. Yu L, Hu X, Kaplan D, Cebe P (2010) Biomacromol 11:2766–2775
28. Huang W, Krishnaji S, Tokareva OR, Kaplan D, Cebe P (2014) Macromolecules 47:8098–8106
29. Huang W, Krishnaji S, Tokareva OR, Kaplan D, Cebe P (2014) Macromolecules 47:8107–8114
30. Mieda S, Aida M (2013) Chem Lett 42:473–475
31. Papadopoulos P, Floudas G, Schnell I, Klok H-A, Aliferis T, Iatrou H, Hadjichristidis N (2005) J Chem Phys 122:224906
32. Floudas G, Spiess HW (2009) Macromol Rapid Commun 30:278–298

33. Papadopoulos P, Floudas G, Klok H-A, Schnell I, Pakula T (2004) Biomacromol 5:81–90
34. Floudas G, Papadopoulos P, Klok HA, Vandermeulen GWM, Rodriguez-Hernandez J (2003) Macromolecules 36:3676–3683
35. Guan X, Stark RE (2010) A general protocol for temperature calibration of MAS NMR probes at arbitrary spinning speeds. Solid State Nucl Magn Reson 38:74–76
36. Saalwächter K, Schnell I (2002) Solid State Nucl Magn Reson 22:154–187
37. Saalwächter K, Spiess HW (2001) J Chem Phys 114:5707–5728
38. Southern SA, Perras FA (2024) J Magn Reson 364:107710
39. Stockmayer WH (1967) Pure Appl Chem 15:539–546
40. Papamokos G, Wudarczy J, Graf R, Schollmeyer D, Baumgarten M, Müllen K, Floudas G (2018) Macromolecules 51:3330–3339
41. Ngai KL, Roland CM (1993) Macromolecules 26:6824–6830
42. Paluch M, Patkowski A, Fischer EW (2000) Phys Rev Lett 85:2140–2143
43. Floudas G, Paluch M, Grzybowski A, Ngai KL (2011) In: Molecular dynamics of glass-forming systems. Effects of pressure, Springer, Berlin, Heidelberg, pp 1–174
44. Mpoukouvalas K, Floudas G, Williams G (2009) Macromolecules 42:4690–4700
45. Wagner WR, Sakiyama-Elbert SE, Zhang G, Yaszemski MJ (2020) Biomaterials science: an introduction to materials in medicine, 4th edn. Academic Press, Elsevier, San Diego, US, pp 1–1586
46. Thomasy SM, Raghunathan VK, Winkler M, Reilly CM, Sadeli AR, Russell TP, Jester JV, Murphy CJ (2014) Acta Biomater 10:785–791
47. Uversky VN (2019) Front Phys 7:1–19
48. Movileanu L, Cheley S, Bayley H (2003) Biophys J 85:897–910
49. Aksimentiev A, Heng JB, Timp G, Schulten K (2004) Biophys J 87:2086–2097
50. Blobel G, Dobberstein B (1975) J Cell Biol 67:835–851
51. Yong H, Molcrette B, Sperling M, Montel F, Sommer J (2021) Macromolecules 54:4432–4442
52. EL-Mahdy AFM, Yu TC, Mohamed MG, Kuo S (2021) Macromolecules 54:1030–1042
53. Nikoubashman A (2021) J Chem Phys 154:090901
54. Spyridakou M, Tsimenidis K, Gkikas M, Steinhart M, Graf R, Floudas G (2022) Macromolecules 55:2615–2626
55. Duran H, Gitsas A, Floudas G, Mondeshki M, Steinhart M, Knoll W (2009) Macromolecules 42:2881–2885
56. Kardasis P, Sakellariou G, Steinhart M, Floudas G (2022) J Phys Chem B 126:5570–5581
57. Dunker AK (2001) J Mol Graph Model 19:26–59
58. Wright P, Dyson HJ (1999) J Mol Biol 293:321–331

Dynamics of Peptides at Low Temperatures: From Monomer to Polymer and the Universal Nature of Slaving Behavior

Silvina Cerveny

Abstract This chapter investigates the coupling between hydration dynamics and peptide conformational behavior using a series of lysine-based model systems. The study spans molecular sizes from the amino acid L-lysine to short oligomers (n-lysine with $n = 3, 4, 5$, and 10) and the polypeptide ε-polylysine (ε-PLL, $n \approx 32$). Dielectric spectroscopy measurements, supported by complementary experimental approaches, reveal a close correspondence between peptide conformational fluctuations near the glass transition and the dynamical properties of the surrounding water. The results indicate that peptide motions are governed by solvent dynamics, consistent with the concept of dynamical "slaving" previously reported for hydrated proteins. Here, this behavior is shown to persist across smaller peptides and selected non-biological polymers within defined thermodynamic regimes. The analysis further differentiates this coupling mechanism from conventional plasticization effects. Overall, the findings underscore the central role of hydration in controlling macromolecular dynamics and provide a general physical perspective relevant to both biological and synthetic polymer systems.

Keywords Dielectric spectroscopy · Water dynamics · Peptide dynamics · Hydration water · Amino acids · Peptides · L-lysine · ε-polylysine · Glass transition temperature · Slaving behavior · Bio- and non-biological aqueous solutions

S. Cerveny (✉)
Centro de Física de Materiales (CFM—CSIC/EHU)—Material Physics Centre (MPC), Paseo Manuel de Lardizabal 5, 20018 Donostia-San Sebastian, Spain
e-mail: silvina.cerveny@ehu.es

Donostia International Physics Centre (DIPC), Donostia-San Sebastian, Spain

S. Cerveny and J. Swenson (eds.), *Dynamics of Biomolecules Studied by Dielectric Spectroscopy*, Advances in Dielectrics, https://doi.org/10.1007/978-3-032-30588-6_3

1 Introduction

1.1 Protein Dynamics

Proteins are the molecular workhorses of living systems, catalyzing biochemical reactions, transmitting signals, providing structural support, and performing many other vital functions. Since the pioneering determination of John Kendrew and coworkers [1] of the three-dimensional structure of myoglobin in 1958, it has been understood that the biological functions of proteins are closely tied to their three-dimensional shape. However, structure alone does not fully explain protein function; equally important are the dynamic motions that enable proteins to transition between different conformational states, bind substrates, and respond to their environment.

The dynamics of proteins cover a remarkable range of timescales, from femtosecond bond vibrations to millisecond or slower conformational changes associated with catalytic activity [2–4]. A key question for understanding these motions is the following: what role does water play in enabling and regulating protein dynamics? Proteins in their native, functional state exist in aqueous environments, and the close association between proteins and their surrounding water molecules has long been recognized as essential for biological activity [5–7]. Protein functionality generally begins at hydration levels around 0.2 g water per gram of dry protein, indicating that water is not just a passive solvent but also an active participant in biomolecular function [6].

In Chap. 1, we explored the behavior of proteins and amino acids (AAs) at room temperature, which closely resembles the conditions where proteins are biologically active. The dynamics of proteins in solution at room temperature are extremely fast and can be studied using terahertz (THz) spectroscopy and dielectric spectroscopy in the gigahertz (GHz) frequency range [8–12]. However, protein motions can also be effectively investigated below room temperature, within the supercooled regime of water, as discussed in Chaps. 4–7. In this low-temperature region, the slowing of water dynamics allows a more precise separation between motions arising from the solvent and those intrinsic to the protein [13–16].

The study of protein dynamics in aqueous solutions at low temperatures has been limited by water crystallization, which has led to the use of hydrated powders instead of biologically realistic aqueous solutions [17, 18] and hindered our understanding of protein–water interactions under physiologically relevant conditions. However, recent advances in studying smaller biomolecules, especially peptides, have opened new opportunities for exploring these fundamental interactions in proper aqueous solutions while preventing water crystallization [19–21]. This chapter provides a detailed analysis of peptide dynamics at low temperatures by examining lysine-based model systems, ranging from the simplest amino acid monomer (L-lysine) to oligomeric peptides (n-lysine, where n = 1–10) and the polypeptide ε-polylysine (ε-PLL, n = 32). Through this systematic investigation, we demonstrate how water dynamics influence the conformational fluctuations of these biomolecular systems and show that the coupling between water and solute dynamics (referred to as “slaving”) is not unique to proteins but is a more universal phenomenon that occurs under certain thermodynamic conditions, with implications extending from biological to non-biological systems.

The experimental methods used in these studies combine several complementary techniques, mainly broadband dielectric spectroscopy (BDS), which examines the reorientational behavior of dipolar molecules over extremely broad frequency ranges (from millihertz to gigahertz), differential scanning calorimetry (DSC), Fourier-transform infrared spectroscopy (FTIR), and small-angle X-ray scattering (SAXS). As already mentioned, conducting experiments at supercooled temperatures, below the freezing point of bulk water, offers important advantages because different dynamic processes become well separated in time, allowing them to be analyzed. At the same time, the importance of studying lysine-based molecules lies in their high solubility, which prevents crystallization in water.

The organization of this chapter is as follows. After discussing the theoretical framework of energy landscapes and the slaving concept, we examine the experimental evidence for water–peptide coupling within the lysine family. Next, we differentiate between slaving and plasticization phenomena and generalize the idea of “slaving behavior” to other types of molecules, not just proteins.

2 Theoretical Framework: Energy Landscapes and Conformational Dynamics

2.1 The Energy Landscape

The conceptual framework for understanding protein dynamics rests on the idea of a rugged energy landscape: a multidimensional surface where the coordinates represent all possible configurations of the protein and its environment, and the heights correspond to the free energies of each configuration [22]. The landscape is a hypersurface in the high-dimensional space of all protein and hydration-shell coordinates. It displays a hierarchical organization (“wells within wells within wells”), in which substates are grouped into tiers of decreasing free-energy barriers, a structure revealed by cryogenic experiments that identified distinct kinetic regimes associated with different tiers [23].

The barriers between conformational substates arise from both enthalpic contributions (associated with changes in bond energies, electrostatic interactions, and van der Waals contacts) and entropic factors, which reflect changes in the accessible configurational space [23, 24]. For a protein to transition between these substates, it must overcome these barriers, a process that requires both enough thermal energy and appropriate fluctuations in the local environment. It is in this context that water plays its crucial role.

Water molecules in the hydration shell and in the bulk solvent undergo continuous rotational and translational motion. These solvent fluctuations can lower the barriers between protein conformational substates by providing entropy contributions that facilitate barrier crossing. According to this view, the large-scale conformational motions of proteins are not solely properties of the protein but result from the combined dynamics of the protein and its aqueous environment [25, 26].

2.2 The Slaving Concept

The term "slaving" was introduced by Frauenfelder [25, 27, 28] and coworkers to describe the observation that some protein motions have the same temperature dependence as solvent dynamics but occur on slower timescales. In the slaving picture [25, 27, 29, 30] developed primarily through studies of myoglobin (MYO), protein motions are classified according to their relationship to solvent dynamics. Large-scale conformational fluctuations are "slaved" to the α-fluctuation (structural relaxation) of the bulk solvent, meaning that they exhibit the same non-Arrhenius temperature dependence as the solvent, but occur 10^3–10^6 times more slowly. The slowness indicates that numerous individual solvent fluctuations are needed to trigger a single large-scale conformational change of the protein. Slaved motions are controlled by solvent viscosity and are absent in a solid environment [25].

The unified model of protein dynamics formulated later by Frauenfelder, Swenson, and coworkers [26] further distinguishes the roles of the bulk solvent and the hydration shell. Large-scale protein motions, including shape changes and motions at tier 0 of the energy landscape hierarchy, are slaved to α-fluctuations (structural relaxation) in the bulk solvent. In contrast, more local internal protein motions, such as side-chain rotations and small backbone fluctuations, are coupled to β-fluctuations (local relaxation) in the hydration shell [29]. These hydration-shell fluctuations involve localized reorientations and hydrogen-bond-breaking events that occur on faster timescales and exhibit Arrhenius temperature dependence.

A key experimental finding supporting the slaving model came from comparing rate coefficients for protein processes in different solvents [25]. Although the dielectric relaxation rates of the solvents differed by more than a factor of 1000, the ratio of the protein rate coefficient to the solvent dielectric rate remained essentially unchanged. This result implies that the rate coefficient for a slaved transition can be written as $k(T) = k_{\text{diel}}(T)/n(T)$, where $k_{\text{diel}}(T)$ is the rate of solvent dielectric fluctuations and $n(T)$ is a coefficient indicating how much the protein and its hydration shell slow the effect of solvent fluctuations. Because $k_{\text{diel}}(T)$ varies considerably with temperature while $n(T)$ depends only weakly on temperature, the solvent dominates the activation enthalpy for slaved processes. In contrast, the protein controls the activation entropy via the energy landscape [25].

On the other hand, more localized protein motions, such as side-chain rotations and small backbone fluctuations, are linked to faster, more local motions in the hydration shell. These "β-fluctuations" of the hydration water involve localized reorientations and hydrogen-bond-breaking events that occur on shorter timescales and display Arrhenius temperature dependence.

Experimental evidence for slaving has been obtained through multiple techniques. Studies using broadband dielectric spectroscopy demonstrated that the temperature dependence of the mean-square displacement of protein atoms (measured by Mössbauer effect or neutron scattering) follows the temperature dependence of dielectric fluctuations in the hydration shell [26, 31]. The result is clear: The protein provides the structural machinery necessary for function, but it is dynamically passive; the environment (water) controls the protein motions.

Water molecules in the hydration shell and even outside of it (bulk-like water) undergo continuous rotational and translational motions. These solvent fluctuations can lower the effective barriers between protein conformational substates by providing entropy contributions that facilitate barrier crossing (see Fig. 1). According to this picture, the large-scale conformational motions of proteins are not intrinsic properties of the protein alone but emerge from the coupled dynamics of the protein and its aqueous environment [25, 30].

The slaving model thus provides a hierarchical picture of protein dynamics in which different classes of solvent motions control different classes of protein motions. It is important to note that this coupling has profound implications. If protein function depends on large-scale conformational changes and these changes are enslaved to solvent dynamics, then manipulating solvent properties provides a route to controlling protein activity.

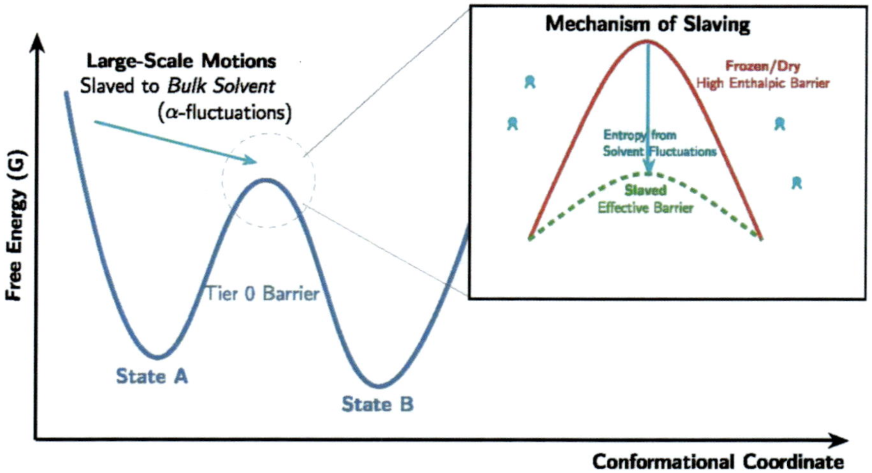

Fig. 1 Hierarchical energy landscape and solvent-slaved dynamics. The main panel shows a schematic one-dimensional projection of the protein energy landscape illustrating the hierarchical organization of conformational states: Tier 0 represents major functional states (A, B) separated by large barriers. The inset zoom illustrates the mechanism of slaving: The free energy barrier (red solid line) is effectively lowered (green dashed line) by entropic contributions from solvent fluctuations (cyan molecules with motion arrows). Large-scale protein motions are controlled by bulk solvent α-relaxation

2.3 Alternative Perspectives: Plasticization

An alternative viewpoint proposes that water mainly acts as a plasticizer for proteins [32–35]. In this picture, water lowers the glass transition temperature (T_g) and reduces internal friction within the protein, enabling motions that would occur even in the dry state, though at much slower rates. According to this idea, water is not the primary factor controlling protein dynamics; instead, it modifies the free-energy barriers the protein must overcome.

The difference between slaving and plasticization has important conceptual and practical consequences. In slaving, the protein and solvent dynamics share the same temperature dependence, differing only in timescale. In plasticization, the solute and solvent dynamics can have independent temperature dependences: Water influences the overall rates of protein motions without fundamentally altering their nature.

Distinguishing these mechanisms (slaving and plasticization) experimentally requires careful analysis of how solute and solvent relaxation times change with temperature across a broad range of conditions. The studies outlined in this chapter explore this question through systematic investigations of model peptide systems where both phenomena can be studied.

3 Experimental Techniques and Methodological Considerations

3.1 Broadband Dielectric Spectroscopy

Broadband dielectric spectroscopy (BDS) is the main technique used in this chapter to examine molecular dynamics. BDS offers the unique ability to measure dielectric properties and molecular dynamics across different time and length scales, especially at low temperatures where conductivity and polarization effects are reduced [36].

BDS measures the complex permittivity $\varepsilon^*(\omega) = \varepsilon'(\omega) - i\varepsilon''(\omega)$ as a function of frequency and temperature. The technique probes reorientational motions of permanent dipoles, which, in aqueous peptide solutions, include both water molecules and the peptide backbone and side chains. The large frequency range accessible to BDS instruments (10^{-2} to 10^9 Hz or beyond) enables characterization of motions spanning more than eleven decades in time. This broad frequency range was obtained by combining two dielectric techniques (10^{-1} to 10^6 Hz: a Novocontrol Alpha Analyzer; 10^6–10^9 Hz: an Agilent rf impedance analyzer 4192B).

To analyze the complex permittivity (ε^*), simultaneous fitting of the real (ε') and imaginary (ε'') parts of the permittivity was performed, using the well-known Cole–Cole equation [37].

$$\varepsilon^*(\omega) = \varepsilon_\infty + \frac{\Delta\varepsilon}{(1 + (i\omega)^\alpha)}, \tag{1}$$

where $\Delta\varepsilon$ is the dielectric strength ($\Delta\varepsilon = (\varepsilon_s - \varepsilon_\infty)$, ε_∞ and ε_s are the unrelaxed and relaxed values of the dielectric constant), τ is the relaxation time, α is the stretching parameter of each relaxation process, and $\omega = \tau^{-1} = 2\pi f$ is the angular frequency (f). In addition, at low frequencies, conductivity effects dominate, so a power-law term was added to account for them.

Additionally, when strong conductivity effects were observed in the dielectric data, the relaxation times were estimated from the temperature derivative of the real part of the permittivity [38].

$$\varepsilon'' \approx -\frac{\pi}{2}\frac{\varepsilon'(\omega)}{\log(\omega)}. \tag{2}$$

For the analysis of the temperature dependences of relaxation times, the Vogel–Fulcher–Tammann (VFT) equation was used for the α-relaxation:

$$\tau = \tau_0 \cdot \exp\left(\frac{DT_0}{T - T_0}\right), \tag{3}$$

where τ_0 is a pre-exponential factor, D is a parameter related to fragility, and T_0 is the temperature at which the relaxation time extrapolates to infinity.

For the analysis of local relaxations, the Arrhenius law was used

$$\tau = \tau_0 \cdot \exp\left(\frac{E_a}{k_B T}\right), \tag{4}$$

where E_a is the activation energy, and k_B is the Boltzmann constant.

3.2 *Complementary Techniques*

Differential scanning calorimetry measures heat flow during controlled temperature programs, enabling the determination of glass transition temperatures, crystallization events, and the width of glass transitions. Small-angle X-ray scattering probes structural inhomogeneities on length scales from nanometers to tens of nanometers. SAXS measurements were done on ε-PLL solutions at pH 10 to confirm the absence of aggregation on these length scales, establishing that the solutions are molecularly homogeneous, which is an important consideration for interpreting the dynamical data. Fourier-transform infrared spectroscopy (FTIR) probes vibrational modes of molecular bonds, and it is used here to monitor changes in the local chemical environment and conformations of *n*-lysine and ε-PLL solutions.

4 Model Systems: From L-Lysine to ε-Polylysine

Studies of protein dynamics at supercooled temperatures face a fundamental experimental challenge: water crystallization. In fact, dilute protein solutions cannot be cooled below the homogeneous nucleation temperature (around 235 K) without forming ice. Traditional approaches avoid this problem by studying hydrated protein powders rather than solutions, but such samples lack the bulk solvent present in physiologically relevant conditions [14, 15, 39, 40].

The lysine-based systems examined in this chapter provide a solution to this challenge [19, 41]. The amino acid L-lysine and its oligomers possess high solubility in water (up to 8 M), enabling the preparation of concentrated solutions that vitrify upon cooling without ice formation. This allows the investigation of the dynamics across the entire temperature range from room temperature through the glass transition temperature, with both the solvent and the solute remaining amorphous.

L-lysine possesses three hydrophilic moieties (two amino groups and one carboxylate group) together with a hydrophobic tail. This chemical structure creates diverse local environments for water molecules, mimicking the heterogeneous surface chemistry found on proteins. Although lysine and its oligomers are far simpler than functional proteins, they retain the essential features needed to explore the relationship between water and peptide dynamics: the ability to adopt multiple conformations, the formation of hydrogen bonds with water, and the presence of both hydrophilic and hydrophobic regions. These secondary structures were verified using Fourier-transform infrared spectroscopy of the amide I band region (1600–1700 cm^{-1}).

4.1 *Properties of n-Lysine Solutions*

Aqueous solutions of *n*-lysine oligomers and ε-PLL were prepared at water concentrations of 40 wt%, a level chosen to avoid crystallization while providing sufficient water to examine hydration-dependent phenomena. The pH of ε-PLL solutions was controlled to investigate conformational effects: at pH 10, ε-PLL adopts a pure β-sheet conformation, whereas at pH 7, the peptide assumes a more disordered structure that mixes β-sheets with β-turns [21].

Figure 2a shows the infrared spectra in the region 1800–1200 cm^{-1}, for *n*-lysine oligomers and ε-PLL solutions at a pH close to 10 and a water content of 40 wt%. The amide I band indicates the peptide conformation [42]. For ε-PLL, the amide I band is at 1629 cm^{-1} and indicates a β-sheet conformation [42], whereas for 10-Lys, it is at 1646 cm^{-1}, indicating an α-helix conformation. However, for the other oligomers, this band shifts to lower frequencies (see blue dotted line), near the bending mode of water at 1642 cm^{-1}, suggesting that they do not adopt this conformation. Additionally, the band at 1402 cm^{-1} for *n*-Lys is assigned to COO– stretching. This band is

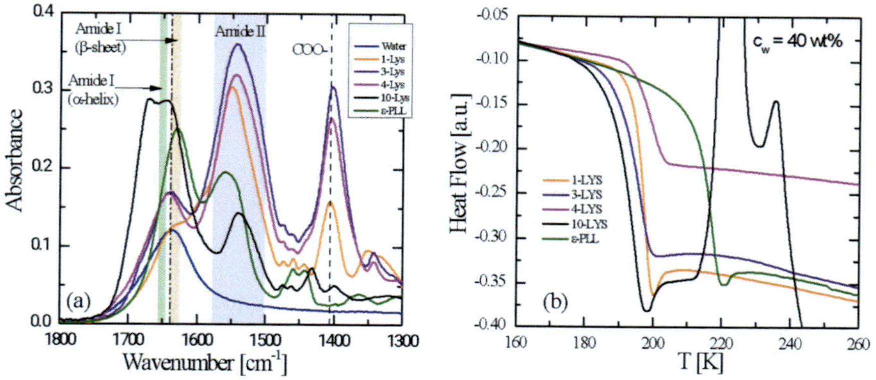

Fig. 2 **a** Infrared spectra of oligomers of L-lysine and ε-PLL aqueous solutions ($c_W = 40$ wt%). Colored regions indicate the Amide I (α-helix or β-sheet bands); black dotted and solid lines represent the bending band of water and the symmetric stretching of COO–, respectively. **b** Heat flow as a function of temperature for oligomers of L-lysine and ε-PLL. Adapted from Ref. [44]. Copyright 2019 with permission from AIP Publishing

absent in ε-PLL because increasing the chain length reduces the number of carboxyl groups.

In addition to the conformation of each solution, differential scanning calorimetry showed that the solutions vitrify when cooled (no crystallization), and on heating, they showed a narrow glass transition (see Fig. 2b). This differs from proteins in solution, which usually display a very broad glass transition over 70 K or more, a consequence of the many different motions occurring on different timescales within these complex molecules [40, 43]. The narrower glass transitions in n-lysine solutions make it easier to interpret dynamical data while still maintaining the key relationship between water and solute dynamics. Table 1 shows the general characteristics of the solutions and T_g values.

Table 1 Molecular weight (M_n), glass transition temperature (T_g), and pH of solutions on n-Lysine (c_W = 40 wt%)

Sample	M_n [g/mol]	T_g [K]	pH
1-Lys	146.2	194.0	10.4
3-Lys	402.5	190.0	10.3
4-Lys	530.7	186.0	10.4
10-Lys	1300	194.6	10.0
ε-PLL (32-Lys)	4090.0	212.5	10.3

5 Evidence for Slaving in *n*-Lysine Aqueous Solutions

5.1 Relaxation Processes in ε-PLL Solutions

Figure 3 shows both components of the complex dielectric permittivity for the ε-PLL-water solution between 160 and 265 K. A single peak is observed in ε'' (see Fig. 3b), but no additional relaxation processes can be seen at higher temperatures (see Fig. 3d). However, the real part of the permittivity (ε', see Fig. 3c) *exhibits* two additional relaxations that are only barely visible in ε''. Moreover, the derivative of ε' (ε_D', see Fig. 4a, b), which is practically identical to $\varepsilon''(\omega)$ but not affected by conductivity contributions [38], confirms the presence of these two relaxations through the corresponding steps in ε'. Isochronal measurements (Fig. 4c) provide consistent results and further demonstrate that weak relaxations can be detected via temperature scans at a fixed frequency.

To analyze the complex permittivity (ε^*) shown in Fig. 3, simultaneous fitting of the real (ε') and imaginary (ε'') parts of the permittivity was performed, using Eq. (1) to describe each relaxation process [37]. In addition, at low frequencies, conductivity

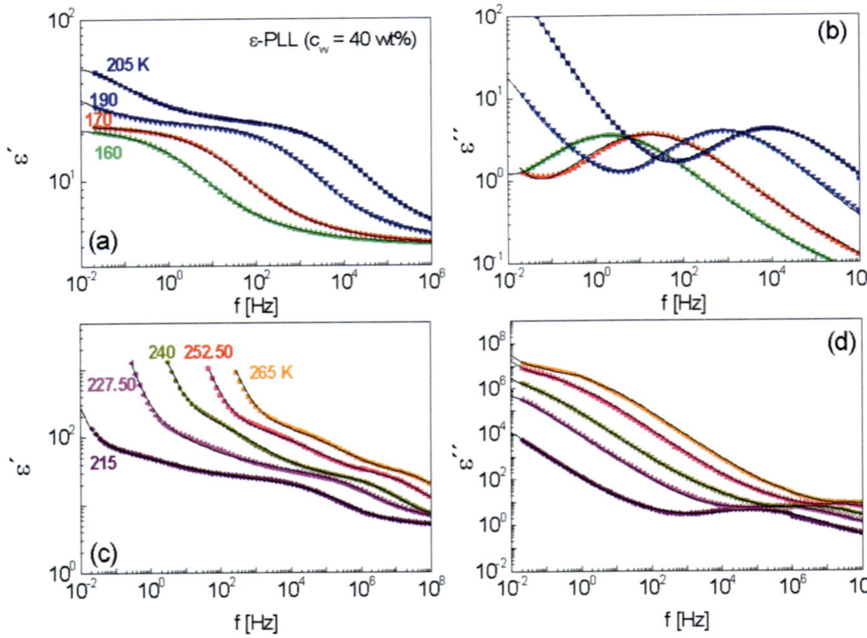

Fig. 3 Real and imaginary parts of the complex dielectric permittivity $[\varepsilon^*(\omega) = \varepsilon'(\omega) - i\varepsilon''(\omega)]$ of the ε-PLL-water solution ($c_W = 40$ wt%) at the different temperatures indicated in the figure. The solid lines through the data points represent fits to the experimental data. $\varepsilon_D''(\omega)$ was calculated from the derivative of $\varepsilon'(\omega)$. Adapted from Ref. [19]. Copyright 2016 with permission from the American Chemical Society

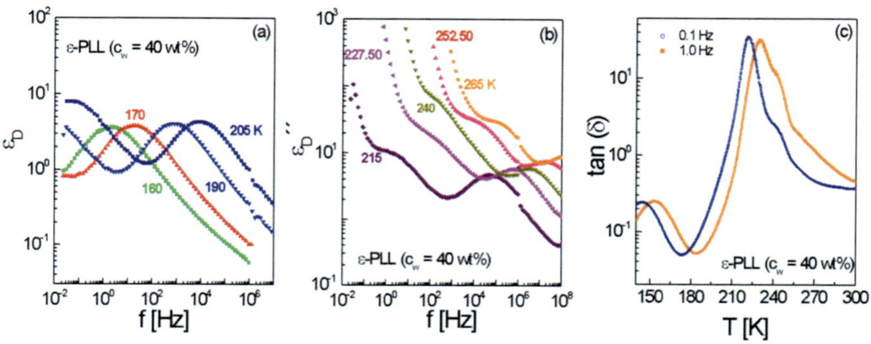

Fig. 4 **a**, **b** Imaginary part of the complex dielectric permittivity obtained through derivative analysis (ε_D'') of ε-polylysine solutions (c_W = 40 wt%) at the temperatures indicated in each plot. **c** Isochronal dielectric measurements of ε-PLL (c_W = 40 wt%). Adapted from Ref. [19]. Copyright 2016 with permission from American Chemical Society

effects dominate, so a power-law term was added to account for them. Over the entire temperature range explored, we found six relaxation processes [19] (only up to three coexist in the frequency window at each temperature), of which the three slowest are related to conductivity effects and are not discussed here.

Figure 5b shows the relaxation map derived from the fitting data in Figs. 3 and 4, which displays the three relevant relaxation processes for analyzing peptide dynamics. In addition, Fig. 5a shows the glass transition temperature for this solution. Below the glass transition temperature, we observe two processes (fast- and slow-water relaxations), whereas above T_g, the dielectric data reveal a new process (α-relaxation). The origin of these processes is interpreted based on the dependence of the relaxation strength ($\Delta\varepsilon$) on water concentration, as shown in Fig. 6. The relaxation strength increases with higher hydration levels for both slow- and fast-water relaxations, indicating that these processes arise from water molecules relaxing in ε-PLL solutions. Conversely, the dielectric strength of α-relaxation grows with increasing ε-PLL concentration, suggesting that this process is related to the reorientation of ε-PLL influenced by water. More information about the origin of these processes can be obtained indirectly via depolarized dynamic light scattering (DDLS), since this technique is insensitive to water molecule relaxation. Therefore, the finding that the two relaxations previously detected by BDS (the slow- and fast-water relaxations) could not be observed with this technique implies that these processes must be related to motions of water molecules [21].

The fast-water relaxation exhibits an Arrhenius temperature dependence with an activation energy of $E_a = 0.53$ eV, similar to that of other water-containing solutions [45–47]. Since this water relaxation occurs below the T_g and exhibits an Arrhenius temperature dependence, it should be considered more localized and labeled here as β-relaxation. This water relaxation, referred to here as "fast-water relaxation", was called ν-relaxation, water relaxation, process II, and process P1, by other authors of this book. In contrast, the relaxation times of α- and slow-water relaxations display

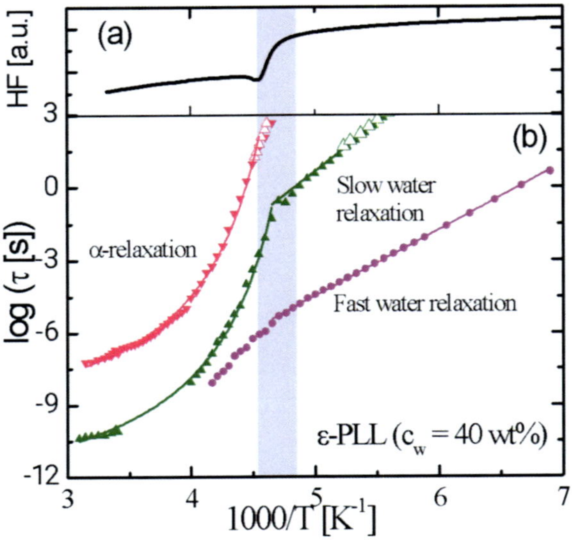

Fig. 5 Temperature dependence of the relaxation times of ε-PLL aqueous solutions at a water concentration of 40 wt%. The panel above shows the calorimetric response of the solution where a glass transition is observed when crossover occurs. Adapted from Ref. [19]. Copyright 2016 with permission from American Chemical Society

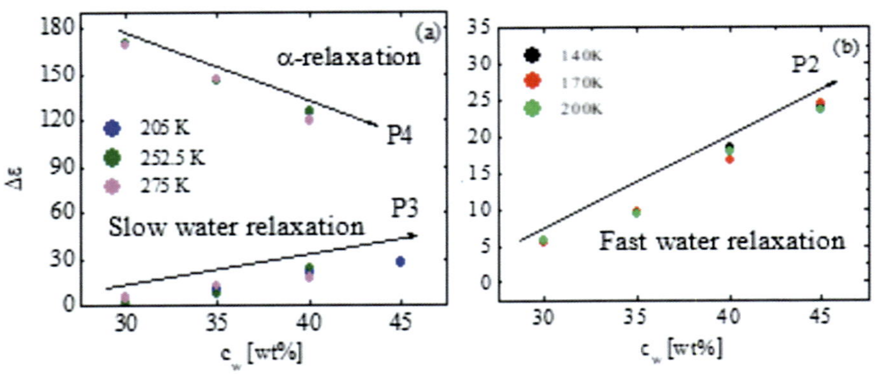

Fig. 6 Water concentration dependence at different temperatures of the relaxation strength of α-relaxation and slow- and fast-water relaxations

the same non-Arrhenius temperature dependence (although the slow-water relaxation is about four decades faster than the α-relaxation). In polymers, these characteristics strongly suggest that such relaxations result from cooperative motions, which are typical of the cooperative and glass transition-related structural (α) relaxation [36]. However, in our case, the cooperative nature of the α-relaxation arises from the cooperativity of the slow-water relaxation [19].

Figure 5a shows the calorimetric response of the ε-PLL-water solution, with the glass transition temperature regions highlighted in blue. The α-relaxation extrapolates to 100 s at a temperature near the calorimetric T_g, confirming its identification as the structural α-relaxation of the peptide. Finally, it is easy to see in Fig. 5 that in both the fast- and slow-water relaxations, a crossover from a low-temperature Arrhenius to a high-temperature non-Arrhenius behavior is produced at T_g. These relaxation processes are similar to those seen in protein solutions [14, 16, 40, 48].

5.2 Relaxation Processes in n-Lysine Solutions

Dielectric spectroscopy measurements of n-lysine aqueous solutions ($n = 1, 3, 4, 10$) at $c_w = 40$ wt% show the same dynamical behavior as ε-PLL aqueous solutions: Three dynamical processes are observed (α-, slow-water, and fast-water relaxations), as depicted in the relaxation map of Fig. 7a for the 3-lysine solution. As already analyzed for ε-PLL, at temperatures below T_g, the two water-related relaxation processes show Arrhenius temperature dependence, with activation energies near 0.5 eV. At temperatures above T_g, the slow-water relaxation and the α-relaxation follow a Vogel–Fulcher–Tammann (VFT) temperature dependence. As in ε-PLL solutions, the dielectric strength of the α-relaxation increases with peptide concentration (decreasing water content) rather than with water concentration, indicating that it arises from peptide motions rather than from water. Furthermore, for all the peptides analyzed, the relaxation time of α-relaxation extrapolates to 100 s at temperatures near the calorimetric T_g, confirming its identification as the structural α-relaxation of the peptide.

The study of water dynamics in peptide solutions has revealed a universal fast-water relaxation process with remarkably consistent features across different systems. This fast relaxation is characterized by an activation energy of about 0.5 eV, consistent with local reorientational motions of water molecules (see Fig. 7d). Its time scale remains nearly independent of the solute molecules, except for the monomer, which is slower but exhibits the same activation energy. This fast-water process has been identified as a local relaxation of water molecules, often interpreted as the Johari–Goldstein β-relaxation of water in solution. This intrinsic local relaxation occurs whenever water content is high enough to form water clusters. The universality of this process across biological and non-biological systems suggests that it reflects a fundamental aspect of water behavior in confined or interfacial environments.

Unlike the fast-water relaxation, the slow-water process (see Fig. 7c) shows a strong dependence on the solute characteristics, indicating that this water relaxation is influenced by the solute. In addition, this relaxation occurs only above a water concentration of about 20 wt% and displays a strong non-Arrhenius temperature dependence above T_g. The underlying mechanism of the slow-water relaxation involves slaving behavior, in which the collective rearrangements of water molecules at the peptide–water interface generate entropy fluctuations that promote peptide conformational dynamics. Note that this dynamics is distinct from those of bulk water and reflects a relaxation process arising from specific interactions between water molecules and peptide functional groups.

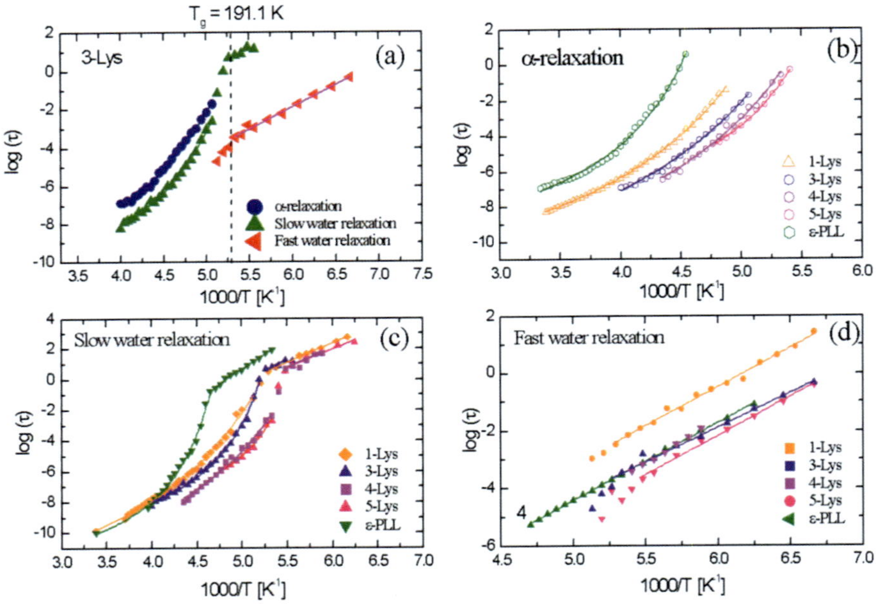

Fig. 7 **a** Temperature dependences of the relaxation times of 3-lysine ($c_W = 40$ wt%). The vertical dashed line represents the onset point of the glass transition. **b–d** Temperature dependences of the relaxation time for α-relaxation and slow and fast-water relaxation, respectively. The straight lines represent fits to the experimental data by the Arrhenius equation, whereas the full lines at temperatures above T_g represent VFT-fits to the experimental data

5.3 *The Slaving Relationship in n-Lysine and ε-PLL Aqueous Solutions*

In the context of this chapter, the key finding (shown in the relaxation map of Figs. 5 and 7) is that the processes α-relaxation and slow-water relaxation exhibit identical temperature dependences above T_g, despite differing by one to three decades in timescale. When plotted as log τ_α versus log τ_w, a linear relationship with unit slope emerges for all oligomer chain lengths examined (see Fig. 8). A slope of unity implies that the activation enthalpy for the peptide conformational change is determined by the solvent dynamics, consistent with the solvent-slaved model.

This observation provides direct evidence for slaving: the dynamics of a peptide follow the same temperature dependence as the cooperative water relaxation but occur on a consistently slower timescale. The result is similar to earlier findings for protein solutions, where the protein structural relaxation was shown to be slaved to the solvent α-relaxation. The present results demonstrate that the same fundamental coupling occurs in peptide solutions, extending the slaving concept from proteins to simpler biomolecular systems.

The slaving relationship persists even though the absolute timescales of water and peptide relaxations vary with chain length. For ε-PLL at low temperatures (below

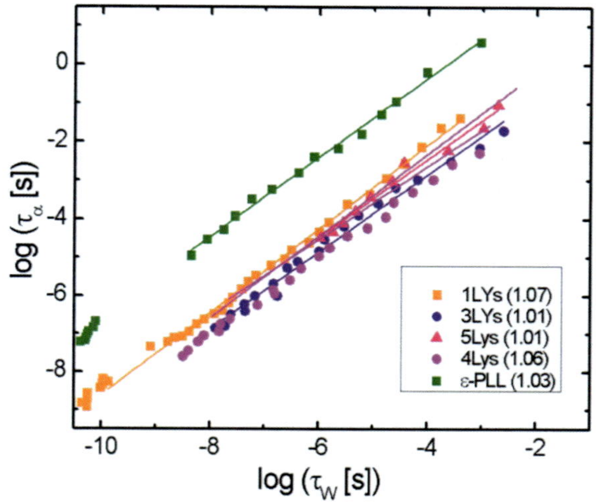

Fig. 8 Relaxation times of α-relaxation (τ_α) as a function of the relaxation times of the slow-water relaxation (τ_W). An almost perfect linear dependence, with a slope of one, is found for all the oligomers

235 K), both water and peptide relaxations are much slower than in shorter oligomers. This indicates that water in the ε-PLL solution forms larger, more bulk-like clusters than in solutions of shorter oligomers, probably because the extended β-sheets of ε-PLL cannot interpenetrate as closely with water as the smaller n-lysine molecules.

6 Slaving Versus Plasticization: Distinguishing Two Mechanisms

In this section, we demonstrate that slaving behavior is not limited to biological systems but can also occur in certain non-biological polymer-water solutions. This challenges the notion that slaving is solely a biological phenomenon and suggests that it is a broader physical principle governing polymer–solvent interactions. The key point is that slaving behavior depends not on the biological origin of the polymer but on specific physical properties, particularly the glass transition temperature of the dry polymer and how T_g changes with water content.

6.1 Concentration Dependence of the Glass Transition Temperature of Aqueous Solutions

Figure 9 shows the concentration dependence of the glass transition temperature, measured by differential scanning calorimetry (DSC), for different aqueous solutions. The T_g-concentration dependence reveals two distinct patterns. For some polymers (see Fig. 9a), including poly(vinyl methyl ether), poly(propylene glycol), and

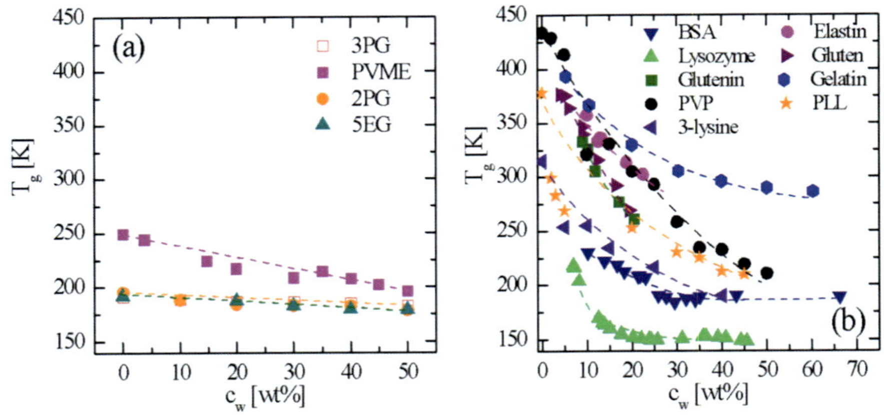

Fig. 9 Concentration dependence of the glass transition temperature (T_g) for different solutions. **a** Poly(vinyl methyl ether) [49], poly(propylene glycol) [50], oligomers of propylene glycol (nPG), and pentaethylene glycol (5EG) [51]. **b** Poly(vinyl pyrrolidone), ε-poly(lysine), dextran, 3-lysine [19], bovine serum albumin (BSA) [14], gluten [17], glutenin [52], gliadin [52], gelatin [52], elastin [53], lysozyme [54], and collagen. The dashed lines are visual guides. Adapted from Ref. [44]. Copyright 2019 with permission of AIP Publishing

oligomers of propylene glycol and ethylene glycol, T_g decreases roughly linearly as water content increases. The decrease in T_g is relatively small; for example, poly(vinyl methyl ether) shows only a 54 K decrease in T_g from the dry state to 50 wt% water.

In contrast, for other materials (see Fig. 9b), including poly(vinyl pyrrolidone) (PVP), ε-polylysine, dextran, proteins, and *n*-lysine oligomers, T_g decreases more sharply and nonlinearly as water concentration increases. PVP exhibits a 223 K decrease from the dry state to 50 wt% water. These two patterns are associated with fundamentally different dynamical behaviors, as revealed by dielectric spectroscopy in the next section.

6.2 *Dynamics of Solutions with Linear T_g Dependence*

For solutes showing linear T_g-concentration dependence (Fig. 9a), dielectric spectroscopy reveals only a single water relaxation at all concentrations. As an example, we show the dynamics [50] of 3PG with $c_w = 50$ wt% in Fig. 10a. For 3PG solutions, we study their behavior at different water contents and observe that, with the addition of water, the relaxation times of the α- and fast-water relaxations become faster, as shown in Fig. 10b, c.

For the water relaxation, the temperature dependence follows an Arrhenius behavior, with $E_a \approx 0.5$ eV at 50 wt%. As this water relaxation occurs below the T_g of the solution and exhibits an Arrhenius temperature dependence, it should be

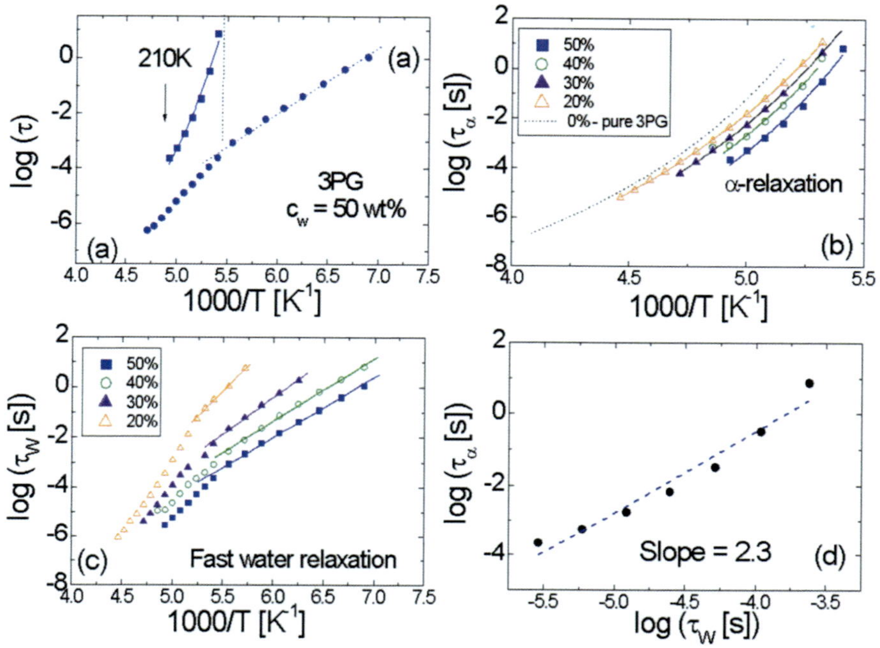

Fig. 10 **a** Temperature dependence of the relaxation times of 3PG water solutions ($c_W = 50$ wt%). **b**, **c** Dependence of relaxation time on concentration for aqueous 3PG solutions (α- and water relaxations, respectively). **d** Relaxation times of the α-process as a function of those of the fast-water relaxation. The slope is 2.3, away from the slope of one, expected for the slaving behavior. Adapted from Ref. [50]. Copyright 2006 with permission of AIP Publishing

more local than the "slow-water relaxation" previously observed for lysine-peptide solutions. By contrast, the α-relaxation follows a VFT temperature dependence and its extrapolation to 100 s agrees with the calorimetric T_g.

However, in the framework of the "slaving behavior", the temperature dependence of the water relaxation differs from that of the solute α-relaxation; when log τ_α is plotted against log τ_W, the relationship deviates substantially from linearity as seen in Fig. 10d. These characteristics are consistent with simple plasticization: Water reduces the glass transition temperature and facilitates solute motions, but the solute and water dynamics remain independent. The solute would exhibit relaxation even in the absence of water (at higher temperatures), and water merely modifies the energy barriers.

6.3 Dynamics of Solutions with Strong T_g Dependence

For solutes showing a strong, nonlinear T_g-concentration pattern, dielectric spectroscopy reveals two distinct water relaxations at sufficiently high water content (above approximately 20 wt%), with dynamics similar to those observed in lysine-based peptide solutions. Figure 11 shows the relaxation map for poly(vinyl pyrrolidone) (PVP) in comparison with the dynamics of ε-PLL at a water content of 40 wt%. For PVP water solutions, three relaxations are observed (α, slow-, and fast-water relaxations), as previously observed in lysine-based solutions. As for lysine solutions, the slow-water relaxations exhibit the same temperature dependence as the solute α-relaxation, consistent with the slaving relationship. This slow-water relaxation is absent [44] at low water concentrations and appears only above a threshold hydration level.

The α-relaxation, in Fig. 11a, follows a VFT temperature dependence, and the extrapolation to 100 s agrees with the calorimetric T_g, whereas the slow-water relaxation (Fig. 11c) follows the same temperature dependence but 5 decades faster. When plotted as log τ_α versus log τ_W, a linear relationship with unit slope emerges for both polymers (see Fig. 12). This implies that the motions of the polymer are determined by the water dynamics, consistent with the solvent-slaved model.

On the other hand, the fast-water relaxation (Fig. 11c) below T_g reflects the local motion of water molecules. Two key observations support this conclusion: (1) the

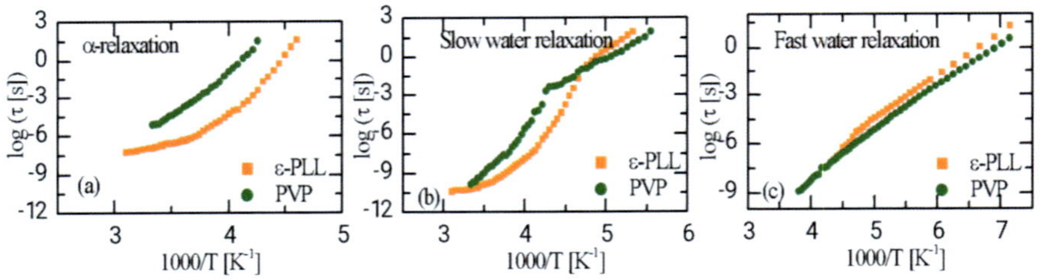

Fig. 11 Temperature dependences of the relaxation times of α- **a**, slow- **b**, and fast **c** water relaxations, respectively, for ε-PLL and PVP. The water content was 40 wt%

Fig. 12 Relaxation times of α-relaxation (τ_α) as a function of the relaxation times of the slow-water relaxation (τ_W). A linear dependence, with a slope of one, is found

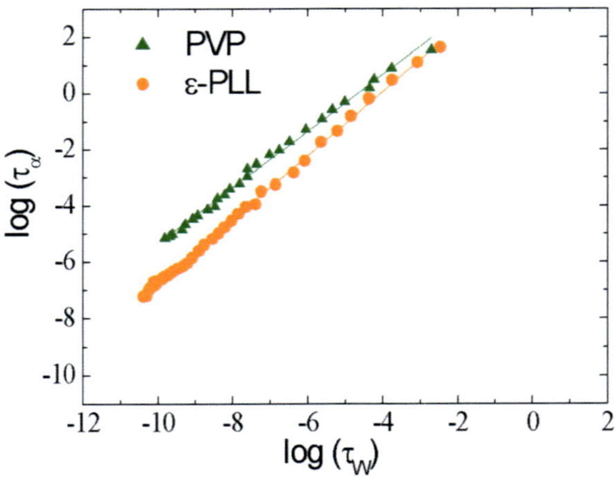

timescale of this relaxation remains independent of the specific solute, and (2) the relaxation is symmetric. As previously discussed, the fast-water relaxation can be classified as a Johari–Goldstein relaxation [55]. In contrast to the fast-water relaxation, the slow-water relaxation strongly depends on the solute type (Fig. 11b). The temperature dependence of the slow-water relaxation times also shows a crossover from non-Arrhenius behavior above T_g to Arrhenius behavior at lower temperatures, similar to that observed for the fast-water relaxation. However, both the timescale and the crossover temperature depend on the specific solute. These observations suggest that the slow-water relaxation involves cooperative interactions, or "cross-terms," between water molecules and solute molecules.

The presence of two distinct water-molecule relaxations is closely linked to the onset of slaving behavior. The fast process reflects a local, Johari–Goldstein-type relaxation of water.In contrast, the slow-water relaxation represents a collective mode involving cooperative motions of water molecules together with specific surface groups of the solute. This coupled water–solute relaxation is the process that actually drives ("slaves") the global conformational dynamics of the macromolecule: Once it becomes active above a threshold hydration level, the structural α-relaxation of the solute follows its temperature dependence with a fixed time offset. Importantly, the emergence and character of this slow and interfacial water relaxation are essentially independent of whether the solute is a protein, a peptide, or a synthetic polymer, indicating that the underlying mechanism does not rely on any particular biological structure or conformation but rather on general thermodynamic conditions and interfacial interactions in hydrated glass-forming systems.

6.4 *Criteria for Slaving in Bio- and Non-biological Aqueous Solutions*

The systematic investigation of various polymer-water systems has revealed specific criteria that determine whether slaving or plasticization dominates the behavior of aqueous solutions.

At low water concentrations (below 5 wt%), all investigated systems exhibit plasticizing effects, in which water molecules reduce the energy barriers to molecular motion without directly controlling polymer dynamics. However, when the water content is increased to 15–20 wt%, the response varies across the systems studied. Materials with high glass transition temperatures in the dry state, including poly(vinylpyrrolidone) (PVP, $T_g = 448$ K) and ε-polylysine (ε-PLL, $T_g = 493$ K), undergo substantial reductions in T_g and, in terms of dynamics, show the onset of solvent-controlled dynamics consistent with slaving behavior. By contrast, polymers with comparatively low dry-state T_g values, such as poly(vinyl methyl ether) (PVME), display only limited T_g shifts and maintain behavior characteristic of conventional plasticization.

The comparative analysis of the dynamics of biological and synthetic molecules showed that, when appropriately hydrated, they exhibit similar dynamic responses. Aqueous solutions of both ε-PLL and PVP show two distinct water relaxation processes: a rapid and localized water relaxation decoupled from the polymer motions, whereas a slower cooperative water relaxation is associated with the polymer dynamics. This slow-water process exhibits slaving behavior and reflects collective motions involving both water molecules and polymer surface groups. The slaving behavior observed across different chemical systems suggests that, under certain thermodynamic conditions, the dynamics of solutions of biological macromolecules and those of synthetic polymers are not significantly different. We believe that these observations provide a comprehensive physical basis for understanding hydration-driven macromolecular behavior.

Finally, these criteria should be considered as empirical guidelines derived from the specific set of peptide, protein, and polymer–water systems studied here. Further work will be needed to test their validity more broadly, for instance, in intrinsically disordered proteins, membrane proteins, and other macromolecular environments.

7 Implications for Protein Function and Biological Activity

7.1 *Onset of Protein Functionality*

Now we discuss the relationship between the onset of the slaving behavior and the protein functionality. It is important to note that the hydration level at which slaving behavior manifests in the dynamics, i.e., 15 or 20 wt% water, closely aligns with the minimum water content required for protein biological activity. This occurrence shows the essential coupling between water dynamics and protein motion (which supports its conformational flexibility).

At hydration levels below this water level, the proteins live in a glass-like state in which large-amplitude conformational rearrangements are strongly suppressed. However, increasing the water content above this threshold allows proteins to explore a broader conformational landscape. This enables the molecular motions associated with catalysis, ligand binding, and related biological processes.

7.2 *Conformational Landscapes and Energy Barriers*

The slaving mechanism offers a physical explanation for how proteins move through their complex conformational landscapes. By studying the slaving mechanism, it is possible to explain how proteins move in their complex conformational landscapes. When no water molecules are present, the energy barriers between conformational states are very high, trapping the protein in local minima. However, in the presence of water, these barriers are easily overcome, as water molecules provide the entropic force required. This enables transitions between the different conformational substates.

This mechanism explains why protein function is so critically dependent on hydration. The water molecules do not act as a passive medium but instead participate in conformational dynamics, providing the thermal fluctuations necessary for barrier crossing. This active role of water in protein function represents a fundamental aspect of biological systems that extends beyond simple chemical interactions.

8 Conclusions

In this chapter, lysine-based systems were used as simple models to explore how hydration water controls peptide dynamics at low temperatures. By combining broadband dielectric spectroscopy with calorimetry and infrared spectroscopy, it was possible to separately follow the motions of water and the peptide backbone and relate them to the glass transition of the solutions.

The main result is that the structural α-relaxation of n-lysine oligomers and ε-polylysine is not an independent peptide process but follows the temperature dependence of the cooperative water relaxation over many decades. This slow-water process appears only above a particular water content, shows non-Arrhenius behavior above T_g, and is strongly affected by the solute, indicating that reflects interfacial water–peptide relaxation rather than bulk water dynamics. Under these conditions, the peptide conformational fluctuations are "slaved" to the dynamics of the surrounding water.

At the same time, all peptide and polymer solutions studied exhibit a fast water relaxation with an activation energy of about 0.5 eV and an almost solute-independent timescale, which can be regarded as a local Johari–Goldstein-type process of water in the mixture. The coexistence of this universal fast process with a slower solute-specific water relaxation is a robust feature, observed for peptides as well as for synthetic polymers such as PVP.

When the analysis is extended to non-biological polymers, two distinct behaviors emerge. Systems with a high glass transition temperature in the dry state and a nonlinear decrease of T_g with water content (for instance, ε-PLL and PVP) show two water relaxations and clear slaving: The α-relaxation of the solute and the slow-water relaxation share the same temperature dependence and differ only by a constant time shift. In contrast, polymers with a lower dry T_g and a nearly linear dependence of T_g (cw) (such as PVME and 3PG) exhibit only one water relaxation that does not track the solute α-process, and their behavior is better described by simple plasticization.

Acknowledgements This work was supported by the Ministerio de Ciencia, Innovación y Universidades (MICIU), the Agencia Estatal de Investigación (AEI), and the European Regional Development Fund (FEDER, UE) under grant [PID2023–146348NB–I00]. We also acknowledge grants

PIBA_2024_1_0011 and IT1820-26 funded by the Department of Education of the Basque Government.

References

1. Kendrew JC, Bodo G, Dintzis HM, Parrish RG, Wyckoff H, Phillips DC (1958) A Three-dimensional model of the myoglobin molecule obtained by X-Ray analysis. Nature 181(4610):662–666
2. McCammon JA, Gelin BR, Karplus M (1977) Dynamics of folded proteins. Nature 267(5612):585–590
3. Karplus M, McCammon JA, Peticolas WL (1981) The internal dynamics of globular protein. Crit Rev Biochem 9(4):293–349
4. Henzler-Wildman K, Kern D (2007) Dynamic personalities of proteins. Nature 450(7172):964–972
5. Kuntz ID, Kauzmann W (1974) Hydration of proteins and polypeptides. In: Anfinsen CB, Edsall JT, Richards FM (eds) Advances in protein chemistry, vol 28. Academic Press, pp 239–345
6. Rupley JA, Careri G (1991) Protein hydration and function. In: Anfinsen CB, Richards FM, Edsall JT, David SE (eds) Advances in protein chemistry, vol 41. Academic Press, pp 37–172
7. Cheng C-Y, Song J, Franck JM, Han S (2015) Mapping out protein hydration dynamics by overhauser dynamic nuclear polarization. In: Berliner L (ed) Protein NMR: modern techniques and biomedical applications. Springer US, Boston, MA, pp 43–74
8. Wolf M, RG, Lunkenheimer P, Loidl A (2012) Relaxation dynamics of a protein solution investigated by dielectric spectroscopy. Biochim Biophys Acta 1824:723–730
9. Emmert S, Lunkenheimer P, Loidl A (2024) THz spectroscopy on the amino acids L-serine and L-cysteine. J Chem Phys 160(14)
10. Sato T, Buchner R, Fernandez S, Chiba A, Kunz W (2005) Dielectric relaxation spectroscopy of aqueous amino acid solutions: dynamics and interactions in aqueous glycine. J Mol Liq 117(1–3):93–98
11. Friesen S, Fedotova MV, Kruchinin SE, Buchner R (2021) Hydration and dynamics of l-glutamate ion in aqueous solution. Phys Chem Chem Phys 23(2):1590–1600
12. Rodriguez-Arteche I, Cerveny S, Alegria A, Colmenero J (2012) Dielectric spectroscopy in the GHz region on fully hydrated zwitterionic amino acids. Phys Chem Chem Phys 14:11352–11362
13. Swenson J, Jansson H, Bergman R (2006) Relaxation processes in supercooled confined water and implications for protein dynamics. Phys Rev Lett 96:247802
14. Ngai KL, Capaccioli S, Shinyashiki N (2008) The protein "glass" transition and the role of the solvent. J Phys Chem B 112(12):3826–3832
15. Shinyashiki N, Yamamoto W, Yokoyama A, Yoshinari T, Yagihara S, Kita R, Ngai KL, Capaccioli S (2009) Glass transitions in aqueous solutions of protein (bovine serum albumin). J Phys Chem B 113:14448–14456
16. Panagopoulou A, Kyritsis A, Shinyashiki N, Pissis P (2012) Protein and water dynamics in bovine serum albumin-water mixtures over wide ranges of composition. J Phys Chem B 116:4593–4602
17. Panagopoulou A, Kyritsis A, Sabater i Serra R, Gómez Ribelles JL, Shinyashiki N, Pissis P (2011) Glass transition and dynamics in BSA–water mixtures over wide ranges of composition studied by thermal and dielectric techniques. Biochim Biophys Acta (BBA) Proteins Proteomics 1814(12):1984–1996
18. Sasaki K, Kita R, Shinyashiki N, Yagihara S (2014) Glass transition of partially crystallized gelatin-water mixtures studied by broadband dielectric spectroscopy. J Chem Phys 140(12):124506 (7 pp.)

19. Cerveny S, Combarro-Palacios I, Swenson J (2016) Evidence of coupling between the motions of water and peptides. J Phys Chem Lett 7(20):4093–4098
20. Weigler M, Combarro-Palacios I, Cerveny S, Vogel M (2020) On the microscopic origins of relaxation processes in aqueous peptide solutions undergoing a glass transition. J Chem Phys 152(23):234503–234503
21. Melillo JH, Gabriel JP, Pabst F, Blochowicz T, Cerveny S (2021) Dynamics of aqueous peptide solutions in folded and disordered states examined by dynamic light scattering and dielectric spectroscopy. Phys Chem Chem Phys 23(28):15020–15029
22. Frauenfelder H, McMahon BH (2001) The energy landscape. In: Rigler R, Orrit M, Basché T (eds) Single molecule spectroscopy: nobel conference lectures. Springer, Berlin, Heidelberg, pp 257–276
23. Frauenfelder H, Sligar SG, Wolynes PG (1991) The energy landscapes and motions of proteins. Science 254(5038):1598–1603
24. Levy Y, Onuchic JN (2006) Water mediation in protein folding and molecular recognition. Ann Rev Biophys 35:389–415
25. Fenimore PW, Frauenfelder H, McMahon BH, Parak FG (2002) Slaving: solvent fluctuations dominate protein dynamics and functions. Proc Natl Acad Sci USA 99(25):16047–16051
26. Frauenfelder H, Chen G, Berendzen J, Fenimore PW, Jansson H, McMahon BH, Stroe IR, Swenson J, Young RD (2009) A unified model of protein dynamics. Proc Natl Acad Sci USA 106:5129–5134
27. Frauenfelder H, Fenimore PW, McMahon BH (2002) Hydration, slaving and protein function. Biophys Chem 98(1–2):35–48
28. Frauenfelder H, McMahon BH, Fenimore PW (2003) Myoglobin: the hydrogen atom of biology and a paradigm of complexity. Proc Natl Acad Sci USA 100(15):8615–8617
29. Fenimore PW, Frauenfelder H, McMahon BH, Young RD (2004) Bulk-solvent and hydration-shell fluctuations, similar to alpha- and beta-fluctuations in glasses, control protein motions and functions. Proc Natl Acad Sci USA 101:14408–14413
30. Frauenfelder H, Fenimore PW, Chen G, McMahon BH (2006) Protein folding is slaved to solvent motions. Proc Natl Acad Sci USA 103:15469–15472
31. Chen G, Fenimore PW, Frauenfelder H, Mezei F, Swenson J, Young RD (2008) Protein fluctuations explored by inelastic neutron scattering and dielectric relaxation spectroscopy. Philos Mag 88(33–35):3877–3883
32. Caliskan G, Mechtani D, Roh JH, Kisliuk A, Sokolov AP, Azzam S, Cicerone MT, Lin-Gibson S, Peral I (2004) Protein and solvent dynamics: How strongly are they coupled? J Chem Phys 121(4):1978–1983
33. Khodadadi S, Roh JH, Kisliuk A, Mamontov E, Tyagi M, Woodson SA, Briber RM, Sokolov AP (2010) Dynamics of biological macromolecules: not a simple slaving by hydration water. Biophys J 98:1321–1326
34. Doster W (2006) Dynamical structural distributions in proteins. Physica B 385–386:831–834
35. Khodadadi S, Sokolov AP (2017) Atomistic details of protein dynamics and the role of hydration water. Biochim et Biophys Acta (BBA) General Subjects 1861(1, Part B):3546–3552
36. Kremer F, Schönhals A (2003) Broadband dielectric spectroscopy. Springer, Berlin
37. Cole KS, Cole RH (1942) Dispersion and absorption in dielectrics—direct current characteristics. J Chem Phys 10:98–105
38. Wübbenhorst M, van Turnhout J (2002) Analysis of complex dielectric spectra. I. One-dimensional derivative techniques and three-dimensional modelling. J Non-Cryst Solids 305:40–49
39. Khodadadi S, Sokolov AP (2015) Protein dynamics: from rattling in a cage to structural relaxation. Soft Matter 11:4984–4998
40. Khodadadi S, Malkovskiy A, Kisliuk A, Sokolov AP (2010) A broad glass transition in hydrated proteins. BBA-Proteins Proteomics 1804:15–19
41. Cerveny S, Swenson J (2014) Dynamics of supercooled water in a biological model system of the amino acid l-lysine. Phys Chem Chem Phys 16(40):22382–22390

42. Mauerer A, Lee G (2006) Changes in the amide I FT-IR bands of poly-l-lysine on spray-drying from α-helix, β-sheet or random coil conformations. Eur J Pharm Biopharm 62:131–142
43. Jansson H, Swenson J (2010) The protein glass transition as measured by dielectric spectroscopy and differential scanning calorimetry. BBA-Proteins Proteomics 1804:20–26
44. Combarro Palacios I, Olsson C, Kamma-Lorger CS, Swenson J, Cerveny S (2019) Motions of water and solutes—Slaving versus plasticization phenomena. J Chem Phys 150(12):124902
45. Cerveny S, Colmenero J, Alegria A (2007) Dielectric properties of water in amorphous mixtures of polymers and other glass forming materials. J Non-Cryst Solids 353(47–51):4523–4527
46. Cerveny S, Colmenero J, Alegria A (2007) Dynamics of confined water in different environments. Eur Phys J Spec Topics 141:49–52
47. Cerveny S, Alegria A, Colmenero J (2008) Universal features of water dynamics in solutions of hydrophilic polymers, biopolymers, and small glass-forming materials. Phys Rev E Stat Nonlinear Soft Matter Phys 77:031803
48. Jansson H, Bergman R, Swenson J (2011) Role of solvent for the dynamics and the glass transition of proteins. J Phys Chem B 115:4099–4109
49. Cerveny S, Colmenero J, Alegría A (2005) Dielectric investigation of the low-temperature water dynamics in the poly(vinyl methyl ether)/H_2O system. Macromolecules 38(16):7056–7063
50. Cerveny S, Schwartz GA, Alegria A, Bergman R, Swenson J (2006) Water dynamics in n-propylene glycol aqueous solutions. J Chem Phys 124(19):194501
51. Sudo S, Tsubotani S, Shimomura M, Shinyashiki N, Yagihara S (2004) Dielectric study of the α and β processes in supercooled ethylene glycol oligomer–water mixtures. J Chem Phys 121(15):7332–7340
52. Kalichevsky MT, Jaroszkiewicz EM, Blanshard JMV (1992) Glass transition of gluten. 1: Gluten and gluten—sugar mixtures. Int J Biol Macromol 14(5):257–266
53. Batzer H, Kreibich UT (1981) Influence of water on thermal transitions in natural polymers and synthetic polyamides. Polym Bull 5(11):585–590
54. Miyazaki Y, Matsuo T, Suga H (2000) Low-temperature heat capacity and glassy behavior of lysozyme crystal. J Phys Chem B 104(33):8044–8052
55. Johari GP, Goldstein M (1970) Viscous liquids and glass transition. 2. Secondary relaxations in glasses of rigid molecules. J Chem Phys 53(6):2372–2388

Atomistic Mechanisms of Protein Dynamics and the Essential Role of Hydration Water

Silvina Cerveny, Jan Swenson, and Alexei P. Sokolov

Abstract This chapter reviews the current knowledge on the dynamics of biological macromolecules, with emphasis on globular proteins. Based on results from different experimental techniques (mainly neutron scattering and broadband dielectric spectroscopy) and supported by molecular dynamics simulations, we show how the protein motions can be organized in a hierarchy extending from picosecond-scale local fluctuations to microsecond-scale structural rearrangements. Within the broader framework of soft-matter physics, this chapter examines dynamical features, such as the boson peak and secondary relaxations, as well as constraints imposed by the protein fold. Particular attention is given to the active role of hydration water and to the coupled protein-solvent dynamics, including the sub-nanosecond localized diffusion process that controls slower protein motions. Moreover, we present evidence that microsecond backbone relaxation is the primary structural relaxation in hydrated proteins and is directly linked to their glass transition behavior. By combining experimental and simulation perspectives, the chapter develops an atomistic picture of the motions underlying biological function, highlighting the contrast of the dynamical behavior of proteins with that of DNA and RNA.

Keywords Protein dynamics · Solvent dynamics · Hydration water · Glass transition temperature · Coupled motions · Dielectric spectroscopy · Neutron scattering · Molecular dynamic simulations

---

S. Cerveny
Centro de Física de Materiales (CFM, CSIC-UPV/EHU), San Sebastián, Spain

Donostia International Physics Center (DIPC), San Sebastián, Spain

J. Swenson
Department of Physics and Astronomy, Chalmers University of Technology, Göteborg, Sweden

A. P. Sokolov (✉)
Department of Chemistry, University of Tennessee, Knoxville, TN, USA
e-mail: sokolov@utk.edu

S. Cerveny and J. Swenson (eds.), *Dynamics of Biomolecules Studied by Dielectric Spectroscopy*, Advances in Dielectrics, https://doi.org/10.1007/978-3-032-30588-6_4

1 Introduction

Biological macromolecules such as proteins, DNA, and RNA exhibit a rich spectrum of internal motions that are essential for their biological activity [1–6]. In contrast to an earlier view of macromolecules as rigid structures, it is now well established that they populate ensembles of conformations and undergo continuous thermally driven fluctuations over a broad range of time and length scales [1–5]. High-resolution structural techniques have provided detailed static descriptions of many proteins and nucleic acids. However, an understanding of their internal dynamics and how these dynamics are related to catalysis, molecular recognition, transport, and regulation remains incomplete [2–4, 7–9].

In this context, it is helpful to analyze the dynamics of proteins using concepts developed for soft condensed-matter systems. Biological macromolecules share several generic features with synthetic polymers, glass-forming liquids, and other soft materials, including rugged free-energy landscapes, dynamic heterogeneity, and sensitivity to external control parameters such as temperature, pressure, and hydration [5, 6, 10–12]. In soft condensed-matter systems, the potential energy surface is characterized by large metastable basins separated by barriers, giving rise to a hierarchy of relaxation processes with distinct characteristic times and length scales. However, globular proteins differ from soft materials because they adopt folded conformation that are stabilized by specific non-covalent interactions. These interactions impose additional steric and energetic constraints, resulting in characteristic dynamical modes that simple polymers do not have.

The internal dynamics of folded proteins can be described as a hierarchical energy landscape comprising different types of motions: fast, small-amplitude vibrations; localized side-chain and backbone motions; larger conformational transitions involving loop or domain rearrangements; and very slow structural relaxations [3, 5, 6, 13–15]. These processes occur over timescales ranging from picoseconds to microseconds and even longer times. Moreover, the surrounding environment of the folded protein strongly influences the rates and amplitudes of these processes. In particular, the solvent (i.e., water) plays a central role in shaping the equilibrium conformational distributions and the dynamical properties. Different experimental and computational studies have shown that hydration water is a critical component of the folded protein, with motions in the first hydration shell being strongly coupled to protein fluctuations over a wide range of timescales [5, 6, 14–18]. Changes of hydration levels and solvent composition affect the lubrication experienced by protein segments during motion. The effective energy barriers separating conformational sub-states are also affected, thereby modulating both local and global dynamics.

The aim of this chapter is to summarize and organize current knowledge on intra-molecular protein dynamics, with particular emphasis on atomistic mechanisms and on the role of hydration water in controlling different classes of motion. The discussion is presented in terms of concepts borrowed from soft-matter physics, but it also highlights those features that distinguish proteins and other biological macromolecules from simpler model systems. By combining insights from neutron scattering (NS), broadband dielectric spectroscopy (BDS), nuclear magnetic resonance (NMR), and molecular dynamics (MD) simulations, a coherent hierarchical picture emerges that relates fast picosecond motions, localized sub-nanosecond processes, nanosecond conformational jumps and microsecond structural relaxation to specific regions of the protein energy landscape and to the dynamical behavior of the surrounding solvent [5, 6, 14–18].

2 Methodology

The investigation of protein dynamics over a broad range of time and length scales requires a multi-modal approach that combines experimental and computational methods. Therefore, this chapter is based on both NS and BDS as well as MD simulations.

2.1 *Neutron Scattering*

Quasielastic neutron scattering (QENS) is a unique and powerful tool for probing dynamics on the atomic scale (Ångströms) and on picosecond-to-nanosecond timescales (10^{-12} to 10^{-9} s). In particular, an incoherent scattering signal arises from the exceptionally high incoherent scattering cross-section of the hydrogen atom compared to other elements (including deuterium). Since proteins are hydrogen-rich, the scattering signal is dominated by incoherent scattering from protein protons, which reflects their self-motion.

Another important advantage is the possibility of using contrast variation. Using deuterium oxide (D_2O) as the solvent, it is possible to mask the signal of water (H_2O), because the total scattering cross-section of deuterium is substantially smaller. This allows the selective observation of the internal motions of proteins. Conversely, deuterated proteins in H_2O allow for the isolation of hydration water dynamics.

QENS measures the broadening of the elastic peak, which is directly related to the timescale of diffusive or relaxational motions. Inelastic scattering is used to probe vibrational modes, such as the boson peak. The energy of the cold neutrons used in these experiments (in the meV range) is non-destructive, making them ideal for delicate biological samples.

2.2 *Broadband Dielectric Spectroscopy*

While neutron scattering provides spatial resolution, it covers very limited time window, and also accuracy is often low due to limited amounts of scattered neutrons. Therefore, using BDS, an extraordinary frequency range of more than 15 orders of magnitude (from 10^{-6} to 10^{12} Hz, i.e., from hours to ps in time) can be covered. BDS measures the complex permittivity of a sample, and it is sensitive to the reorientation of permanent dipoles and the translation of charge carriers. In protein solutions and hydrated powders, BDS can detect different relaxation processes arising from the solvent, protein side chains, and the global protein dipole moment. Using BDS, very high-precision measurements of relaxation times and activation energies can be obtained, though it lacks the microscopic spatial information provided by neutron scattering techniques.

2.3 *Molecular Dynamics Simulations*

MD simulations provide atomistic trajectories that can be directly compared with experimental observables to validate physical models and allow direct visualization of atomic motions, distinguishing between side-chain rotations, backbone fluctuations, and solvent displacements [19–22]. However, these results depend on the force fields and the type of approximation used; therefore, it is essential to combine MD simulations with experimental results to validate them. For instance, by calculating the dynamic structure factor $S(Q, \omega)$ or the intermediate scattering function $I(Q, t)$ from simulated trajectories, computational results can be directly validated against NS data. This cooperation between MD and NS is critical for interpreting the complex, averaged signals obtained from experiments.

3 Dynamics of Soft Materials

As already noted above, the dynamics of biological macromolecules covers a huge temporal range. To organize this complexity, in the following we adopt the classification scheme used in soft-matter physics (in this case, we have used polyisoprene as an example), distinguishing between fast vibrations, picosecond and secondary relaxations, and primary structural relaxations [23] (see Fig. 1).

In the following, we briefly explain each relaxation process shown in Fig. 1, from the slowest to the fastest dynamics. This will serve as a reference for the later comparison with protein dynamics in the next section.

The slowest process (called "chain" in Fig. 1) represents end-to-end relaxation, i.e., the slowest chain relaxation mode associated with the motion of the entire macromolecular chain [23]. This process determines the long-time diffusion of the polymer. It is well described by classical Rouse and reptation models [24], with a strong dependence on chain length (molecular weight).

The main structural relaxation (α-relaxation, called "segmental" in Fig. 1) is the key process that governs the macroscopic viscosity and the transition from a glassy (solid) state to a flowing liquid state. It is a cooperative motion involving many structural units. In polymers, this α-process is often called segmental relaxation,

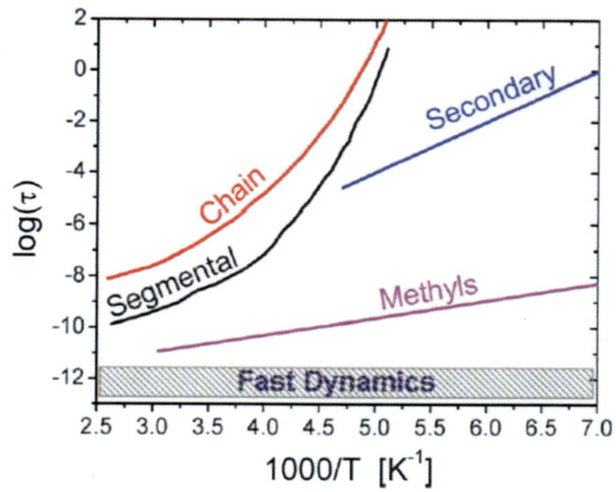

Fig. 1 Traditional relaxation map of a soft material (poly(isoprene) [23]). It depicts the temperature dependence of chain relaxation, segmental relaxation (α), secondary relaxation (β), methyl rotation, and fast picosecond relaxation. This figure establishes the baseline hierarchy of processes that applies to proteins. This figure is reproduced from Ref. [12]. Copyright (2015) with permission from RSC (Royal Society of Chemistry)

since it corresponds to the motion of chain segments. The relaxation time exhibits a non-Arrhenius temperature dependence, described by the Vogel–Fulcher–Tammann (VFT) law, and its sharp slowing down on cooling is linked to growing cooperativity and dynamic heterogeneity.

Faster than the α-relaxation, we can find the secondary relaxations and methyl group dynamics [25–29] in Fig. 1, which are more local and chemically specific. These relaxations can involve intramolecular motions and side-group rotations (for instance, the methyl group rotation), or the more general Johari–Goldstein β-relaxation. These processes exhibit Arrhenius temperature dependences with well-defined activation energies. They are usually easiest to separate from the main structural relaxation at temperatures close to the glass transition temperature, T_g.

Finally, all the fastest processes are shown in Fig. 1 as "fast dynamics". These processes include low-frequency collective vibrations (the boson peak) [30, 31] and fast picosecond fluctuations [32, 33]. The boson peak corresponds to an excess of vibrational modes compared with the Debye model expected for a crystal and is a generic feature of disordered solids. The picosecond motions are usually interpreted as "rattling" of a structural unit inside a cage formed by its neighbors and are considered precursors of the slower structural relaxation [34, 35]. Their times-scale depends only weakly on temperature, indicating very low energy barriers [32, 33].

4 Dynamics of Biological Macromolecules

As already mentioned in the introduction, there are similarities between how biological macromolecules, such as proteins, DNA, and RNA, move and how other soft materials behave. At the same time, proteins also show important differences because they have a globular structure, which is not typically found in soft matter. One common way to describe biomolecular motions is through a hierarchical energy landscape, in which different types of motions occur on different levels [2–4, 6]. This idea is similar to energy landscape models used for other soft materials.

Therefore, in this section, we use the same classification of dynamics commonly applied to soft materials, following the example of polymers discussed in Sect. 3. The studies presented here span a wide range of temperatures, well beyond normal biological conditions. This approach is widely used in physics because it helps to separate different types of motions and distinguish energetic from entropic effects.

4.1 Fast Dynamics: The Boson Peak and Cage Rattling

On the picosecond timescale ($t < 1$ ps) and at frequencies around 1 THz, all disordered materials, including proteins, DNA, and RNA, exhibit a boson peak [30, 31]. This appears in the dynamic structure factor $S(Q, \omega)$ as an excess of inelastic scattering intensity relative to the Debye model prediction for acoustic phonons (where the density of states $g(\omega) \propto \omega^2$). As mentioned above, this feature is universal to disordered solids and arises from heterogeneity in elastic constants [36–39] or the presence of soft vibrational modes [40].

Together with the boson peak, we can find the anharmonic fast picosecond fluctuations, often described as "rattling in a cage." Residues are transiently trapped by steric barriers from neighboring residues and by the solvent cage [41]. The amplitude of these motions, measured as the mean-squared displacement (MSD), is typically small, $\langle r^2 \rangle \approx 0.15 - 0.4$ Å^2 at ambient conditions [32, 33].

A crucial finding is the strong dependence of these fast dynamics on the solvent environment. As shown in Fig. 2, the amplitude of the boson peak and the quasielastic broadening (fast relaxation) track the solvent dynamics. In high-viscosity solvents such as glycerol or trehalose, these rapid protein fluctuations are significantly suppressed compared to water. This suppression is believed to be a key mechanism in the biopreservation of proteins, effectively arresting the degradative pathways that require structural flexibility [42].

4.2 Methyl Group Dynamics: The Internal Plasticizers

Proteins can contain a high density of methyl groups ($-CH_3$), which are essential for forming their hydrophobic core. The dynamics of methyl groups are unique and distinct from those of the rest of the protein matrix. The methyl group rotations, studied mainly by NMR and MD simulations, are rapid (picosecond-to-nanosecond scale), governed by relatively low energy barriers ($E_a \approx 10$–20 kJ/mol), and remain fast even at low hydration and low temperatures [25–29]. Furthermore, the methyl group dynamics is insensitivity to hydration. Unlike backbone motions and other types of slower protein motions, which freeze upon dehydration, methyl rotations remain active even in dry powders and at cryogenic temperatures ($T < 100$ K) [23,

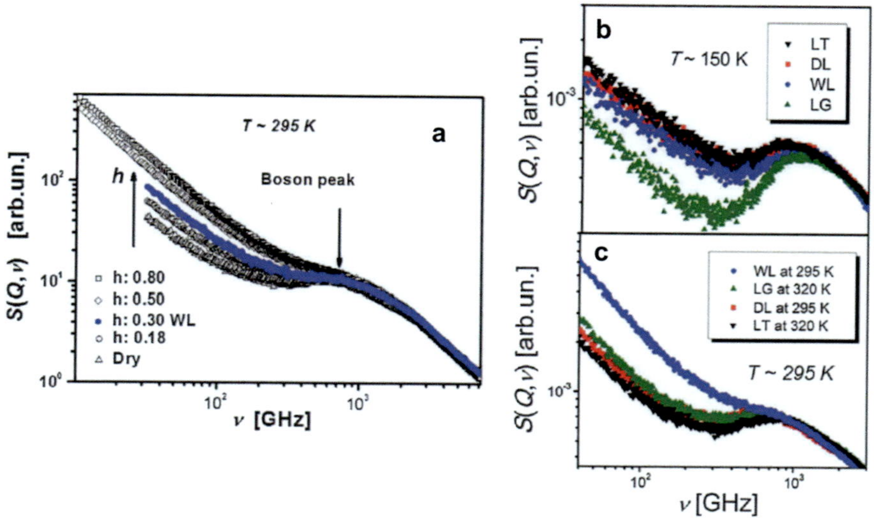

Fig. 2 Dynamic structure factor $S(Q, \upsilon)$ of lysozyme. **a** Effect of hydration: Boson peak shifts and quasielastic broadening increases with hydration. H presents g of water per g of protein. **b**, **c** Effect of solvent (glycerol, trehalose) and temperature on the fast relaxation process in lysozyme. WL—lysozyme in water; DL—dry (dehydrated) lysozyme; LG—lysozyme in glycerol and LT—lysozyme in trehalose. The strong coupling between fast protein dynamics and solvent dynamics is evident from the similarity of temperature dependence across different samples. This figure is reproduced from Ref. [12]. Copyright (2015) with permission from RSC (Royal Society of Chemistry)

24]. In fact, methyl group dynamics may facilitate these other types of slower protein motions, although it is important to note that also hydration water is needed for these other protein motions to occur. By maintaining entropic freedom within the densely packed core, the methyl groups prevent the protein from becoming a completely brittle glass, even under extreme environmental stress. It has been observed in lysozyme and myoglobin that the fastest methyl groups are often located near active sites [41, 43, 44], suggesting a functional role in providing local flexibility together with water.

4.3 Comparison Between Nucleic Acids

The methyl groups differentiate proteins from DNA and RNA. It is well-known that RNA lacks methyl groups entirely in its primary structure, and DNA possesses them only on the thymine bases (see Fig. 3c). Consequently, DNA and RNA lack the internal plasticization mechanism discussed above.

This contributes to their greater dependence on hydration for flexibility and to their more rigid behavior at low temperatures. The onset of methyl group rotation at

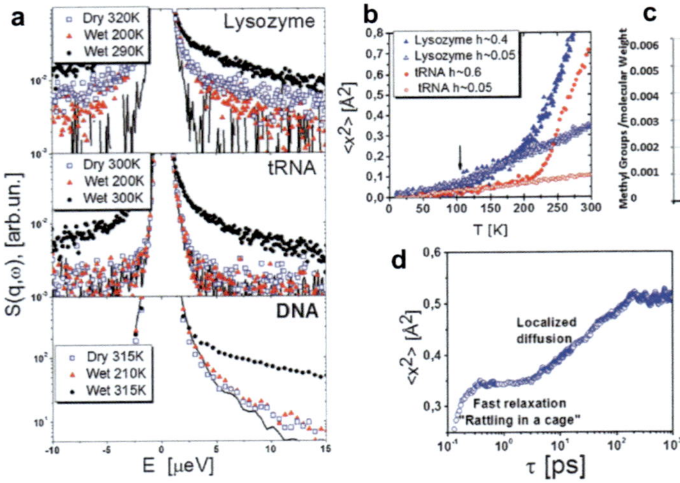

Fig. 3 **a** Spectra of Lysozyme show significant quasielastic contribution even at low temperatures and hydrations in contrast to spectra of tRNA/DNA. This contribution comes from methyl group rotation in lysozyme. **b** MSD shows onset of methyl rotation at T ~ 100 K (arrow) in proteins but not in nucleic acids. **c** Methyl group density comparison across biological macromolecules. **d** MSD time trace from MD simulations showing "caging" at short times followed by "localized diffusion" at longer times with saturation indicating confinement. This figure is reproduced from Ref. [12]. Copyright (2015) with permission from RSC (Royal Society of Chemistry)

$T \approx 100$ K is visible in the MSD of most proteins but absent in tRNA. As illustrated in Fig. 3.

4.4 *Sub-nanosecond Process: Coupled Protein-Solvent Relaxation*

Neutron scattering experiments on hydrated proteins have revealed a highly stretched relaxation process with characteristic times of 10–30 ps at room temperature [6, 8, 12, 45–48]. Using hydrogen/deuterium contrast variation, it is possible to separate the dynamics of the protein from the motions in the surrounding hydration water [41, 49–51]. Both components exhibit dynamical processes occurring on similar timescales with similar temperature dependences, indicating strong coupling of these processes (see Fig. 4a).

However, an analysis of the wave-vector (Q) dependence of these relaxations reveals a significant difference. Hydration water exhibits diffusive translational motion, though slowed by approximately 2–4 times relative to bulk water [26, 58]. In contrast, these protein motions are localized rather than diffusive. This is evident from the MSD obtained from molecular dynamics simulations: the displacement increases slowly but saturates at timescales of approximately 100–200 ps or longer

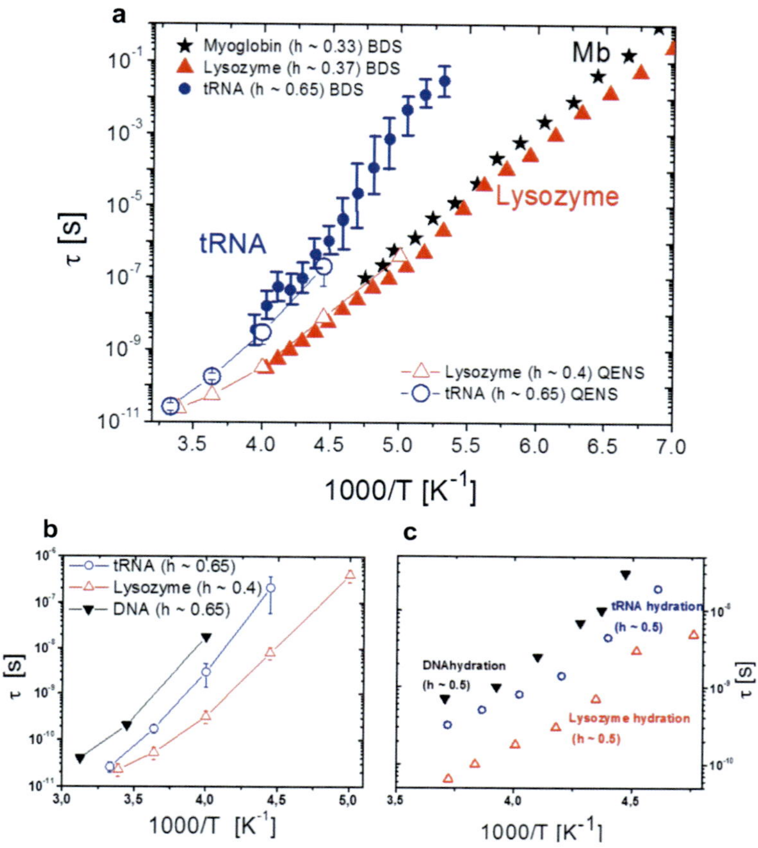

Fig. 4 **a** Temperature dependence of characteristic biomolecular relaxation times of hydrated lysozyme [49, 52], Myoglobin [53] and tRNA [49, 52], as obtained by BDS (closed symbols) and quasielastic neutron scattering (QENS) (open symbols. All relaxation times were obtained from the susceptibility maximum (see Fig. 5) or dielectric loss maximum, $(2\upsilon_{max})^{-1}$). **b** Temperature dependence of characteristic biomolecular relaxation times from QENS spectra of tRNA [49], lysozyme [49] and DNA [54] hydrated with D_2O; **c** Characteristic relaxation times of hydration water of RNA [55], lysozyme [56], and DNA [57], also obtained from QENS measurements. This figure is reproduced from Ref. [12]. Copyright (2015) with permission from RSC (Royal Society of Chemistry)

(Fig. 3d). Analysis of atomic trajectories from simulations reveals Brownian-like motion confined to a spatial region with a radius of roughly [8] 1–1.5 Å. For this reason, the process is referred to as "localized diffusion" [8].

The relaxation spectra are often analyzed in a susceptibility representation, analogous to dielectric or mechanical loss spectra, which offers several advantages over the traditional dynamic structure factor. The susceptibility spectra of this protein process exhibit a symmetric, strongly stretched shape, well described by a Cole–Cole distribution function with a stretching parameter $\alpha \approx 0.25$–0.4 (see Fig. 5) [41,

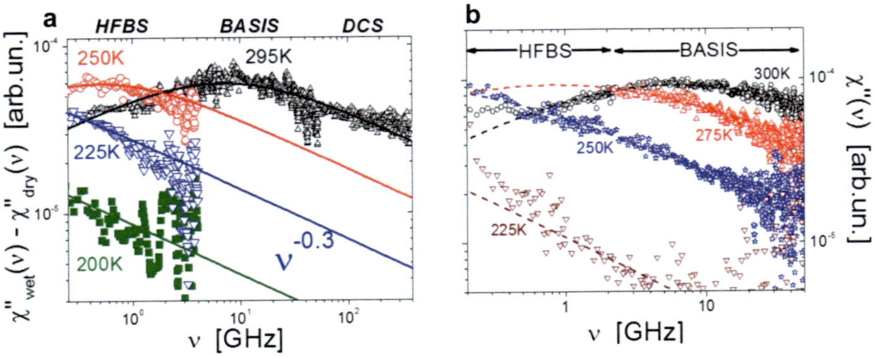

Fig. 5 **a** Susceptibility spectra of the localized motion in hydrated lysozyme ($h \sim 0.35$) at different temperatures after correction for methyl group rotation, as obtained by combining data from the three neutron scattering spectrometers HFBS, BASIS, and DCS [48]. **b** Corresponding susceptibility spectra of the localized motion in hydrated tRNA ($h \sim 0.65$) at different temperatures [49]. The experimental data (symbols) have been fitted by Cole–Cole distribution functions (lines). This figure is reproduced from Ref. [12]. Copyright (2015) with permission from RSC (Royal Society of Chemistry)

59]. Such extreme stretching indicates significant dynamic heterogeneity, reflecting differences in the chemical structure and local environment of different residues within the protein. The appearance and characteristic time of this process depend strongly on protein hydration [41] and, more generally, on the choice of solvent [60, 61].

The temperature dependence of this localized diffusion is slightly non-Arrhenius (Fig. 4). It is associated with a sharp increase in the MSD near 200–230 K for hydrated proteins [25, 41, 45, 47, 62–64]. This experimental observation is often misleadingly denoted as a "dynamic transition". However, it appears due to the timescale of the dynamic process becoming comparable to the resolution of the neutron scattering spectrometer. Neutron scattering and simulations estimate the localization length of this motion to be approximately 1–3 Å [11, 65]. The symmetric relaxation spectrum and the localized nature of the motion suggest that this process is a secondary relaxation in protein dynamics. However, it appears to involve most, if not all, of the residues [66]. Molecular dynamics simulations show that both surface and core residues participate in this motion. According to various simulation studies [9], the process predominantly involves side-group motions, with relatively minor contributions from backbone fluctuations. Detailed analysis of lysozyme and green fluorescent protein suggests that the process can be described as overdamped diffusive motion in a harmonic potential well [67]. In this picture, the motions are not jumps between well-defined conformational states but rather relatively continuous diffusion within a restricted spatial region, characterized by a particular friction coefficient.

This sub-nanosecond process also appears in dielectric relaxation spectra and has traditionally been attributed to the dynamics of hydration water [68–70]. However, NMR, neutron scattering, and simulations all provide evidence for protein dynamics on the same 10–50 ps timescale [8, 17, 41, 47, 65, 71]. The coincidence of characteristic timescales and the similarity of the temperature dependencies of hydration water and protein dynamics suggest strong coupling between the two processes (see Fig. 4a). Thus, the translational motion of hydration water molecules appears tightly coupled to the localized motion of the protein. This coupling has been directly demonstrated in MD simulations in which the translational motion of water molecules was artificially restricted; such restrictions strongly suppressed protein relaxation on the sub-nanosecond timescale [22, 37].

A similar relaxation process has also been observed in RNA and DNA dynamics, but significant differences exist. In tRNA, the process has a significantly larger localization length of approximately 7 Å compared with proteins [72]. Nucleic acids also exhibit stronger temperature dependence, with tRNA dynamics slower than those of proteins such as lysozyme and myoglobin, and DNA showing even slower dynamics than tRNA (see Fig. 4b). In all cases, the dynamics of the corresponding hydration water exhibit the same relative differences (see Fig. 4c) [56, 57], suggesting that biomolecular dynamics are not only simply "slaved" by dynamics of hydration water but also influence water dynamics. Thus, coupled hydration water—biomolecule dynamics is also affected by interactions with the biomolecule, which, in turn, depend on its chemical composition and three-dimensional structure.

4.5 *Nanosecond Dynamics: Conformational Jumps*

Detailed analysis of MD simulations has identified jump-like motions in proteins occurring on the nanosecond timescale [7, 8]. QENS experiments on protein solutions have also detected relaxation processes in the 10–30 ns range [10, 73, 74]. These motions involve relatively large atomic displacements of approximately 8 Å and have been ascribed to domain-like motions [10, 73]. The amplitude and timescale of this process are sensitive to ligand binding [13, 45, 73]. A similar process has recently been reported for the intrinsically disordered protein myelin [11], suggesting that this nanosecond process may not be limited to domain motions but rather reflects more general conformational changes in proteins. In contrast to the "localized diffusion" process observed on sub-nanosecond timescales, the nanosecond process corresponds to jumps between well-defined conformational states [8, 10, 73]. Protein relaxations in the nanosecond range have also been detected in numerous NMR studies [27, 75].

Detailed MD simulations indicate a broad distribution of energy barriers for side-group conformational motions. These barriers range from approximately 10 to 35 kJ mol^{-1} and higher [76]. Methyl groups tend to dominate motions with lower energy barriers, whereas aromatic side groups typically exhibit higher energy barriers [47, 76]. There is also heterogeneity in the behavior of hydrophobic versus hydrophilic groups [47].

Dielectric spectroscopy has revealed nanosecond relaxation processes in both protein solutions and hydrated powders [77–79] (see Fig. 6). Although this dielectric process is often ascribed to strongly bound hydration water to proteins, its large dielectric amplitude would require hundreds of tightly bound hydration water molecules per protein, moving 100–1000 times slower than bulk water. Such behavior contradicts current findings from neutron scattering, NMR, and simulations [8, 10, 27, 73]. Thus, the nanosecond relaxation process should be attributed to internal protein dynamics.

The temperature dependence of the nanosecond process has been analyzed primarily using dielectric relaxation spectroscopy [52, 78–80]. It exhibits slightly non-Arrhenius behavior, similar to the sub-nanosecond relaxation process, but is approximately 100–1000 times slower (Fig. 6). In dielectric spectra, this process slows down significantly as the hydration level decreases [52, 78, 80]. The process is also strongly stretched, and its characteristic relaxation time is comparable across different globular proteins regardless of their size [52]. This observation is consistent with a somewhat universal distribution of energy barriers for side-group motions across different proteins, as suggested by MD simulations [76].

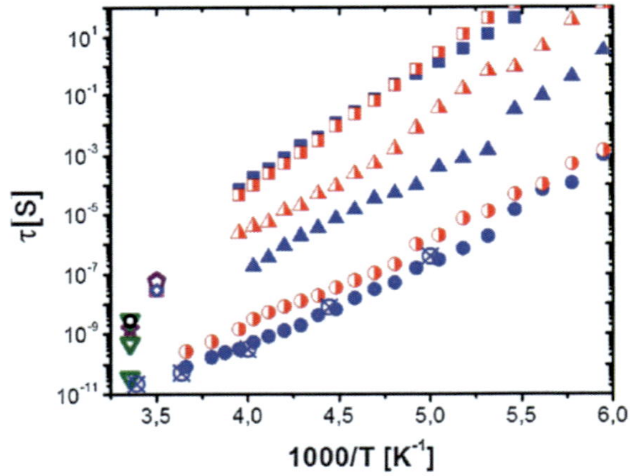

Fig. 6 Temperature dependence of relaxation times obtained by BDS for hydrated lysozyme ($h \approx 0.4$; filled blue symbols) and myoglobin ($h \approx 0.33$; half-filled red symbols). QENS data for hydrated lysozyme are shown as crossed circles. Three different relaxation processes are observed for each hydrated protein: (i) the fastest one (circles) represents a coupled protein-solvent relaxation involving localized diffusion of residues, side chains and backbone segments; (ii) the intermediate process (triangles) is due to domain-like motions within a basin, and (iii) the slowest one (squares) represents glass transition related segmental relaxation ascribed to transitions between different basins. The figure demonstrates the wide range of timescales of protein dynamics. This figure is reproduced from Ref. [12]. Copyright (2015) with permission from RSC (Royal Society of Chemistry)

This process clearly represents conformational jumps between well-defined states. Analysis of MD simulations indicates that the nanosecond process involves side-group motions more than backbone motions [7, 47]. It is likely dominated by conformational jumps of side groups with relatively large amplitudes, and the energy barriers are governed by chemical structure, interactions, and steric constraints. This process involves both the protein surface and its core [47]. According to neutron scattering, motions of secondary structures (such as domain-like motions in multi-domain proteins) also occur in the nanosecond time range. The temperature dependence of this process is similar to that of the coupled protein-solvent relaxation, which involves the sub-nanosecond localized protein diffusion, and therefore it is likely that it is governed by this faster protein-solvent dynamics. More detailed studies using MD simulations and neutron scattering would be valuable for unraveling the microscopic details of nanosecond protein relaxation.

4.6 Microsecond Dynamics: Main Structural Relaxation

Previous dielectric relaxation studies on hydrated protein powders (i.e., no solutions) have revealed a relaxation process in the microsecond time range with an anomalously large dielectric amplitude ($\Delta\varepsilon \approx 10^3$–$10^4$) [12, 48, 49, 52]. This process exhibits a somewhat non-Arrhenius temperature dependence and is strongly dependent on the hydration level. By extrapolating the temperature dependence of its relaxation time to 100 s, an estimation of the calorimetric T_g can be obtained. The dielectric T_g obtained in this way is approximately between 170 and 200 K, depending on the hydration level. These temperatures agree with those detected by calorimetric measurements. Taken together, these observations provide strong evidence that the microsecond process represents the main structural relaxation of hydrated proteins, analogous to the α-relaxation in conventional glass-forming systems [15, 16, 81].

Direct experimental characterization of this process is challenging because its timescale is longer than the rotational tumbling time of proteins in solution ($\sim$ 10–100 ns) and lies beyond the typical time window of current neutron spectrometers and most MD simulations. Nevertheless, a few long-time MD simulations extending into the millisecond range, together with NMR relaxation-dispersion experiments, reveal that backbone motions dominate microsecond dynamics: rotations and jumps between well-defined basins in the protein energy landscape [7, 17, 46, 47]. The effective activation barriers for these motions are on the order of 50–100 kJ mol^{-1}, comparable to typical peptide backbone torsional barriers [17, 46, 82]. In this sense, the microsecond process can be viewed as a backbone-dominated segmental relaxation of the folded protein, i.e., an inter-basin structural relaxation strongly governed by hydration and directly linked to the glass-transition behavior of hydrated proteins [12, 17, 46, 52, 82]. Because its characteristic timescale overlaps with that of many biochemical processes, this relaxation is likely the most biologically relevant internal motion, potentially associated with partial unfolding, large-scale rearrangements of

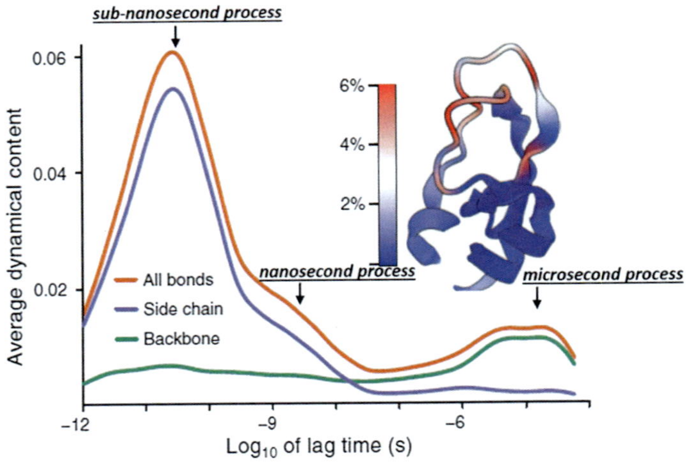

Fig. 7 Schematic representation of the dynamical content of long MD simulations of the small globular protein BPTI illustrating sub-nanosecond and nanosecond side-chain motions and the microsecond relaxation dominated by backbone motions. The figure also places the microsecond process in the overall hierarchy of protein dynamics and connects it to the glass transition of hydrated proteins. This figure is reproduced from Ref. [12]. Copyright (2015) with permission from RSC (Royal Society of Chemistry)

secondary structures, and functionally critical conformational transitions [7, 17, 46, 47] (Fig. 7).

5 General Picture of Protein Dynamics and Its Energy Landscape

From all these previous studies, an atomistic picture of protein dynamics emerges where the various relaxation processes identified by neutron scattering, dielectric spectroscopy, NMR, and molecular dynamics (MD) simulations are considered together as a hierarchy of motions spanning from sub-picosecond-to-microsecond timescales (see Fig. 8).

The relaxation map in Fig. 8 is similar to those observed in other soft materials, featuring fast dynamics (vibrations and picosecond relaxation), secondary relaxations, and a main structural (segmental) relaxation, but it is shaped by the specific geometry and chemistry of folded proteins rather than by random-coil polymer statistics.

Under ambient conditions, hydrated proteins exhibit a wide range of conformational sub-states rather than a single native structure, as illustrated by the hierarchical nature of its energy landscape in Fig. 9. The internal dynamics cover more than six orders of magnitude in time, from motions faster than 1 ps to processes slower than 1 μs, with typical atomic displacements ranging from less than 0.5 Å to about 10 Å

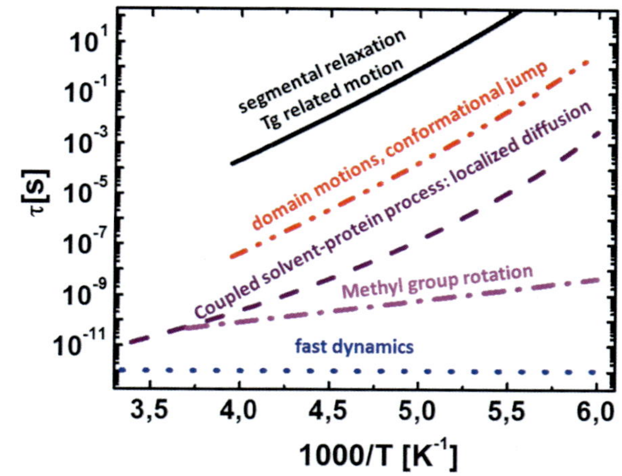

Fig. 8 Schematic illustration of the temperature dependences of the five different types of internal protein motions, divided into (i) the main structural (segmental) relaxation reaching $\tau \sim 100$ s at the glass transition temperature of the protein, (ii) domain and side-group motions on a nanosecond timescale at room temperature, (iii) "localized diffusion" of residues and side-group motions coupled to solvent dynamics with a relaxation time of tens of picoseconds at room temperature, (iv) methyl group rotation with Arrhenius temperature dependence, and (v) a fast picosecond relaxation with only a weak temperature dependence of its relaxation time. This figure is reproduced from Ref. [83]. Copyright (2017) with permission from Elsevier

[6, 8, 84]. This wide range of timescales and amplitudes is one reason why proteins are often described as more "liquid-like" than solid, and why a static crystallographic structure can be a poor representation of the dynamically active state [6, 85, 86].

At the fastest part of the spectrum ($\approx$ 1 ps and below), proteins show small-amplitude conformational fluctuations. These motions, with mean-squared displacements of about 0.2–0.4 Å^2, are associated with very low energy barriers of the order of 2–5 kJ mol^{-1} (see the bottom of Fig. 9) and are considered a generic feature of soft condensed matter [30, 35, 36, 39]. In the same high-frequency regime, proteins exhibit low-frequency collective vibrations known as the boson peak, which appears as an excess in vibrational density of states relative to the Debye model and is also regarded as a generic feature of disordered solids arising from elastic heterogeneity or soft modes [12, 30, 32, 38, 39, 49]. Together, the rattling motion and the boson peak form the lowest level of the hierarchy illustrated in Fig. 9.

On a longer timescale of several to tens of picoseconds, methyl group rotations and other side-chain motions become evident [27, 75, 87, 88]. Methyl group rotations have relatively low barriers ($\approx$ 10–20 kJ mol^{-1}) and show Arrhenius temperature dependence, remaining dynamically active even at low hydration and low temperature [27, 75, 87, 88]. Because proteins often contain a high density of methyl groups in their hydrophobic cores, whereas RNA lacks methyl groups and DNA has them only on thymine bases, these motions together with solvent motions act as internal plasticizers in proteins but not in nucleic acids [26, 27, 75]. This helps to maintain conformational flexibility in the protein core under conditions where the rest of the

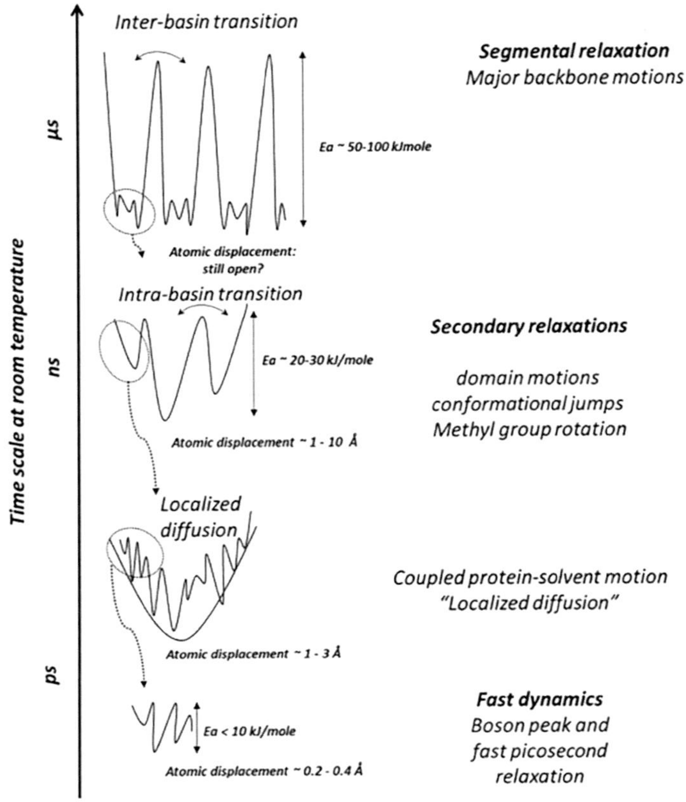

Fig. 9 Hierarchical picture of protein dynamics that includes (i) inter-basin transitions ascribed to structural (segmental) relaxation; (ii) intra-basin conformational jumps mainly related to side-chain motions and some domain motions; (iii); coupled protein-solvent dynamics described as localized diffusion (diffusion in a potential well), and (iv) fast dynamics with relatively low energy barriers (~ 3–5 kJ/mol) and small atomic displacements. This figure is reproduced from Ref. [83]. Copyright (2017) with permission from Elsevier

matrix would otherwise behave as a brittle glass, and may be particularly important near active sites, where the fastest methyl groups are often located [26, 75, 88].

Translational and rotational dynamics of hydration water at ambient temperature occur on timescales of 10–50 ps [49, 55, 58, 89]. NMR, neutron scattering, and MD simulations show that hydration water is slowed by a factor of 2–4 (on average) relative to bulk water, with stronger slowing near polar and charged groups [49, 55, 58, 89]. The activation energy for hydration-water motion is generally 15–18 kJ mol^{-1}, but the temperature dependence becomes non-Arrhenius at high temperature, and is well described by a Vogel–Fulcher–Tammann (VFT) law, reflecting the cooperative nature of these motions [55, 58, 80]. Importantly, both the absolute timescale and temperature dependence of hydration water dynamics differ for proteins, tRNA, and

DNA, underscoring the role of biomolecular surface structure and chemical nature in modulating the interfacial solvent [49, 57, 72].

On a similar timescale (10–50 ps), neutron scattering reveals a strongly stretched relaxation in proteins, interpreted as a coupled protein-solvent process termed "localized diffusion", see Fig. 8 [41, 48, 65, 84]. This motion corresponds to conformational fluctuations of residues, side chains, and backbone segments with amplitudes of about 1–3 Å [8, 12, 26, 49, 83, 90] and appears as a localized Brownian-like motion in a harmonic potential well with a confinement length of the same distance (see Fig. 9) [48, 84]. The relaxation process is well described by a Cole–Cole function with a stretching exponent of 0.25–0.4, indicating pronounced dynamic heterogeneity across different residues and environments [41, 48, 65]. The timescale and temperature dependence of this process closely follow those of hydration water, and both neutron scattering and MD simulations show that artificially restricting water motions suppresses this protein relaxation, demonstrating strong coupling between the two [48, 63]. This localized diffusion is strongly hydration-dependent, disappearing in dry proteins or when proteins are embedded in rigid glassy matrices [48, 60, 65, 91]. It is also associated with the so-called "dynamic transition" observed as a sharp increase in the MSD of hydrated proteins around 200–230 K. Although localized diffusion resembles a secondary relaxation in its amplitude and microscopic character, its VFT-like temperature dependence, persistence as a distinct process down to very short times, and strong coupling to the structural relaxation of hydration water make it a distinct type of secondary relaxation unique to hydrated or solvated biomacromolecules [41, 48, 65, 92].

The second highest tier of the hierarchy illustrated in Fig. 9 consists of conformational jumps, which correspond to larger-scale intra-basin transitions mainly involving side groups and, in some cases, domain-like motions [8, 10–12, 62, 65, 83]. These jumps produce atomic displacements on the order of 1–10 Å and have characteristic activation energies of about 20–40 kJ mol^{-1}, appearing on the nanosecond timescale at ambient temperatures [8, 10–12, 25, 62, 65, 83] and can be detected by neutron scattering, NMR, dielectric spectroscopy, and MD simulations. They are interpreted as transitions between well-defined conformational substates within a broader basin of the protein energy landscape [8, 10–12, 25, 62, 65, 83]. The temperature dependence of these nanosecond processes is thought to be controlled by the coupled protein-solvent relaxation, which involves more local protein motions, as discussed above.

At the slower end of the experimentally and computationally accessible range lies a microsecond relaxation observed in dielectric spectroscopy of hydrated protein powders and in long MD simulations [77, 78, 80, 86, 93]. NMR and long-time MD simulations indicate that this process is dominated by inter-basin transitions due to backbone rearrangements on microsecond and longer timescales [12, 17, 45, 83, 89, 94]. It corresponds to the main structural (segmental) relaxation of the protein and is associated with high torsional barriers for backbone rotations, of order 30–80 kJ mol^{-1}, as illustrated in Fig. 9. Dielectric and calorimetric data relate this microsecond process to the glass transition of hydrated proteins ($\approx$ 170–220 K, see Fig. 8) since its relaxation time reaches $\approx$ 100 s in the same temperature range as the calorimetric glass transition, suggesting that it plays a role analogous to the α-relaxation in glass-forming liquids and synthetic polymers [17, 83, 89, 94]. Furthermore, since its timescale overlaps with many biochemical processes, this microsecond structural relaxation is likely to be the most directly relevant to function, folding, and large-scale conformational transitions [30, 80, 89].

Taken together, these observations motivate the hierarchical picture of the protein energy landscape and its associated dynamics, as summarized schematically in Fig. 9. At the bottom of the hierarchy are the fast picosecond "rattling" motions, with small amplitudes (≈ 0.3–0.5 Å) and low barriers (≈2–5 kJ mol^{-1}), typical to all soft materials [30, 35, 36, 39]. On slightly longer timescales, methyl group rotations and other side-chain motions appear as chemically specific secondary relaxations with Arrhenius-like temperature dependence [27, 75, 87, 88]. Hydration-water dynamics and the coupled sub-nanosecond localized diffusion form a distinct tier of strongly cooperative, VFT-like motions with amplitudes of 1–3 Å; they are highly sensitive to hydration and solvent conditions and found in proteins, DNA, and RNA, with biomolecule-specific characteristics [41, 48, 49, 55, 58, 62, 63, 65, 89].

Above this level, nanosecond conformational jumps of side chains, loops, and domains, with amplitudes of 2–8 Å and barriers of 20–40 kJ mol^{-1}, constitute intra-basin secondary relaxations whose temperature and hydration dependence are controlled mainly by the underlying localized diffusion [11, 21, 47, 62, 73, 76, 95]. Finally, the microsecond backbone relaxation appears as an inter-basin structural (segmental) relaxation with barriers above 50 kJ mol^{-1}, closely tied to the glass transition of hydrated proteins and likely to be directly involved in biologically relevant conformational changes [30, 80, 89].

Obviously, there are even slower relaxation processes in protein dynamics, e.g., the unfolding process. Here it is important to comment on the paper by Smith and coworkers [96] suggesting that protein seems to remain non-ergodic (non-equilibrated) throughout its entire life. Using MD simulations, the authors clearly demonstrated that the protein dynamics change with the simulations waiting time all the way to microseconds. Considering that the main structural relaxation in proteins happens on timescale ~ 10 microsecond and longer (Figs. 7 and 8), this observation is consistent with the expected aging process. Usually, to reach equilibrium in any glass-forming liquid (including synthetic polymers) requires aging (annealing) time ~ 10 times the equilibrium structural relaxation time. This analogy suggests that protein dynamics are expected to become ergodic only on ms timescales. However, according to the analysis presented in [96], even experimental studies on milliseconds–seconds timescale provide estimates of protein relaxation times in good agreement with extrapolations from simulations. Thus, it is possible that proteins indeed remain non-ergodic for a very long time.

6 Conclusion

In this chapter, we summarized the different types of protein motions existing on timescales from sub-picoseconds to several microseconds. We provided atomistic details and the physical nature of the different motions. The fastest motions are also the most local with the lowest activation energies. With increasing timescale the motions become less restricted and need to overcome higher energy barriers. It is also evident from the experimental data that certain protein motions (those occurring on a 10–50 ps timescale) are strongly coupled to the motions in the hydration water. Thus, this rather local diffusion of residues, side chains, and backbone segments also involves interacting water molecules, which are thereby slowed down compared to bulk water. On the other hand, this coupled relaxation process exhibits a non-Arrhenius temperature dependence, which would not be expected for a protein relaxation of non-cooperative character, and, furthermore, is not present without any solvent, suggesting a symbiotic relationship between the two components.

The slower relaxation processes, occurring on a nanosecond-microsecond timescale, are also strongly dependent on the hydration level and/or the nature of the solvent, verifying the important role of the solvent for slower protein dynamics of functional importance. The entropy contribution of the solvent helps the protein overcome enthalpic energy barriers in the protein energy landscape and thereby facilitates these motions. In fact, the solvent also plays an important role in even slower protein motions, such as the unfolding process, but these processes have not been discussed in this chapter.

In this chapter, we have not discussed how protein dynamics is affected by replacing water with other solvents, such as glycerol and sugar solutions. This is another important topic of high relevance for cryopreservation of food and biomaterials, as well as for the stabilization of proteins and nucleic acids in pharmaceuticals. However, the mechanism behind the stabilization by such co-solvents remains elusive, mainly due to the complex dynamical coupling of different motions in both the protein and its surrounding solvent. Therefore, to gain further understanding of this dynamical coupling and to relate the different types of protein motions to biological activities are essential for understanding both life and how properties of proteins can be tailored for different applications.

Acknowledgements A.P.S. thanks NSF Polymer program (award DMR-2515834) for partial financial support.

References

1. Köhler W, Friedrich J (1989) Probing of conformational relaxation processes of proteins by frequency labeling of optical states. J Chem Phys 90(2):1270–1273
2. Frauenfelder H, Sligar SG, Wolynes PG (1991) The energy landscapes and motions of proteins. Science 254(5038):1598–1603

3. Henzler-Wildman K, Kern D (2007) Dynamic personalities of proteins. Nature 450(7172):964–972
4. Henzler-Wildman KA, Lei M, Thai V, Kerns SJ, Karplus M, Kern D (2007) A hierarchy of timescales in protein dynamics is linked to enzyme catalysis. Nature 450(7171):913–916
5. Lange OF, Lakomek N-A, Farès C, Schröder GF, Walter KFA, Becker S, Meiler J, Grubmüller H, Griesinger C, de Groot BL (2008) Recognition dynamics up to microseconds revealed from an RDC-derived ubiquitin ensemble in solution. Science 320(5882):1471–1475
6. Frauenfelder H, Chen G, Berendzen J, Fenimore PW, Jansson H, McMahon BH, Stroe IR, Swenson J, Young RD (2009) A unified model of protein dynamics. Proc Natl Acad Sci USA 106:5129–5134
7. Shaw DE, Maragakis P, Lindorff-Larsen K, Piana S, Dror RO, Eastwood MP, Bank JA, Jumper JM, Salmon JK, Shan Y, Wriggers W (2010) Atomic-level characterization of the structural dynamics of proteins. Science 330(6002):341–346
8. Hong L, Smolin N, Lindner B, Sokolov AP, Smith JC (2011) Three classes of motion in the dynamic neutron-scattering susceptibility of a globular protein. Phys Rev Lett 107(14):148102
9. Baldwin AJ, Kay LE (2012) Dynamic binding. Nature 488(7410):165–166
10. Smolin N, Biehl R, Kneller GR, Richter D, Smith JC (2012) Functional domain motions in proteins on the ns timescale: comparison of neutron spin-echo spectroscopy of phosphoglycerate kinase with molecular-dynamics simulation. Biophys J 102(5):1108–1117
11. Stadler AM, Stingaciu L, Radulescu A, Holderer O, Monkenbusch M, Biehl R, Richter D (2014) Internal nanosecond dynamics in the intrinsically disordered myelin basic protein. J Am Chem Soc 136(19):6987–6994
12. Khodadadi S, Sokolov AP (2015) Protein dynamics: from rattling in a cage to structural relaxation. Soft Matter 11:4984–4998
13. Kay LE (2005) NMR studies of protein structure and dynamics. J Magn Reson 173(2):193–207
14. Ball P (2011) More than a bystander. Nature 478(7370):467–468
15. Vitkup D, Ringe D, Petsko GA, Karplus M (2000) Solvent mobility and the protein 'glass' transition. Nat Struct Mol Biol 7(1):34–38
16. Fenimore PW, Frauenfelder H, McMahon BH, Young RD (2004) Bulk-solvent and hydration-shell fluctuations, similar to alpha- and beta-fluctuations in glasses, control protein motions and functions. Proc Natl Acad Sci USA 101:14408–14413
17. Lewandowski JR, Halse ME, Blackledge M, Emsley L (2015) Direct observation of hierarchical protein dynamics. Science 348:578–581
18. Schirò G, Fichou Y, Gallat F-X, Wood K, Gabel F, Moulin M, Härtlein M, Heyden M, Colletier J-P, Orecchini A, Paciaroni A, Wuttke J, Tobias DJ, Weik M (2015) Translational diffusion of hydration water correlates with functional motions in folded and intrinsically disordered proteins. Nat Commun 6:6490
19. Glass DC, Krishnan M, Nutt DR, Smith JC (2010) Temperature dependence of protein dynamics simulated with three different water models. J Chem Theory Comput 6(4):1390–1400
20. Khodadadi S, Curtis JE, Sokolov AP (2011) Nanosecond relaxation dynamics of hydrated proteins: water versus protein contributions. J Phys Chem B 115(19):6222–6226
21. Krishnan M, Kurkal-Siebert V, Smith JC (2008) Methyl group dynamics and the onset of anharmonicity in myoglobin. J Phys Chem B 112(17):5522–5533
22. Smith JC (1991) Protein dynamics: comparison of simulations with inelastic neutron scattering experiments. Q Rev Biophys 24(3):227–291
23. Sokolov AP, Garcia Sakai V (2012) Dynamics of soft matter: neutron applications, Chap. 1. Springer Science & Business Media, New York, p 1–23
24. Rubinstein M, Colby RH (2003) Polymer physics. Oxford University Press
25. Doster W, Settles M (2005) Protein-water displacement distributions. BBA-Proteins Proteomics 1749(2):173–186
26. Nickels JD, O'Neill H, Hong L, Tyagi M, Ehlers G, Weiss KL, Zhang Q, Yi Z, Mamontov E, Smith JC, Sokolov AP (2012) Dynamics of protein and its hydration water: neutron scattering studies on fully deuterated GFP. Biophys J 103(7):1566–1575

27. Buck M, Boyd J, Redfield C, MacKenzie DA, Jeenes DJ, Archer DB, Dobson CM (1995) Structural determinants of protein dynamics: analysis of 15N NMR relaxation measurements for main-chain and side-chain nuclei of hen egg white lysozyme. Biochemistry 34(12):4041–4055
28. Krushelnitsky A, Reichert D (2005) Solid-state NMR and protein dynamics. Prog Nucl Magn Reson Spectrosc 47(1):1–25
29. Nickels JD, Curtis JE, O'Neill H, Sokolov AP (2012) Role of methyl groups in dynamics and evolution of biomolecules. J Biol Phys 38(3):497–505
30. Sokolov AP, Rössler E, Kisliuk A, Quitmann D (1993) Dynamics of strong and fragile glass formers: differences and correlation with low-temperature properties. Phys Rev Lett 71(13):2062–2065
31. Jäckle J (1981) Amorphous solids—low-temperature properties. Springer-Verlag, Berlin, Heidelberg, New York
32. Parshin DA, Schober HR, Gurevich VL (2007) Vibrational instability, two-level systems, and the boson peak in glasses. Phys Rev B 76(6):064206
33. Surovtsev NV, Wiedersich JAH, Novikov VN, Rössler E, Sokolov AP (1998) Light-scattering spectra of fast relaxation in glasses. Phys Rev B 58(22):14888–14891
34. Gotze W, Sjogren L (1989) Logarithmic decay laws in glassy systems. J Phys: Condens Matter 1(26):4203
35. Gotze W, Sjogren L (1992) Relaxation processes in supercooled liquids. Rep Prog Phys 55(3):241
36. Malinovsky VK, Novikov VN, Parshin PP, Sokolov AP, Zemlyanov MG (1990) Universal form of the low-energy (2 to 10 meV) vibrational spectrum of glasses. Europhys Lett 11(1):43
37. Sokolov AP, Calemczuk R, Salce B, Kisliuk A, Quitmann D, Duval E (1997) Low-temperature anomalies in strong and fragile glass formers. Phys Rev Lett 78(12):2405–2408
38. Schirmacher W, Ruocco G, Scopigno T (2007) Acoustic attenuation in glasses and its relation with the boson peak. Phys Rev Lett 98(2):025501
39. Buchenau U, Nücker N, Dianoux AJ (1984) Neutron scattering study of the low-frequency vibrations in vitreous silica. Phys Rev Lett 53(24):2316–2319
40. Gurevich VL, Parshin DA, Pelous J, Schober HR (1993) Theory of low-energy Raman scattering in glasses. Phys Rev B 48(22):16318–16331
41. Roh JH, Curtis JE, Azzam S, Novikov VN, Peral I, Chowdhuri Z, Gregory RB, Sokolov AP (2006) Influence of hydration on the dynamics of lysozyme. Biophys J 91(7):2573–2588
42. Sakai VG, Khodadadi S, Cicerone MT, Curtis JE, Sokolov AP, Roh JH (2013) Solvent effects on protein fast dynamics: implications for biopreservation. Soft Matter 9(22):5336–5340
43. Bajaj VS, van der Wel PCA, Griffin RG (2009) Observation of a low-temperature, dynamically driven structural transition in a polypeptide by solid-state NMR spectroscopy. J Am Chem Soc 131(1):118–128
44. Wong K-B, Daggett V (1998) Barstar has a highly dynamic hydrophobic core: evidence from molecular dynamics simulations and nuclear magnetic resonance relaxation data. Biochemistry 37(32):11182–11192
45. Palmer AG III (2004) NMR characterization of the dynamics of biomacromolecules. Chem Rev 104(8):3623–3640
46. Markwick PRL, Malliavin T, Nilges M (2008) Structural biology by NMR: structure, dynamics, and interactions. PLoS Comput Biol 4(9):e1000168
47. Miao Y, Yi Z, Glass DC, Hong L, Tyagi M, Baudry J, Jain N, Smith JC (2012) Temperature-dependent dynamical transitions of different classes of amino acid residue in a globular protein. J Am Chem Soc 134(48):19576–19579
48. Khodadadi S, Pawlus S, Roh JH, Sakai VG, Mamontov E, Sokolov AP (2008) The origin of the dynamic transition in proteins. J Chem Phys 128:195106
49. Khodadadi S, Roh JH, Kisliuk A, Mamontov E, Tyagi M, Woodson SA, Briber RM, Sokolov AP (2010) Dynamics of biological macromolecules: not a simple slaving by hydration water. Biophys J 98:1321–1326

50. Gabel F, Bicout D, Lehnert U, Tehei M, Weik M, Zaccai G (2002) Protein dynamics studied by neutron scattering. Q Rev Biophys 35(4):327–367
51. Gallat FX, Laganowsky A, Wood K, Gabel F, van Eijck L, Wuttke J, Moulin M, Härtlein M, Eisenberg D, Colletier JP, Zaccai G, Weik M (2012) Dynamical coupling of intrinsically disordered proteins and their hydration water: comparison with folded soluble and membrane proteins. Biophys J 103(1):129–136
52. Khodadadi S, Pawlus S, Sokolov AP (2008) Influence of hydration on protein dynamics: combining dielectric and neutron scattering spectroscopy data. J Phys Chem B 112(45):14273–14280
53. Swenson J, Jansson H, Bergman R (2006) Relaxation processes in supercooled confined water and implications for protein dynamics. Phys Rev Lett 96:247802/1–247802/4
54. Sokolov AP, Gregory RB (2006) Internal dynamics of proteins and DNA: analogy to glass-forming systems. In: Fitter J, Gutberlet T, Katsaras J (eds) Neutron scattering in biology: techniques and applications. Springer Science & Business Media, New York, pp 485–502
55. Chu X-Q, Fratini E, Baglioni P, Faraone A, Chen S-H (2008) Observation of a dynamic crossover in RNA hydration water which triggers a dynamic transition in the biopolymer. Phys Rev E 77:011908/1–011908/6
56. Chen S-H, Liu L, Fratini E, Baglioni P, Faraone A, Mamontov E (2006) Observation of fragile-to-strong dynamic crossover in protein hydration water. Proc Natl Acad Sci USA 103:9012–9016
57. Chen S-H, Liu L, Chu X, Zhang Y, Fratini E, Baglioni P, Faraone A, Mamontov E (2006) Experimental evidence of fragile-to-strong dynamic crossover in DNA hydration water. J Chem Phys 125(17):171103/1–171103/4
58. Chen SH, Mallamace F, Mou CY, Broccio M, Corsaro C, Faraone A, Liu L (2006) The violation of the Stokes-Einstein relation in supercooled water. Proc Natl Acad Sci USA 103(35):12974–12978
59. Doster W, Cusack S, Petry W (1990) Dynamic instability of liquid like motions in a globular protein observed by inelastic neutron scattering. Phys Rev Lett 65(8):1080–1083
60. Tsai AM, Neumann DA, Bell LN (2000) Molecular dynamics of solid-state lysozyme as affected by glycerol and water: a neutron scattering study. Biophys J 79(5):2728–2732
61. Paciaroni A, Cinelli S, Onori G (2002) Effect of the environment on the protein dynamical transition: a neutron scattering study. Biophys J 83(2):1157–1164
62. Biehl R, Hoffmann B, Monkenbusch M, Falus P, Préost S, Merkel R, Richter D (2008) Direct observation of correlated interdomain motion in alcohol dehydrogenase. Phys Rev Lett 101(13):138102
63. Tarek M, Tobias DJ (2002) Role of protein-water hydrogen bond dynamics in the protein dynamical transition. Phys Rev Lett 88:138101
64. Daniel RM, Finney JL, Stoneham M, Finney JL (2004) Water? What's so special about it? Philos Trans R Soc Lond Ser B: Biol Sci 359(1448):1145–1165
65. Nickels JD, García Sakai V, Sokolov AP (2013) Dynamics in protein powders on the nanosecond-picosecond time scale are dominated by localized motions. J Phys Chem B 117(39):11548–11555
66. Doster W, Cusack S, Petry W (1989) Dynamical transition of myoglobin revealed by inelastic neutron scattering. Nature 337:754–756
67. Hong L, Cheng X, Glass DC, Smith JC (2012) Surface hydration amplifies single-well protein atom diffusion propagating into the macromolecular core. Phys Rev Lett 108(23):238102
68. Nandi N, Bhattacharyya K, Bagchi B (2000) Dielectric relaxation and solvation dynamics of water in complex chemical and biological systems. Chem Rev 100(6):2013–2045
69. Bone S, Pethig R (1982) Dielectric studies of the binding of water to lysozyme. J Mol Biol 157(3):571–575
70. Bone S, Pethig R (1985) Dielectric studies of protein hydration and hydration-induced flexibility. J Mol Biol 181(2):323–326
71. Biehl R, Monkenbusch M, Richter D (2011) Exploring internal protein dynamics by neutron spin echo spectroscopy. Soft Matter 7(4):1299–1307

72. Roh JH, Briber RM, Damjanovic A, Thirumalai D, Woodson SA, Sokolov AP (2009) Dynamics of tRNA at different levels of hydration. Biophys J 96(7):2755–2762
73. Inoue R, Biehl R, Rosenkranz T, Fitter J, Monkenbusch M, Radulescu A, Farago B, Richter D (2010) Large domain fluctuations on 50-ns timescale enable catalytic activity in phosphoglycerate kinase. Biophys J 99(7):2309–2317
74. Sterpone F, Stirnemann G, Laage D (2012) Magnitude and molecular origin of water slowdown next to a protein. J Am Chem Soc 134(9):4116–4119
75. Krushelnitsky A, Reichert D, Saalwächter K (2013) Solid-state NMR approaches to internal dynamics of proteins: from picoseconds to microseconds and seconds. Acc Chem Res 46(9):2028–2036
76. Rajitha T, Smith JC, Krishnan M (2014) Hidden regularity and universal classification of fast side chain motions in proteins. J Am Chem Soc 136(24):8590–8605
77. Wolf M, Gulich R, Lunkenheimer P, Loidl A (2012) Relaxation dynamics of a protein solution investigated by dielectric spectroscopy. Biochimica et Biophysica Acta 1824:723–730
78. Shinyashiki N, Yamamoto W, Yokoyama A, Yoshinari T, Yagihara S, Kita R, Ngai KL, Capaccioli S (2009) Glass transitions in aqueous solutions of protein (bovine serum albumin). J Phys Chem B 113:14448–14456
79. Jansson H, Bergman R, Swenson J (2011) Role of solvent for the dynamics and the glass transition of proteins. J Phys Chem B 115:4099–4109
80. Nakanishi M, Sokolov AP (2015) Protein dynamics in a broad frequency range: dielectric spectroscopy studies. J Non-Cryst Solids 407:478–485
81. Arrondo JLR, Goñi FM (1999) Structure and dynamics of membrane proteins as studied by infrared spectroscopy. Prog Biophys Mol Biol 72(4):367–405
82. Forman-Kay JD, Mittag T (2013) From sequence and forces to structure, function, and evolution of intrinsically disordered proteins. Structure 21(9):1492–1499
83. Khodadadi S, Sokolov AP (2017) Atomistic details of protein dynamics and the role of hydration water. Biochimica et Biophysica Acta (BBA)—General Subjects 1861(1, Part B):3546–3552
84. Hong L, Glass DC, Nickels JD, Perticaroli S, Yi Z, Tyagi M, O'Neill H, Zhang Q, Sokolov AP, Smith JC (2013) Elastic and conformational softness of a globular protein. Phys Rev Lett 110(2):028104
85. Lindorff-Larsen K, Best RB, DePristo MA, Dobson CM, Vendruscolo M (2005) Simultaneous determination of protein structure and dynamics. Nature 433(7022):128–132
86. Zhou Y, Vitkup D, Karplus M (1999) Native proteins are surface-molten solids: application of the Lindemann criterion for the solid versus liquid state. J Molec Biol 285(4):1371–1375
87. Roh JH, Novikov VN, Gregory RB, Curtis JE, Chowdhuri Z, Sokolov AP (2005) Onsets of anharmonicity in protein dynamics. Phys Rev Lett 95(3):038101
88. Kasinath V, Sharp KA, Wand AJ (2013) Microscopic insights into the NMR relaxation-based protein conformational entropy meter. J Am Chem Soc 135(40):15092–15100
89. Maragakis P, Lindorff-Larsen K, Eastwood MP, Dror RO, Klepeis JL, Arkin IT, Jensen MØ, Xu H, Trbovic N, Friesner RA, Palmer AG, Shaw DE (2008) Microsecond molecular dynamics simulation shows effect of slow loop dynamics on backbone amide order parameters of proteins. J Phys Chem B 112(19):6155–6158
90. Madhusudan Tyagi AA, Colmenero J (2005) Heterogeneous dynamics of polyvinyl acetatefar above Tg—a combined study by dielectric spectroscopy and quasielastic neutron scattering. J Chem Phys 122:244909
91. Cornicchi E, Marconi M, Onori G, Paciaroni A (2006) Controlling the protein dynamical transition with sugar-based bioprotectant matrices: a neutron scattering study. Biophys J 91(1):289–297
92. Ngai KL, Capaccioli S, Ancherbak S, Shinyashiki N (2011) Resolving the ambiguity of the dynamics of water and clarifying its role in hydrated proteins. Philos Mag 91(13–15):1809–1835
93. Haller JD, Schanda P (2013) Amplitudes and time scales of picosecond-to-microsecond motion in proteins studied by solid-state NMR: a critical evaluation of experimental approaches and application to crystalline ubiquitin. J Biomol NMR 57(3):263–280

94. Mittermaier A, Kay LE (2006) New tools provide new insights in NMR studies of protein dynamics. Science 312(5771):224–228
95. Oleinikova A, Smolin N, Brovchenko I (2007) Influence of water clustering on the dynamics of hydration water at the surface of a lysozyme. Biophys J 93(9):2986–3000
96. Hu X, Hong L, Dean Smith M, Neusius T, Cheng X, Smith, Jeremy C (2016) The dynamics of single protein molecules is non-equilibrium and self-similar over thirteen decades in time. Nat Phys 12(2):171–174

Studies of the Protein Glass Transition and Its Related Dynamics—Comparison with Various Aqueous Solutions

K. P. Safna Hussan, Shin Yagihara, Rio Kita, Kaito Sasaki, and Naoki Shinyashiki

Abstract Water plays a pivotal role in determining the structural and dynamical properties of biomolecules and polymers. This chapter explores the molecular dynamics of unfrozen and frozen water mixtures, emphasizing their dielectric relaxation behavior and relevance to protein dynamics. In unfrozen (supercooled) water mixtures, small molecular liquids such as alcohol–water and fructose–water systems typically exhibit a single cooperative relaxation, reflecting collective motion of water and solute molecules. As the temperature decreases, multiple relaxation processes emerge, including local ν-relaxations and cooperative α-relaxations, which indicate stronger intermolecular interactions. Comparative studies on small molecular liquids and polymers reveal distinct cooperativity of water and solute molecules, governed by intermolecular interactions that depend on hydrogen bonding, chemical structure, and molecular size. In frozen water mixtures, crystalline and amorphous phases coexist, allowing for the observation of the dynamics of ice in the crystalline phase and the dynamics of water and solute molecules in the amorphous phase. Proteins such as BSA, gelatin, and lysozyme exhibit similar coupling behaviors, providing insights into hydration-dependent glass transitions and restricted mobility at low temperatures. By establishing a unified framework linking hydrogen-bond disorder, defect-driven dynamics, and segmental mobility, this chapter elucidates how water governs the flexibility, stability, and functionality of macromolecules. The findings have profound implications for understanding protein dynamics, cryopreservation, and the design of amorphous formulations in pharmaceutical and biophysical research.

Keywords Dielectric relaxation · Hydrogen bonding · Unfrozen water mixture · Frozen water mixture · Cooperative α-relaxation · Local ν-relaxation · Polymer–water dynamics · Protein hydration · Glass transition · Cryopreservation

K. P. Safna Hussan · R. Kita · K. Sasaki · N. Shinyashiki (✉)
Micro/Nano Technology Center, Tokai University, Hiratsuka-Shi, Kanagawa, Japan
e-mail: ns@tokai.ac.jp

S. Yagihara · R. Kita · K. Sasaki · N. Shinyashiki
Department of Physics, School of Science, Tokai University, Hiratsuka-Shi, Kanagawa, Japan

S. Cerveny and J. Swenson (eds.), *Dynamics of Biomolecules Studied by Dielectric Spectroscopy*, Advances in Dielectrics, https://doi.org/10.1007/978-3-032-30588-6_5

1 Introduction

Proteins are dynamic macromolecules whose functional behavior depends critically on their internal motions—ranging from rapid side-chain fluctuations to slower, large-scale conformational rearrangements. These dynamics, occurring over a broad timescale from picoseconds to seconds, underpin essential biological processes, including ligand binding, enzymatic catalysis, allosteric modulation, and signal transduction. However, these internal motions do not occur in isolation; they are intricately coupled to the surrounding hydration shell—a thin layer of water molecules dynamically interacting with the protein surface [1].

This hydration water, often termed "biological water," differs fundamentally from bulk water in both structure and dynamics. It facilitates conformational flexibility and acts as a dynamic lubricant for internal protein motions [2]. Importantly, protein dynamics cannot be meaningfully studied in the absence of water, especially under thermal stress. Proteins are highly sensitive to dehydration and heat; when studied in the dry state or at elevated temperatures without water, they rapidly undergo structural degradation, rendering any dynamical measurements physiologically irrelevant. Therefore, the hydration shell is not only essential for function but also for maintaining structural integrity during thermal or spectroscopic analysis. The mobility of these interfacial water molecules is crucial for maintaining the protein's dynamic landscape, particularly near the dynamic transition temperature (~ 200 K), below which the arrest of hydration water motion results in a dramatic suppression of protein flexibility and biological activity. Consequently, hydration dynamics are not merely ancillary but are fundamental to protein stability and function.

The dynamics of hydrated proteins are more complex, unlike small molecule or polymer glass-formers, proteins possess a wide variety of amino acid side chains, secondary and tertiary structural elements, and specific intramolecular interactions that create a rugged energy landscape. This structural complexity leads to multiple overlapping relaxation processes spanning a broad range of timescales and spatial scales. However, recent studies showcase that draw parallels between the dynamic transitions in hydrated proteins and those observed in simpler model systems such as aqueous mixtures of hydrophilic glass-formers and polymers. In these systems, two major relaxation processes have been consistently identified: the structural α-relaxation associated with the collective motion of the hydrogen-bonded matrix, and the β-relaxation corresponding to the reorientation of water molecules. Strikingly, analogous relaxation phenomena occur in hydrated proteins, where the α-relaxation reflects conformational fluctuations of the protein backbone, while the β-relaxation is attributed to the mobility of the hydration water [3].

Below the glass transition temperature, the relaxation time of β-process of hydration water exhibits Arrhenius-type temperature dependence, with an activation energy (E_a) closely matching that of water in the glassy state of aqueous mixtures. This behavior underscores a universal aspect of water dynamics across diverse systems. As temperature increases and surpasses the protein-specific glass transition temperature, the deviation from Arrhenius behavior of the relaxation time of water signals the onset of cooperative structural motions—a hallmark of the α-relaxation process.

To fully understand these phenomena, it is essential first to have a fundamental grasp of the unfrozen dynamics of water in simpler mixtures—for example, water combined with hydrophilic polymers or small glass-formers—before extending these insights to the more intricate hydrated protein systems.

2 Unfrozen Water Mixtures

Numerous studies on unfrozen aqueous mixtures have followed two major research approaches to elucidate their dynamic behavior. The first approach focuses on the behavior of aqueous mixtures at or near room temperature, where they remain in the liquid state. In this case, the primary variable is solute concentration rather than temperature. Studies employing molecular dynamics simulations, nuclear magnetic resonance (NMR) spectroscopy, and broadband dielectric spectroscopy (BDS) have provided detailed insights into composition-dependent dynamics. The second approach involves temperature-dependent investigations over a broad frequency and temperature range, encompassing both the liquid and glassy states. These studies aim to understand the dynamics of the glass transition, focusing on the primary (α) relaxation associated with cooperative structural rearrangements, as well as secondary relaxations such as β or v, which are related to the local motion of water. In this article, the terminology varies among the references of the cited figures; however, the water relaxation is referred to as the β or v process.

2.1 Relaxation Behavior at Room Temperature

At room temperature (T_R), the dielectric spectra of water solutions generally display a single relaxation process as shown in Fig. 1. For small molecular liquid–water mixtures, this relaxation follows a stretched exponential, i.e., Kohlrausch–Williams–Watts (KWW) function, reflecting a cooperative motion of small solute and water molecules [4–6]. The KWW function is used as the asymmetric broadening parameter of frequency dependence of dielectric relaxation given by the following equation.

$$\varepsilon^*(\omega)_{\mathrm{KWW}} = \Delta\varepsilon \int_0^\infty \left[-\frac{\mathrm{d}\varphi(t)}{\mathrm{d}t} \right] \exp(-j\omega t)\mathrm{d}t$$
$$\text{for } \left(\varphi(t) = \exp\left[-\left(\frac{t}{\tau} \right)^{\beta_{\mathrm{KWW}}} \right] \right), \quad (0 < \beta_{\mathrm{KWW}} \leq 1), \tag{1}$$

where t represents the time, ω represents the angular frequency, $\varepsilon^*(\omega)$ is the complex permittivity, τ represents the relaxation time, j represents the imaginary unit given by $j^2 = -1$, $\Delta\varepsilon$ represents the relaxation strength, β_{KWW} value smaller than 1 stretches the exponential function in time domain and cause asymmetric broadening of the relaxation function in the frequency domain. In contrast, for water mixtures containing relatively larger solute molecules, the symmetrically broadened single relaxation process appears and is well described by the Cole–Cole equation given below, reflecting a local motion of water molecules constrained by the presence of large solute molecules

$$\varepsilon^*(\omega)_{\mathrm{CC}} = \frac{\Delta\varepsilon}{1 + (j\omega\tau)^{\beta_{\mathrm{CC}}}}, \tag{2}$$

where β_{CC} represents the symmetric broadening parameter ($0 < \beta_{\mathrm{CC}} \leq 1$). At higher solute concentrations at T_{R}, the spectra generally reveal one relaxation process, except for fructose-water mixtures, which exhibit two distinct relaxations [7]. In polymer-water mixtures, dielectric relaxation associated with polymer and water can be clearly resolved [8–10]. Taken together, these two complementary approaches—temperature-dependent and composition-dependent studies—provide a comprehensive understanding of the cooperative dynamics, intermolecular interactions, and relaxation behavior in unfrozen aqueous mixtures.

2.1.1 Molecular Size Dependence of Relaxation Processes at T_{R}

In alcohol–water mixtures as exemplified in Fig. 1a, although the intrinsic relaxation times of alcohol and water differ by nearly two orders of magnitude, the formation of strong hydrogen-bonded interactions allows the two components to move cooperatively. As a result, their individual relaxation processes merge into a single relaxation mode, reflecting collective molecular dynamics [4]. For such small molecule-water mixtures, only a single relaxation process is typically observed, as water and solute molecules can move cooperatively, leading to a stretched exponential (KWW) type relaxation.

A similar trend is observed in ethylene glycol oligomer (EGO) [5]. The single relaxation process arising from the combined contribution of water and EGO was observed for each mixture. The shape of the relaxation process depends on both solute concentration and molecular size as shown in Fig. 1b. For the small EGO molecules (one and two repeat units), the relaxation shows a broad and asymmetric, and it is best described by the KWW function, reflecting cooperative molecular motion between solute and water. However, as the number of repeating units are larger than two, the relaxation exhibits a broad and symmetric and it can be well described by the Cole–Cole function, indicating a shift toward localized relaxation and it reflects the variations in local structure in mixtures with larger EGO molecules. Larger EGO molecules cannot move cooperatively with water molecules and instead act as geometrical constraints on the motion of water molecules.

In contrast, in polymer–water systems, the relaxation process of water becomes symmetric and can be described by the Cole–Cole function (Fig. 1c). This behavior

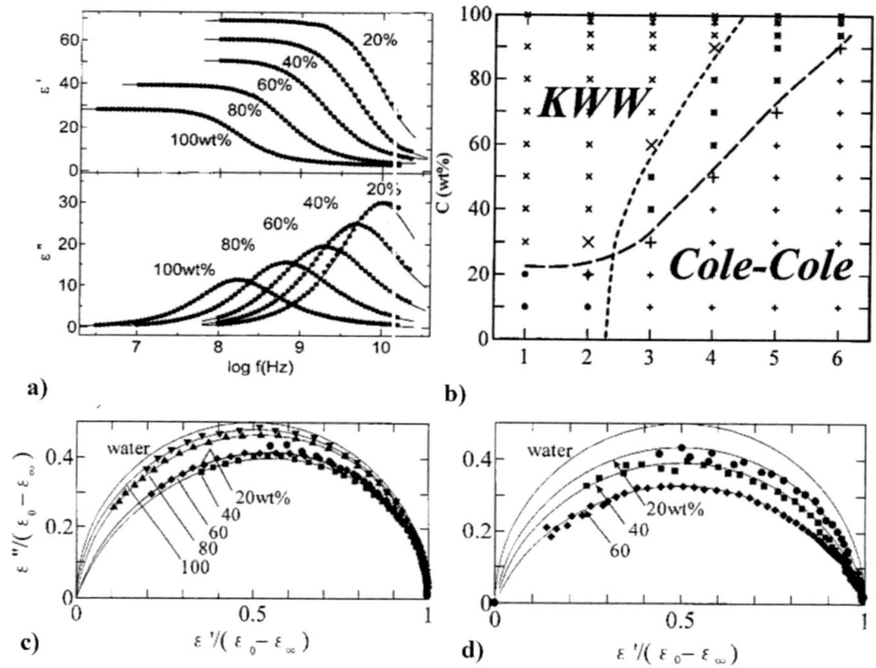

Fig. 1 **a** Dielectric constants and losses for water–1,3-butanediol mixtures at various concentrations at 25 °C. The solid lines represent fits using the Kohlrausch–Williams–Watts (KWW) function. **b** Map showing the applicable relaxation functions as a function of ethylene glycol oligomer (EGO) concentration (*C*, wt%) and number of repeat units (*N*). The region on the upper left (×) corresponds to systems well described by the KWW function, while the lower right (+) region corresponds to systems fitting the Cole–Cole equation. Filled circles (●) indicate mixtures described by both functions, and asterisks (*) denote cases where neither model provides a satisfactory fit. Cole–Cole plots for **c** VP–water and **d** PVP–water mixtures at various concentrations at 25 °C. Solid lines represent fits using KWW and Cole–Cole Equations. For PVP–water systems, the contribution from the lower-frequency relaxation peak arising from the micro-Brownian motion of PVP chains has been subtracted. These figures are reproduced from Ref. (a) [4], (b) [5], and (c) [6] with permission from the American Chemical Society

indicates localized motion—where water molecules experience spatial confinement within the polymer matrix. Consequently, the relaxation reflects local fluctuations of water molecules embedded in the polymer network rather than the global cooperative dynamics observed in small-molecule –water systems [6, 11–14].

Further investigation into the concentration dependence of fructose–water mixtures with various fructose concentrations, C_f (wt%), ranging from 5 to 94.6 wt%, over a frequency range of 2 mHz to 50 GHz at 25 and 0 °C revealed distinct relaxation behaviors. For water-rich mixtures with C_f below 45 wt%, a single relaxation process was predominant at higher temperatures, while two relaxation processes emerged for C_f above 45 wt%, indicating the presence of both α and ν relaxations (Fig. 2) [7]. This behavior is closely linked to the intermolecular interactions between

Fig. 2 Frequency-dependent plots of **a** the real and **b** the imaginary part of the dielectric function for fructose–water mixtures of various concentrations ranging from 5 to 94.6 wt% at 25 °C. For comparison, the imaginary part of pure fructose at 24 °C was taken from literature [15]. To improve clarity, the low-frequency region influenced solely by DC conductivity near the loss peak was excluded. Reproduced from Ref. [7] with permission from Elsevier

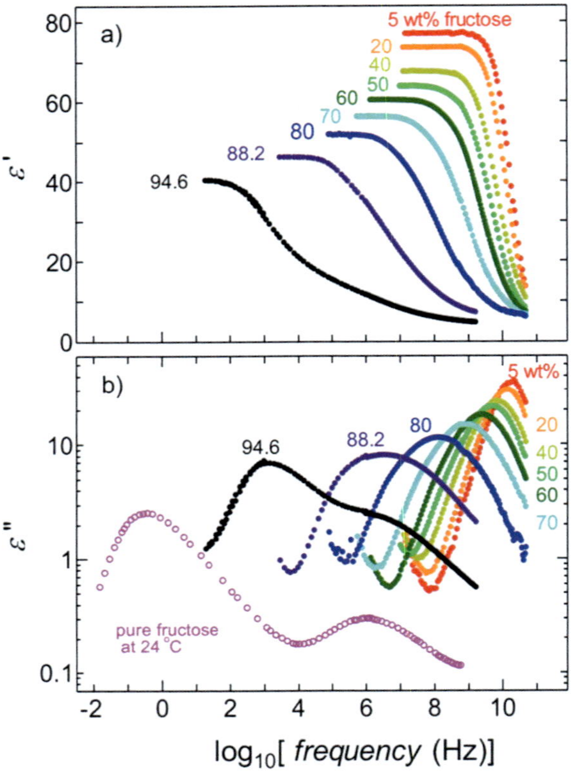

fructose molecules. In water-rich mixtures ($C_f < 45$ wt%), where there are more than 12 water molecules per fructose molecule, the α relaxation strength ($\Delta\varepsilon_\alpha$) becomes negligible at 25 °C, indicating the absence of cooperative molecular motion. In such dilute systems, fructose molecules are surrounded by an excess of water, which prevents direct intermolecular interactions. Furthermore, the strong thermal fluctuations at room temperature inhibit significant cooperative dynamics between fructose and water molecules. Conversely, in mixtures with lower water content ($C_f \geq 45$ wt%), direct intermolecular interactions between fructose molecules become feasible, enabling the emergence of cooperative α relaxation.

2.2 *Temperature Dependence of Relaxation Behavior*

2.2.1 Behavior of Relaxation Process from T_R to T_g

Figure 3 shows the temperature dependence of the dielectric relaxation parameters for the 65 wt% pentaethylene glycol (5EG)–water mixture, plotted against the reciprocal temperature. Two relaxations, the α and β processes were observed. Panel (a) presents the logarithm of the relaxation times ($\log_{10}\tau$) for the α and β processes. At high temperatures (low 1000/T), a single relaxation process dominates, characterized by a strong temperature dependence of τ. As the temperature decreases toward the critical temperature, T_C, this single process separates into two distinct relaxation processes: the α and β processes. Below T_C, the single process continuously developed into the β process as a symmetrically broadened relaxation at lower temperatures, exhibiting weaker temperature dependence, while the α process appears and it shows a more pronounced change in τ with temperature.

Panel Fig. 3b illustrates the relaxation strengths, $\Delta\varepsilon_\alpha$ and $\Delta\varepsilon_\beta$, as well as their sum ($\Delta\varepsilon_\alpha + \Delta\varepsilon_\beta$). The relaxation strength of the α process ($\Delta\varepsilon_\alpha$) decreases gradually with increasing temperature below T_C and vanishes at T_C. In contrast, the relaxation strength of the β process ($\Delta\varepsilon_\beta$) increases with temperature and smoothly connects to the strength of the single relaxation observed above T_C. The total relaxation strength ($\Delta\varepsilon_\alpha + \Delta\varepsilon_\beta$) increases linearly with decreasing temperature and agrees well with the extrapolated value from that above T_C. This indicates that the single relaxation observed at higher temperatures simply splits into the α and β components below T_C, without loss of the total dipole contribution.

Panel Fig. 3c shows the KWW asymmetric shape parameter, β_{KWW}, for the α process and the Cole–Cole symmetric broadening parameter, β_{CC}, for the β process. As the temperature decreases below T_C, β_{KWW} decreases, reflecting an increase in asymmetry and broadening of the α relaxation. Meanwhile, β_{CC} remains relatively high at low temperatures but gradually decreases with decreasing temperature, indicating changes in the symmetry of the β relaxation. Just below T_C, the dipole moments contribute predominantly to the β process, as $\Delta\varepsilon_\beta$ is much larger than $\Delta\varepsilon_\alpha$.

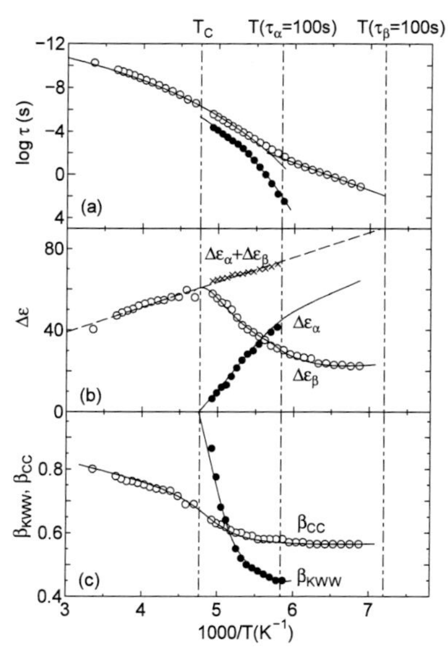

Fig. 3 Temperature dependences of dielectric relaxation parameters for the 65 wt% pentaethylene glycol (5EG)–water mixture plotted against reciprocal temperature: **a** logarithm of relaxation times, **b** relaxation strengths $\Delta\varepsilon_\alpha$, $\Delta\varepsilon_\beta$, and their sum, and **c** shape parameters β_{KWW} (α process) and β_{CC} (β process). This figure is reprinted from the Ref. [16] with permission from American Institute of Physics

With further cooling, a fraction of dipole moments gradually shifts from the β to the α process, eventually leading to $\Delta\varepsilon_\alpha$ exceeding $\Delta\varepsilon_\beta$ at lower temperatures.

The relaxation dynamics of the EGO–water mixtures exhibit an evolution as temperature decreases from T_R toward the T_g, as depicted in Fig. 4 [17]. At T_R, the dielectric response is characterized by a single relaxation process, as shown in Fig. 4 (a typical trend for small-molecule water mixtures). This single relaxation reflects the dynamics of both water and EGO, in which the entire system behaves as one homogeneous relaxation entity. However, as the temperature decreases and approaches the separation temperature (T_S), this single process distinctly separates into two components: the α-process and the ν-process. The α-process corresponds to the structural relaxation associated with the cooperative motion of water and EGO molecules. At the same time, the ν-process originates from the more localized, faster motion of water molecules. The change from one to two distinct processes at T_S highlights a dynamic heterogeneity that becomes evident upon cooling.

Above T_g, the ν-process deviates from simple thermally activated (Arrhenius) behavior and follows a non-Arrhenius temperature dependence, implying that the system requires increasing cooperative motion for molecular rearrangements as temperature decreases. Below T_g, however, the ν-process transitions to Arrhenius behavior, indicating that the molecular motions are confined to local environments

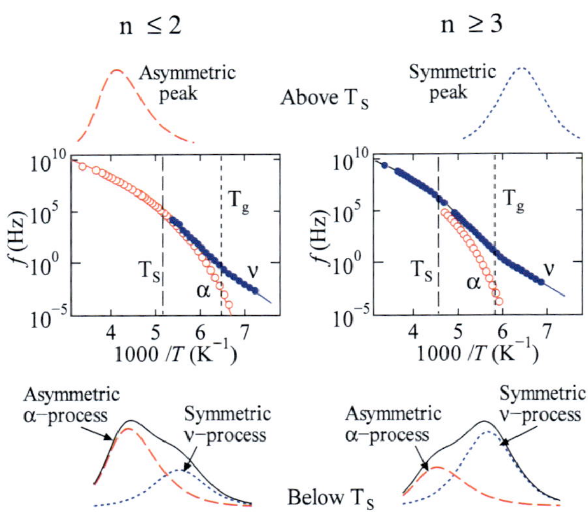

Fig. 4 Schematic representation of temperature-dependent relaxation behavior in ethylene glycol oligomer–water mixtures (65 wt%) for small ($n \leq 2$) and large ($n \geq 3$) numbers of repeat units. The middle panels show the variation of relaxation frequencies (f) with 1000/T, highlighting the separation of the single relaxation process at above T_S into two distinct processes (α and ν) near T_S. The top and bottom panels illustrate the shape of dielectric loss peaks above and below T_S, respectively, where the α-process exhibits asymmetric broadening while the ν-process remains more symmetric. The change in symmetry from the left ($n \leq 2$) to right ($n \geq 3$) above T_S indicates a transition from cooperative to more independent molecular motion as molecular size increases. This figure is reprinted from Ref. [17] with permission from the American Chemical Society

and follow a simple thermally activated process with reduced cooperativity. From the dielectric loss patterns (top and bottom panels of Fig. 4), it is evident that the asymmetric α-process becomes more prominent at lower temperatures, while the ν-process maintains a more symmetric peak shape. The comparison between small ($n \leq 2$) and large ($n \geq 3$) EGOs further reveals that increasing molecular size leads to more symmetric relaxation behavior, signifying a loss of cooperativity at larger molecular size. Despite the trend toward less cooperative motion at higher temperatures, the relaxation process in small EGO-water mixtures remains non-Debye, as reflected by the KWW function fit, confirming the persistence of partial cooperative even in the liquid regime.

In general, this behavior is typical of systems whose T_g lies far below T_R, as seen in many small-molecule water mixtures. In such systems, the structural (α) relaxation process is observed at ambient conditions, leaving only one dominant relaxation mode that reflects the cooperative or confined dynamics of the liquid, depending on the solute molecular size. Upon cooling, however, the molecular motion becomes increasingly constrained, and the emergence of two distinct processes—α and ν—marks the onset of dynamic heterogeneity and the gradual transition toward the glassy state. This trend underscores the temperature-dependent nature of molecular cooperativity, which diminishes at higher temperatures and becomes more pronounced as the system approaches T_g.

2.2.2 Common Temperature-Dependent Dynamics Across Materials

(a) Universal Property of the Relaxation of Water in the Mixtures of Small Molecules and Polymer

Extensive studies on aqueous mixtures of hydrophilic polymers (polyvinyl pyrrolidone (PVP) [18], poly (ethylene glycol) (PEG) 400/600 [16], Poly(vinyl methyl ether) (PVME) [19–21], Polyethylenimine (PEI)) and small-molecular glass formers (ethylene glycol, tripropylene glycol propanol, glycerol, xylitol, sorbitol) [17, 22] have revealed characteristic relaxation processes that determine the behavior of water in constrained environments. In all systems, besides the structural α-relaxation associated with the glass transition, a faster water-related relaxation—the ν-process—appears universally with consistent characteristics, independent of chemical composition or water content. The relaxation time τ_ν follows an Arrhenius dependence below T_g but crosses over to a stronger temperature dependence above T_g. In mixtures such as 35 wt% water with PEG 400, PEG 600, PVP, PVME (Fig. 5a) and 2EG–6EG (Fig. 5b), the activation energies (E_a) cluster around ~ 50 kJ/mol [23].

Then the water mixtures of 3EG at different concentration (65, 70, 80, 90, 94, and 100 wt%) were measured using broadband dielectric spectroscopies in the frequency range 10 μHz–20 GHz and in the temperature range 300–80 K without water crystallization was measured. The results were compared to data on pure glycerol and 60, 65, 70 and 80 wt% glycerol–water mixtures. [24], as well as on 60, 70, and 80

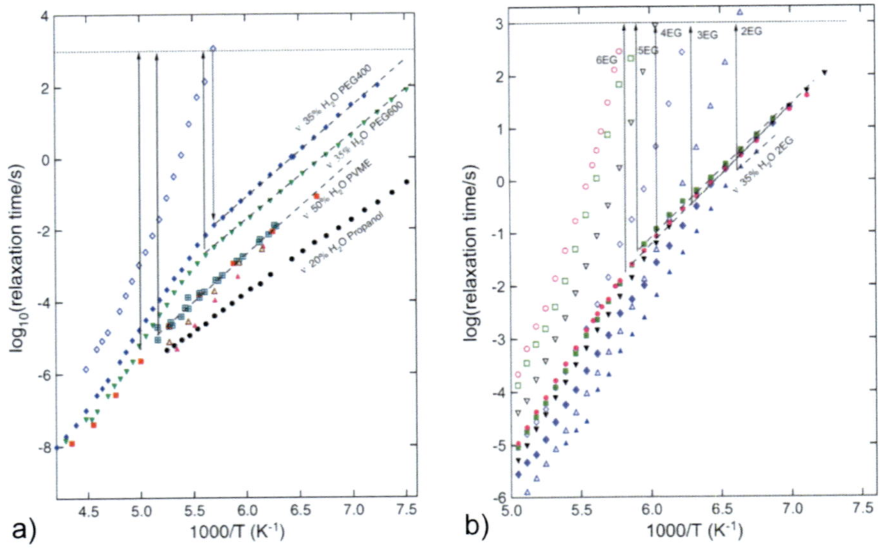

Fig. 5 Temperature dependence of the dielectric relaxation time, **a** τ_ν (closed diamond) and τ_a (open diamond) of mixtures of 35 wt% of water with PEG400, the rest of the symbols are τ_ν of various aqueous mixtures. Closed squares are for mixtures of 50 wt% of water with PVME. Closed circles are for 20 wt% water with propanol. Squares with cross inside are for 50 wt% of water with poly(vinylpyrro1idone) (PVP) of molecular weight 40,000. Open upward-pointing triangles for 38.6 wt% water containing poly(2-hydroxyethyl methacrylate) (PHEMA). Closed upward-pointing triangles for 48.4 wt% water containing PHEMA. **b** 35 wt% of water with various ethylene glycol oligomers as indicated. Circles for 6EG. Squares for 5EG. Downward-pointing triangles for 4EG. Diamond for 3EG. Upward-pointing triangles for 2EG. This figure is reproduced from Ref. [23] with permission from the American Chemical Society

wt% ethylene glycol–water mixtures [25], pure 1-propanol and 80 wt% propanol–water mixture [26], 65 wt% one to six repeat unit of EGO–water mixtures [27, 28], 65 wt% PEG400- and PEG600-water mixtures [16] and 94.6, 88.2, 86.5, 77.8 wt% fructose–water mixtures [29]. These investigations revealed two kinds of relaxation processes for water: ν-the first process behaves as its inherent mobility and appears as a separate relaxation process, and another α-process, water moves cooperatively with solute molecules.

In further studies of the fructose–water mixtures, the same ν process is observed alongside the α process, which originates from the cooperative motions of water and fructose molecules. The relaxation time (τ_ν) and relaxation strength ($\Delta\varepsilon_\nu$) of this β process exhibit a change in temperature dependence across T_g: below T_g, they display relatively weak temperature dependence, while above T_g, the temperature dependence becomes markedly stronger. Near and above T_g, the ratio of τ_α to τ_ν agrees well with the prediction of the coupling model, indicating that the β process is dynamically coupled to the α process in this temperature regime [29].

(b) **For Aqueous Solutions of Polymer**

In aqueous solutions of sugars and alcohols, the α- and ν-processes become distinguishable with decreasing temperature or increasing solute concentration. In contrast, for aqueous polymer solutions, the α- and ν-processes are observed separately over the entire temperature range studied [18, 20, 30, 31]. In protein–water mixtures with a higher water content, the α process cannot be observed because of the significant contributions from DC conductivity and electrode polarization, which originated from the relatively larger amount of ions in the mixtures. For example, in a high T_g polymer PVP, with concentrations of 60, 65, and 70 wt% PVP in a temperature range of 123 − 298 K (Fig. 6c, d) [30], two distinct relaxation processes are identified in the PVP–water mixture; one originated from the segmental chain motion of PVP, α process (in Ref. [18], it is called as l process) and the other corresponds to the primary relaxation of water, ν process (h-process in Ref. [18]). It is observed that $\Delta\varepsilon_\alpha$ increases with decreasing temperature from 298 to T_g. This may be due to the attribution of the cooperative segmental motion of PVP, not both PVP and water, unlike in sugar/alcohol mixtures (Fig. 6a, b), where the α-process involves cooperative motion of solute and water. However, $\Delta\varepsilon_\nu$ (from local water motion) typically decreases with decreasing temperature above T_g of the α-process. The relaxation times of the ν processes exhibited a crossover from Vogel–Fulcher–Tammann (VFT) to Arrhenius behavior at the fragile-to-strong transition (FST) of water, indicating a change in the dynamical regime of water and depicted Johari–Goldstein β-process [18]. Later, the systems of PVP-water mixtures were studied from another perspective and reported that the large difference in T_g between PVP and water, their mixture resembles Highly Asymmetric Mixtures (HAM): where two components have very different dynamics/T_g and Highly Asymmetric Polymer Blends (HAPB): a special case where both components are polymers with a large ΔT_g. The dynamic behavior of PVP–water mixtures shows remarkable similarities to that observed in HAM and HAPB systems, though both systems appear to be governed by the same underlying physics [30]. In our previous study [18], the slow water relaxation process (m in this article, α_2 in publication [30] in PVP–water mixtures was either not observed or inadvertently overlooked. However, in subsequent work [30], we re-evaluate this aspect in light of more recent findings. Specifically, Palacios et al. [32] investigated aqueous solutions of tri-lysine, ε-polylysine, PVP, and dextran and presented an Arrhenius plot that identified three distinct relaxation processes in a 60 wt% PVP–water mixture. These were termed the α-relaxation, slow water relaxation, and fast water relaxation, corresponding to the α_1, $m(\alpha_2)$, and ν processes, respectively, in our current analysis. While Palacios et al. interpreted these dynamics within the framework of the slaving phenomenon—highlighting the coupling between water and biomolecular motion—our aim is distinct: we seek to establish that the α_1, α_2, and ν processes in PVP–water mixtures exhibit dynamical features analogous to those found in HAM and HAPB.

Not only for high T_g polymer, but also for low T_g polymers PVME-water mixtures with PVME concentration: 60–98.7 wt%, a slow water-related process (m process) was observed in addition to the α-process that originated from segmental chain motion of PVME and ν-process from local motion of water (primary relaxation of water) as depicted in Fig. 6e, f [20]. The m process is a small relaxation that appears on the high-frequency side of the α process, and it is observed in a narrow temperature

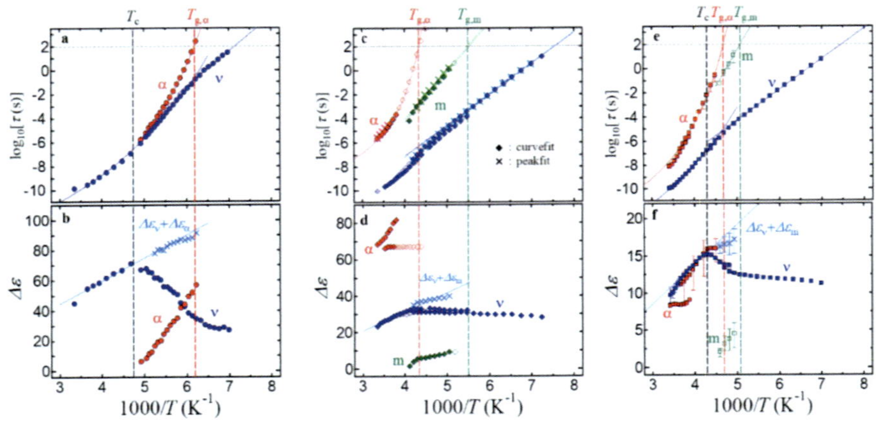

Fig. 6 Plots of log[τ (s)] and $\Delta\varepsilon$ versus reciprocal temperature for the α (red), v (blue), and m (green) relaxation processes in **a, b** 65 wt% 3EG–water (Ref. [17]), **c, d** 65 wt% PVP–water (Ref. [30]), and **e, f** 75 wt% PVME–water mixtures. Panels (b), (d), and (f) show summed relaxation strengths ($\Delta\varepsilon_\alpha + \Delta\varepsilon v$ and $\Delta\varepsilon v + \Delta\varepsilon m$) as blue crosses. Open symbols indicate processes needed to reproduce data but difficult to observe. Red and green dashed lines mark T_g for α and m processes; black dashed lines mark T_c. v-process strengths in (d) and (f) are fitted by least-squares lines. Error bars in (e) and (f) show experimental variations; others are within symbol size. This figure is reprinted from the Ref. [20] with permission from American Chemical Society

range below 218 K, with its strength $\Delta\varepsilon_m$ increasing with increasing water content. This result, along with the comparison with relaxation processes observed in dry PVME and a 98.7 wt% PVME-water mixture, indicates that the m process is closely associated with water. The change of the relaxation time of the α process plasticized by water shows no effect on the relaxation time of the m process of water. At very low water concentrations, the m-process manifests as a combined relaxation process of slow water and the h_α-process, which is Johari–Goldstein relaxation of PVME [20]. The strength of the h_α process, $\Delta\varepsilon_{h\alpha}$ (0.3–0.5), is significantly lower than that of the m process, $\Delta\varepsilon_m$ (2–5) [20].

However, in PEI–water mixtures, the m-process is absent; instead, the α-process appears as a broad and unresolved relaxation, suggesting a different dynamic behavior compared to other polymer–water systems with strong overlap or coupling of motions [31]. The v-process (water relaxation) in PEI–water mixtures at 200 K is ~ 100 × slower than in PVP or PVME–water mixtures, indicating stronger interaction and reduced mobility of water. $\Delta\varepsilon_v$ increases with cooling down to T_c, then decreases below T_c, and levels off near T_g—similar to small molecular liquids and PVME. More interestingly, only below 70 wt% PEI (i.e., with a higher water content), water starts to plasticize PEI, thereby decreasing its T_g. In contrast, PVP and PVME show earlier plasticization at much lower water contents. PEI shows a strong concentration dependence of T_g of water relaxation, unlike PVP and PVME, where T_g of the water process is independent of polymer content. This indicates the restricted mobility of water molecules in the PEI matrix and the prevention of ice crystallization even at higher water concentrations [33]. This may be due to the intermolecular environment around water in PEI-water mixtures, which is more like a polymer than water-like due to intermolecular hydrogen bonding [31, 34].

3 Frozen Water Mixtures

3.1 Frozen Versus Uncrystallized Water Dynamics

Frozen water dynamics encompasses the study of molecular motions and interactions of water in its solid or vitrified states, typically at temperatures below its freezing point. At temperatures below the melting temperature T_m and at low solute concentrations, in the system containing water, the water crystallizes, and the system separates into two distinct regions:

1. **Ice Phase**

This region consists of pure crystalline water molecules arranged in an ordered ice lattice. The area shown by blue snowflakes in Fig. 7. Solute molecules (represented as elephants) cannot enter this phase because their size, polarity, or chemical nature disrupts the hydrogen-bonded ice network. Hence, ice is almost pure water with negligible solute content.

2. **Non-Crystalline (Uncrystallized) Solution Phase**

Fig. 7 Schematic representation of the phase-separated state of an aqueous solution below the melting temperature (T_m) at low solute concentration. The blue snowflakes represent the crystalline ice phase, composed of water molecules arranged in an ordered lattice. The orange elephants symbolize solute molecules, which are excluded from the ice lattice and remain dispersed within the non-crystalline solution phase (depicted by blue droplets)

This is the residual solution that remains after part of the water freezes. The uncrystallized portion contains solute molecules (represented as elephants) surrounded by uncrystallized water molecules (blue drops) as shown in Fig. 7. These water molecules remain unfrozen because the presence of solute depresses the freezing point and disturbs the cooperative hydrogen bonding required for crystallization. This region remains as a supercooled liquid or glass, depending on how far below T_{m} the temperature is.

Normally, the molecular mobility of water in the ice phase is extremely slow due to the rigid hydrogen-bonded lattice. In protein and polymer–water systems, the dielectric relaxation associated with water molecules crystallized in the ice phase appears in the kHz range (~ 10 kHz), indicating restricted dipolar motion. In contrast, the relaxation process corresponding to uncrystallized (non-frozen) water occurs in the GHz region, reflecting much faster molecular dynamics and higher local mobility of water molecules [33, 35–39].

3.2 Protein–Water Systems

Extending these investigations, we examined 20 and 40 wt% bovine serum albumin (BSA)–water mixture, [37] a heterogeneous system comprising protein, water in hydration shell, and bulk water as ice. Temperature dependence of relaxation times (τ) for 20 wt% and 40 wt% BSA–water mixtures were depicted in Fig. 8. Three major dielectric relaxation processes were identified:

1. ν(I) process—the fastest process, observed from 100 K and shifting to higher frequencies up to 270 K. I_c is originated from liquid water, but as described later, I_a and I_b processes were identified to be associated with ice, not with water.
2. Ice(II) process—a slower process appearing on the low-frequency side of the ν process, observed between 140 and 270 K. II_a and II_b processes were interpreted to be due to ice.
3. α(III) process—the slowest process, appearing between 190 and 260 K, partially masked by DC conductivity. The III process was interpreted as being due to hydrated protein; however, it seems to be accompanied by interfacial polarization, since its strength is too large to originate solely from fluctuations of the dipole moments of protein and water molecules.

The characteristic relaxation times $\tau(T)$ for the ice processes in both 20 wt% and 40 wt% BSA-water mixtures (II_a and II_b) closely match the relaxation time of pure ice, $\tau_{\mathrm{ice}}(T)$, and they exhibit comparable E_{a}. As demonstrated in Fig. 8, these findings correlate well with the calorimetric observations of Kawai et al. [40], who reported broad enthalpy relaxations in a 20% (w/w) aqueous BSA solution using adiabatic calorimetry [37].

Panagopoulou et al. investigated the dynamics of water in highly concentrated BSA–water mixtures (40–98 wt% of BSA) using differential scanning calorimetry

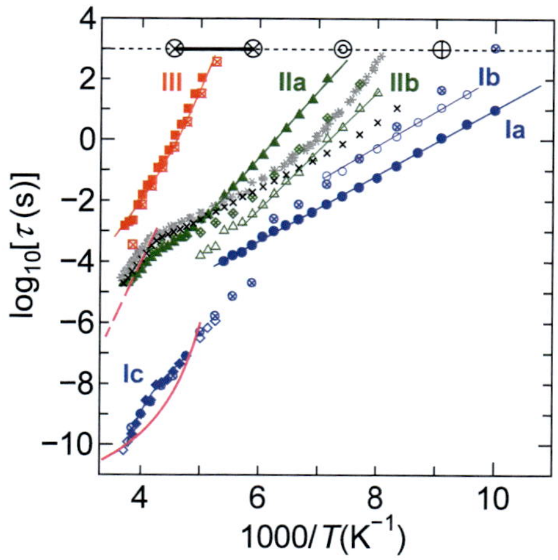

Fig. 8 Temperature dependence of relaxation times (τ) for 20 wt% BSA–water mixtures. Symbols represent different relaxation processes: I_a, I_b, I_c, II_a, II_b, and III. Open symbols indicate processes with weak or indistinct loss peaks. Crosses and asterisks correspond to relaxation times of pure ice, the magenta curve shows τ_c of D_2O reorientation near the myoglobin surface, while the magenta dashed line denotes τ_{HN} for α-fluctuations in hydrated myoglobin embedded in polyvinyl alcohol. Enthalpy relaxations with a relaxation time of 10^3 s originating from different processes found by Kawai et al. [40] are indicated by the thick line bounded on both sides by circles with a cross in the broad range from 170 to 220 K, a double circle at 135 K, and a circle with a plus at 110 K. This figure is reprinted from Ref. [37] with permission from American Chemical Society

(DSC), thermally stimulated depolarization current (TSDC), and dielectric relaxation spectroscopy (DRS) [41]. These complementary thermal and dielectric analyses enabled the identification of multiple relaxation processes arising from both the protein matrix and hydration water. DRS provides the most comprehensive view of the relaxation dynamics due to its broad time–frequency window (10^{-1}–10^6 Hz, − 150 to 20 °C), whereas TSDC covers a narrower range (10^{-4}–10^{-2} Hz). However, the superior resolution of the relaxation peak provides valuable insight. At very low hydration ($h_w = 0.02$), a broad dielectric loss peak near − 75 °C corresponded to motions of small polar protein side groups plasticized by residual water. With increasing hydration ($h_w = 0.07$–0.18), two relaxation modes were observed: a low-temperature process (≈ − 100 °C at $h_w = 0.07$) from side-chain motions and a higher-temperature α-relaxation associated with the glass transition of the water-plasticized protein. The temperature positions of these peaks aligned well with the TSDC data, with minor shifts due to the techniques' different frequencies.

At higher hydration ($h_w = 0.40$), three distinct modes appeared—a low-temperature process at ~ − 130 °C from polar protein groups and two α-relaxations near − 90 °C ($T_{g,1}$) and − 60 °C ($T_{g,2}$) attributed to the glassy dynamics of the BSA–water matrix. Isothermal $\varepsilon''(f)$ spectra showed a clear hydration-dependent plasticization: at − 90 °C, the main relaxation shifted from ~ 30 Hz ($h_w = 0.07$) to ~ 500 Hz ($h_w = 0.13$) and ~ 50 kHz ($h_w = 0.18$), saturating around 10 kHz for

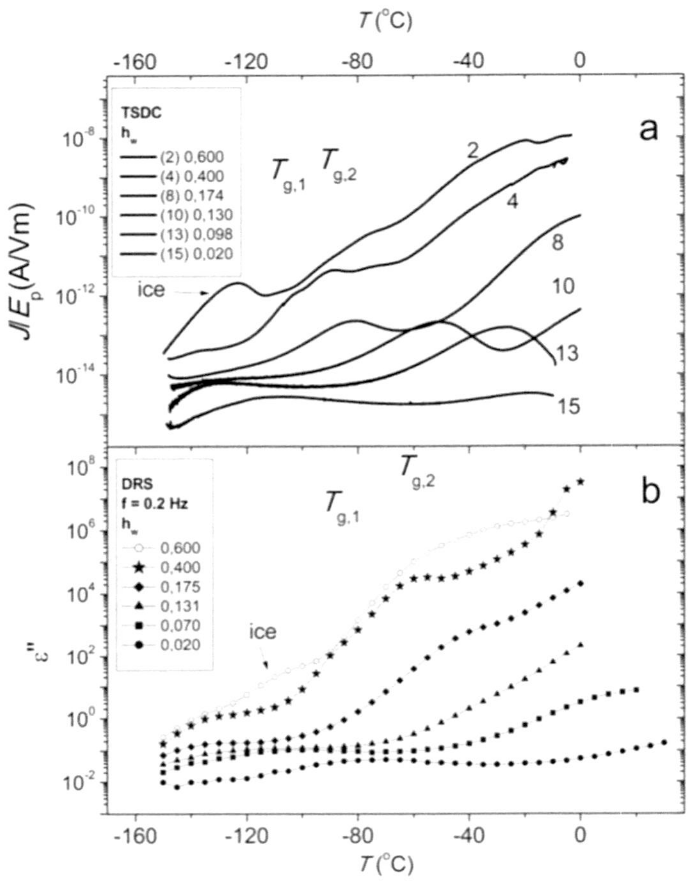

Fig. 9 Comparative TSDC thermograms and isochronal dielectric loss plots for BSA–water mixtures at different water fractions (h_n). **a** Normalized TSDC thermograms (J/E_p vs. T) for solutions and compressed pellets. **b** Isochronal dielectric loss at 0.2 Hz for the same samples, showing hydration-dependent relaxation behavior. This figure is reprinted from the Ref. [41] with permission from Elsevier

$h_w \geq 0.28$. Additional low-frequency (< 1 kHz) α-relaxations and ice-related losses appeared at high hydration ($h_w \geq 0.60$). The close correspondence between DRS and TSDC results confirms that both probe the same molecular mechanisms, indicating that hydration water in protein-rich environments exhibits restricted mobility, similar to supercooled or vitrified water, with relaxation times showing a slower temperature dependence and higher E_a than that of bulk water [41, 42].

Building on the insights from high-concentration BSA–water mixtures, comparative studies involving both BSA and lysozyme at similar hydration levels further elucidate how protein-specific structural features modulate frozen-state water dynamics. Using a combination of TSDC and DRS, the hydration-dependent relaxation behavior of these globular proteins has been systematically characterized.

Normalized TSDC thermograms reveal that at low water fractions (e.g., $h_w = 0.020$), both proteins exhibit a broad relaxation peak—located near –100 °C for lysozyme and – 110 °C for BSA—attributed to the motion of polar side groups plasticized by residual water. Increasing hydration to $h_w \approx 0.15$ this peak shifts to lower temperatures (~ – 125 °C) and increases its intensity, consistent with the progressive plasticization effect of water [43].

At moderate hydration ($h_w \approx 0.085$–0.098), an α-relaxation associated with the glass transition of the protein–water matrix emerges at – 40 °C for lysozyme and – 30 °C for BSA, with BSA consistently showing a higher glass transition temperature (T_g), suggesting stronger protein–water interactions and greater structural rigidity. This trend persists at higher hydration ($h_w \approx 0.148$–0.153), reinforcing the idea that BSA maintains a more constrained hydration shell compared to lysozyme.

When partial crystallization occurs upon cooling, both proteins exhibit splitting of the α-relaxation peak, albeit via different mechanisms—defective ice crystal populations in lysozyme versus probable microphase separation in BSA. Complementary DRS measurements, particularly isochronal dielectric loss plots at 0.2 Hz, confirm these assignments: polar side group relaxations appear around – 100 °C ($h_w = 0.070$), while α-relaxations shift toward higher temperatures (~ 0 °C for BSA and ~ – 10 °C for lysozyme). At high hydration, an additional peak at – 110 °C signals bulk ice formation, accompanied by increased low-frequency dielectric loss indicative of conductivity and electrode polarization effects [43].

3.3 Gelatin–Water Systems

Multiple relaxation processes are identified in 10–45 wt% gelatin-water mixtures. Each relaxation process exhibits distinct temperature and frequency dependencies, as depicted in Fig. 10 [38, 39]. A total of four major relaxation processes and two auxiliary hump features were observed:

(1) **ν-Relaxation**: appears at 123 K at low frequency (~ 10 mHz), which is associated with reorientational motion of water molecules (Process I_a). At temperatures below 181 K, τ_{Ia} follows Arrhenius-type behavior, while above 181 K, the temperature dependence changes to be the VFT-type., A deviations from the VFT-type occur at 260 K and 270 K due to cold crystallization and melting of ice. The relaxation strength of the ν process, $\Delta\varepsilon_\nu$ is nearly constant from 123 to 260 K, decreases slightly between 263 and 267 K (cold crystallization), then rises sharply above 270 K due to ice melting. Another relaxation associated with the water process was observed at temperatures above 183 K as a small step up to 260 K. It is clearly visible in the real part, suggesting a molecular origin rather than an artifact, which might be associated with the water process.

(2) **Ice-Related:** A slower relaxation on the lower frequency side of the ν process is visible at 153 K, then dominates around 193 K but is obscured by the α process above 263 K. $\Delta\varepsilon$ is comparable to pure ice, the τ shows good agreement with ice

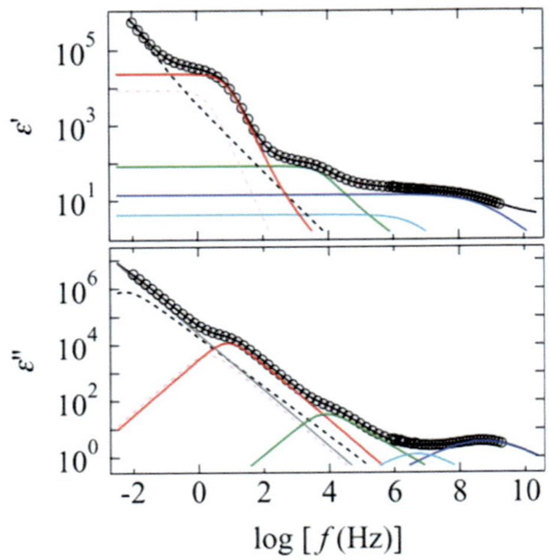

Fig. 10 Real and imaginary parts of dielectric functions for 40 wt% gelatin–water mixture at 258 K. The plots were obtained experimentally. The black solid curves were the sum of all relaxation processes. The blue and light blue are relaxation processes of water. Green, red, gray lines, and pink and black dashed lines are the relaxation processes of ice, hydrated gelatin, dc conductivity, and electrode polarization, respectively. This figure is reprinted from Ref. [38] with permission from American Institute of Physics

above 180 K but deviates below 190 K. A hump is observed at lower frequency at 173–193 K. Another shoulder like hump is observed on the low frequency side of the ice process between 223 and 272 K. At that time, it was just taken as artifacts and not considered as genuine molecular process.

(3) ***α* Process**: identified as the slowest relaxation in the 40 wt% gelatin-water mixture, emerges around 193 K, and becomes the dominant process at higher temperatures. This relaxation is strongly associated with the amorphous phase and the glass transition dynamics of hydrated gelatin, showing a T_g of approximately 193 K. This value is in close agreement with T_g values reported for hydrated proteins (~ 200 K), confirming its connection to protein-water interactions. A notable crossover in τ_{IIIa} occurs near 260 K, indicating changes in the underlying molecular mobility with increasing temperature. Moreover, the dielectric strength ($\Delta\varepsilon_{IIIa}$) is exceptionally high—about two orders of magnitude greater than that of typical polymer segmental relaxations—suggesting contributions not only from hydrated gelatin but also the interfacial polarization in heterogeneous phases.

To gain deeper insight into the ice-related process in gelatin-water mixtures, the temperature dependence of the relaxation time for the ice process was analyzed for systems containing 10–45 wt% gelatin-water mixtures and compared with pure ice data (Fig. 11) reported by Johari and Whalley [44] as well as Auty and Cole [45]. The relaxation time of ice in all gelatin-water mixtures exhibits Arrhenius-type. At lower temperatures (higher 1000/T values), the deviation from pure ice behavior becomes

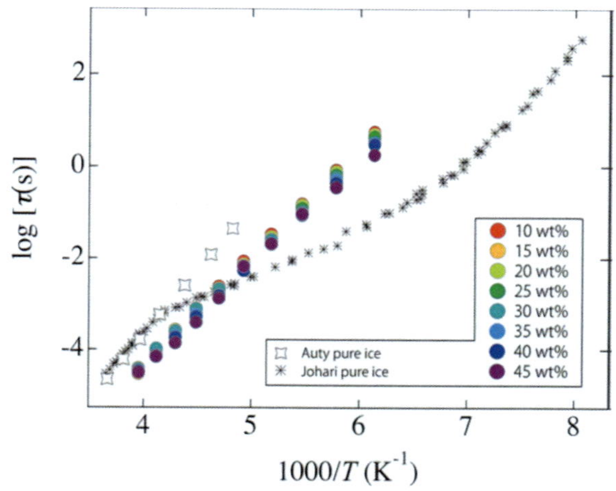

Fig. 11 Temperature dependence of the relaxation time (log τ) for ice relaxation (process II) in gelatin–water mixtures (10–45 wt%) compared with two types of ice relaxation observed in pure ice. Data for pure ice are taken from Johari and Whalley [44] and Auty and Cole [45]. This figure is reprinted with permission from Ref. [39] with permission from the American Chemical Society

more evident, reflecting a suppression of complete crystallization and the presence of amorphous or interfacial water domains. Increasing gelatin concentration from 10 to 45 wt% causes only minor shifts in relaxation times, indicating that beyond ~ 20 wt%, additional gelatin has a negligible effect on ice dynamics. The nearly parallel trends of relaxation time in all mixtures imply comparable E_a. These findings confirm that process II is associated with ice-like relaxation but is substantially influenced by the heterogeneous environment created by the gelatin network, thereby altering freezing behavior compared to pure-water ice.

3.4 *Types of Ice Relaxations*

The dielectric relaxation behavior of ice is a subject of long-standing scientific interest, primarily due to its relevance in understanding molecular dynamics, hydrogen bonding, and proton ordering within the ice lattice. Ice typically shows a single dielectric loss peak in broadband dielectric spectroscopy, which corresponds to the reorientation of water molecules and the movement of ionic or Bjerrum defects (D- and L-defects). The Bjerrum defects arise from orientational disorder in the hydrogen-bond network in ice, where two hydrogens (D-defect) or none (L-defect) occupy a single hydrogen bond. Such orientational defects propagate through molecular reorientation, while ionic defects (H_3O^+ and OH^-) move via proton translation along hydrogen bonds. The migration of these defects underlies the dielectric relaxation process in ice.

However, intriguingly, two distinct temperature dependences of the dielectric relaxation time have been reported, termed Auty-type [45] and Johari-type relaxations [44], It was reported that the different temperature dependence of the relaxation time of ice reflect different dynamic mechanisms and preparation conditions [46]. The Auty-type relaxation, originally reported by Auty and Cole, follows a classic

Arrhenius behavior over the temperature range of 273–208 K. This implies that the relaxation time (τ) of ice increases exponentially with decreasing temperature, described by the Arrhenius equation:

$$\tau = \tau_0 \exp\left(\frac{E_a}{RT}\right), \tag{3}$$

where τ_0 is the pre-exponential factor, E_a is the activation energy (approximately 58 kJ mol^{-1} for Auty-type ice), R is the gas constant, and T is the temperature in Kelvin. This behavior indicates a uniform and thermally activated molecular process, observed in ice crystallized from pure water under stirring at around 265 K [46]. It was interpreted that stirring facilitates uniform nucleation and growth, resulting in fewer impurities and a lower defect concentration. Consequently, fewer mobile Bjerrum or ionic defects are present, leading to longer dielectric relaxation times.

In contrast, the Johari-type relaxation displays a non-Arrhenius behavior, with three distinct E_a regimes depending on the temperature. Specifically, the E_a is 53.2 kJ mol^{-1} above a critical temperature T_{C1} (~ 240 K), drops to 18.8 kJ mol^{-1} between T_{C1} and T_{C2} (~ 170 K), and rises again to 46.4 kJ mol^{-1} below T_{C2}. This deviation from a single Arrhenius trend reflects changes in the dominant relaxation mechanism with temperature—such as a transition between ionic and orientational defect motion. Johari-type ice is formed by slow cooling of water without stirring, resulting in non-uniform crystal growth and a higher concentration of defects. These defects enhance molecular mobility, yielding significantly shorter relaxation times compared to the Auty-type. Thus, the dielectric relaxation of ice is not a singular process but is highly sensitive to preparation conditions, with distinct temperature dependencies reflecting underlying changes in defect populations and dynamics.

BDS measurements on partially crystallized water-rich gelatin−water mixtures (1–5 wt%) were done in the temperature range of 123–298 K, and four distinct relaxation processes of ice were observed below 200 K, despite most of the ice being hexagonal ice [47]. At higher temperatures, a single broad loss peak was observed, which split into four peaks around 225 K. The temperature- and concentration-dependent behavior of these processes confirmed their origin in ice. As shown in Fig. 12, the dominant relaxation process changed with increasing gelatin concentration—from process A at 1 wt% to B and C at 2 wt% and 3 wt%, respectively—while process D was only detected at concentrations below 3 wt%. The relaxation time (τ_{ice}) exhibited three distinct temperature regimes (high-, intermediate-, and low-temperature) characterized by different E_a: the low-temperature range showed the highest E_a (~ 43 kJ·mol^{-1}), the intermediate range had the smallest E_a (~ 20–25 kJ·mol^{-1}). The high-temperature range exhibited moderate E_a values (~ 33 kJ·mol^{-1}). With increasing gelatin concentration, the intermediate range, which exhibits the smallest E_a, narrows and ultimately disappears as gelatin concentration approaches ~ 9 wt%, leading to an Arrhenius-type temperature dependence for mixtures with $\geq$ 10 wt% gelatin. Additionally, processes with larger τ_{ice} consistently correspond to higher E_a, indicating that the formation of a denser gelatin network strongly affects to the ice dynamics. Furthermore, processes with larger τ_{ice} consistently exhibit higher E_a,

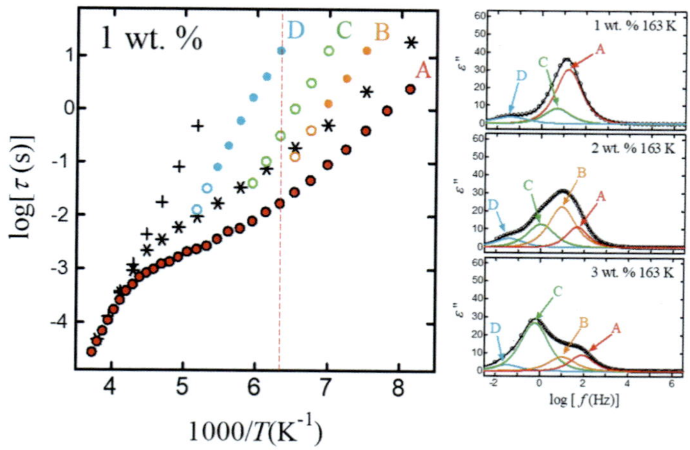

Fig. 12 Temperature dependence of the relaxation time (log τ) for ice-related processes (**a**–**d**) in a 1 wt% gelatin–water mixture (left) and dielectric loss spectra at 163 K for 1 wt%, 2 wt%, and 3 wt% mixtures (right). This figure is reprinted with permission from Ref. [47] with permission from the American Chemical Society

suggesting that slower relaxations are associated with stronger molecular constraints. Analysis of E_a against relaxation time revealed that processes A and B are more affected by impurity-induced orientational defects, while processes C and D are less influenced, indicating that gelatin modulates defect formation by altering ice crystallization kinetics. Overall, these observations demonstrate that gelatin concentration plays a crucial role in controlling ice dynamics by modifying defect densities, restricting crystal growth, and altering the relaxation mechanism without changing the underlying crystal structure of ice.

3.5 Relaxation of Ice in Polymer–Water Systems

Figure 13 compares the temperature dependence of dielectric relaxation time ($\log_{10}[\tau]$, top panel) and dielectric strength ($\Delta\varepsilon$, bottom panel) of ice relaxations in various partially crystallized 10 wt% water mixtures: Poly(N-isopropylacrylamide)(PNIPAM) microgel, PVP, BSA, and gelatin. These behaviors are benchmarked against pure ice data from Auty-Cole [45], Johari et al. [44], and Sasaki et al. [46]. The temperature range is categorized into three distinct regions separated by two crossover temperatures: $T_{C1} \approx 240$ K and $T_{C2} \approx 170$ K. These boundaries correspond to different dominant mechanisms within the trap-controlled proton transport model: orientational defects dominate above T_{C1} (> 240 K), ionic defects between T_{C1} and T_{C2} (240 K–170 K), and trap-controlled proton hopping becomes dominant below T_{C2} (< 170 K) [48–51].

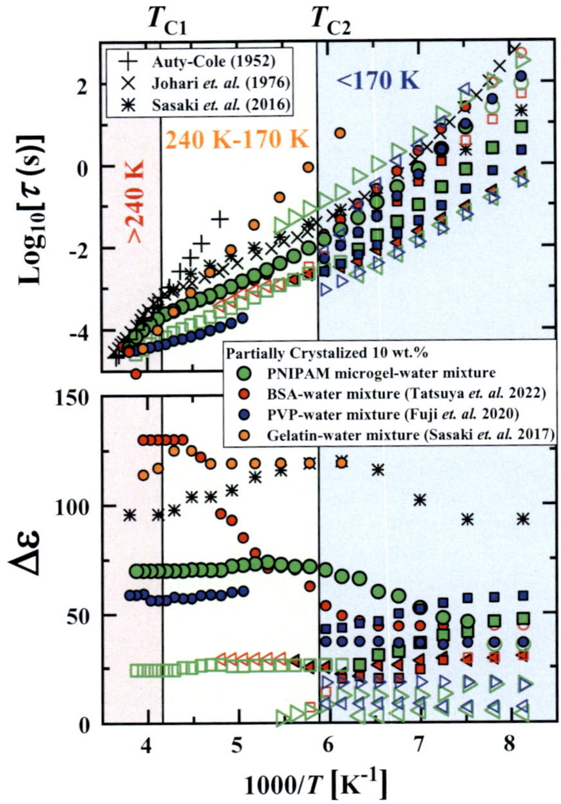

Fig. 13 Temperature-dependent relaxation time and dielectric strength of the ice process in partially crystallized 10 wt% PNIPAM (green), PVP (blue), BSA (red), and gelatin (orange) water mixtures. The black x, *, and + symbols represent Johari ice, Sasaki ice (Ih), and Auty ice, respectively. This figure is reprinted from Ref. [52] with permission from the Royal Society of Chemistry

In the top panel of Fig. 13, the temperature dependences of the relaxation time of ice for the water mixtures of PVP, BSA, and PNIPAM exhibit two or three temperature regions, which are similar to those observed for Johari-type ice relaxation [44]. On the other hand, the relaxation time of ice observed for the gelatin-water mixture shows an Arrhenius temperature dependence similar to that of the Auty type [45]. The relaxation times of ice are significantly shorter in polymer- and protein–water mixtures than in pure ice, due to the suppression of ionic relaxation and the dominance of proton trapping below T_{C2}. The presence of two or more relaxation processes, i.e., the broad distribution of relaxation times and activation energies, E_a, reflects variations in ice conditions within the mixtures. In contrast, gelatin–water mixtures exhibit Arrhenius-type temperature dependence associated with orientational defects [52].

3.6 α-Relaxation

For the first time, Process III, appearing on the lower-frequency side of the ice process observed in dielectric studies of PVP–water mixtures, has been conclusively identified as the α-relaxation process corresponding to the segmental motion of PVP, specifically under conditions below T_m [35]. However, when the relaxation times (τ_{III}) for Process III in 10–50 wt% frozen PVP–water mixtures were found to match those of the α-relaxation (τ_{PVP}) closely observed in 60–70 wt% uncrystallized PVP–water systems. Despite the lower polymer content, τ_{III} values in the crystallized mixture fell within the same range as τ_{PVP}, suggesting that freeze-concentration effects locally increased the effective polymer concentration, giving rise to segmental dynamics characteristic of the α-process.

Further confirmation came from the concentration dependence of the dielectric relaxation strength ($\Delta\varepsilon_{III}$) [35]. At 238 K, the $\Delta\varepsilon_{III}$ values for the frozen 40 and 50 wt% PVP-water mixtures aligned closely with the α-relaxation strength of PVP in non-frozen PVP-water mixtures measured at 288 K. This behavior is illustrated in Fig. 14, which compares dielectric strength at two temperatures—238 K (blue, Process III) and 288 K (red, α-process)—across a concentration range of 10–50 wt% PVP. At 288 K in PVP-water mixtures without crystallization, $\Delta\varepsilon$ increases nearly linearly from ~ 18 (10 wt%) to ~ 78 (50 wt%), reflecting that the dielectric relaxation strength is proportional to the polymer concentration. In contrast, at 238 K in crystallized PVP-water mixtures, $\Delta\varepsilon$ values for process III start high (~ 78) at low concentrations but remain relatively flat or slightly decrease (to ~ 65) with increasing PVP content. For PVP concentrations below 40 wt%, the ice phase occupies a large portion of the mixture, while small uncrystallized phases are dispersed within it. Thus, interfacial polarization occurs in the small uncrystallized phase and contributes to $\Delta\varepsilon_{III}$. The uncrystallized phase is smaller at lower PVP concentrations, and interfacial polarization is expected to increase with decreasing PVP concentration. The increase of τ_{III} with decreasing PVP concentration also seems to be affected by the interfacial polarization [35].

3.7 Comparative Dynamics

The temperature-dependent behavior of relaxation times in aqueous systems containing polymers or proteins reveals two distinct regimes that characterize the dielectric relaxation: one associated with uncrystallized (supercooled) water and the other with ice. At higher temperatures (lower values of $1000/T$), the relaxation times of uncrystallized water exhibit a steep slope. This behavior is due to the melting and/or crystallization of water. Crystallization increases the polymer concentration in non-crystalline phase, and higher concentration leads to larger relaxation times of both water and polymer processes.

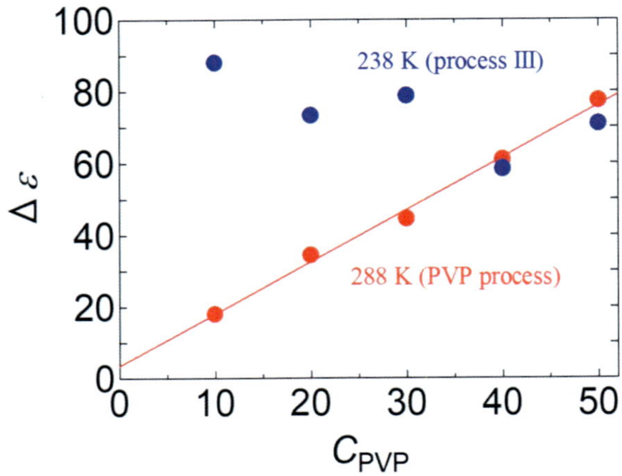

Fig. 14 Concentration dependence of the relaxation strength ($\Delta\varepsilon$) for process III at 238 K (blue) and for the PVP process at 288 K (red) in 10–50 wt% PVP–water mixtures. This figure is reprinted from Ref. [35] with permission from the American Chemical Society

Figure 15 illustrates the temperature dependence of relaxation times ($\log_{10}(\tau)$) as a function of the inverse temperature ($1000/T$) for protein–water systems, clearly showing two distinct dynamical regimes corresponding to the uncrystallized (supercooled) water and ice. At higher temperatures (lower $1000/T$ region), the ν process originated from uncrystallized water with small relaxation times exhibit a steep slope, reflecting the uncrystallized solution phase in which water molecules remain mobile and dynamically coupled to the protein matrix. As the temperature decreases, this ν process continues to move to a larger relaxation time and eventually overlaps with the ice process; the ν process originated from uncrystallized water and ice process remains spatially distinct. This overlap has been extensively reported in biomolecular systems such as BSA–water mixtures [37, 53] and protein hydration layers, as well as in polymer systems like PVP–water mixtures [35]. To elucidate the dynamics of uncrystallized water in the low-temperature region (below 200 K), the uncrystallized water process in water mixtures containing 60 and 72 wt% BSA [41], 64 wt% lysozyme and 66 wt% myoglobin [54] were compared. By introducing the relaxation times obtained from protein–water mixtures with low water content, the relaxation process of uncrystallized water can be clearly distinguished, owing to the small contribution from the ice component in these mixtures. Below 200 K, the temperature dependence of the relaxation time for the uncrystallized water process is larger than that of the ice-related processes. This comparison suggests that the relaxation previously interpreted as originated from uncrystallized water in the 20 and 40 wt% BSA–water mixtures [37] was in fact, due to ice rather than uncrystallized water [53].

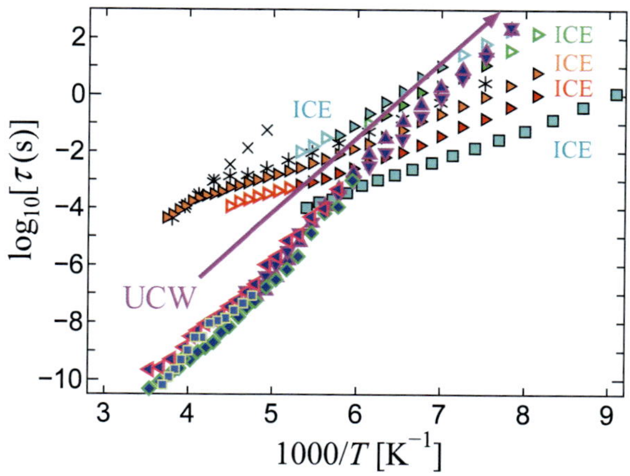

Fig. 15 Temperature dependence of the relaxation times ($\log_{10}(\tau)$) for protein–water systems showing two distinct regimes corresponding to uncrystallized (supercooled) water (UCW) and ice. Symbols indicate the four ice processes in the water mixtures of 2 wt% BSA (right-pointing triangles) [53], 20 wt% BSA (sky blue squares) [37]: it was interpreted as originated from UCW in Ref. [37], but identified to be associated with ice below 200 K in Ref. [53]. Other symbols indicate the UCW process for the water mixtures of 20 wt% BSA (squares above 200 K) [37], 60 wt% (upward triangles) and 72 wt% (downward triangles) BSA [42], 64 wt% Lysozyme (diamonds) and 66 wt% Myoglobin (left-pointing triangles) [54]

4 Conclusion

In small molecular liquids, a single relaxation process is observed at $T > T_c$, and this relaxation splits to two processes at $T < T_c$; (1) α-process: cooperative motion of solute and water and (2) ν-process: local motion of water. And the $\Delta\varepsilon_\nu$ increases as temperature drops to T_c, then decreases from T_c to T_g. However, the combined dielectric strength ($\Delta\varepsilon_\alpha + \Delta\varepsilon_\nu$) continues to increase with cooling, suggesting that water gradually participates more in the cooperative α-process at lower temperatures. But in High T_g Polymer PVP, $\Delta\varepsilon_\nu$ does not decrease between T_c and T_g, unlike in small-molecule–water mixtures, implying that the ν-process (water) in PVP–water is not strongly coupled with PVP's cooperative segmental motion. Instead, a slow water process (sometimes called m-process or α_2) appears between α and ν processes, supporting this uncoupled behavior. Hence, PVP dynamics dominate, with limited plasticization from water, unless at low PVP content. Like PVP, PVME also shows distinct α and ν processes, and in some cases, a slow m-process. The behavior of $\Delta\varepsilon_\nu$ in PVME–water mixtures is more similar to that in small molecular liquids, with a maximum at T_c and a decrease thereafter. Unlike PVP and PVME, PEI exhibited strong water–polymer coupling, no m-process, delayed plasticization, slowed water relaxation, and high water T_g, due to hydrogen bonding with PEI amines. These differences highlight how chemical structure and hydrogen bonding capacity of the polymer strongly influence the dynamics and glass transition behavior in water–polymer systems. These studies have clarified important aspects including the cooperative α-relaxation resulting from the dynamics of hydrogen-bonded networks and the local α-relaxation ascribed to the reorientation of water molecules. These model systems are crucial benchmarks that shed light on how water dynamics affect the flexibility and functionality of macromolecules. A more sophisticated view of the intricate, diverse interactions driving protein-water coupling and the ensuing dynamical behavior is made possible by applying this knowledge to hydrated protein systems.

The study of frozen water mixtures reveals that molecular dynamics in such systems are governed by the coexistence and interplay of crystalline (ice) and non-crystalline (uncrystallized) solution domains. Below the melting temperature, pure ice forms an ordered hydrogen-bonded lattice with extremely slow dipolar motion, while the remaining uncrystallized phase retains liquid-like dynamics characterized by fast relaxation in the GHz region. Two distinct relaxation types of ice—Auty-type (Arrhenius) and Johari-type (non-Arrhenius)—emerge, reflecting differences in defect concentration and crystallization pathways. Comparative dielectric analyses across protein–water (BSA, gelatin, lysozyme) and polymer–water (PVP, PNIPAM) systems demonstrate that solutes profoundly influence ice nucleation, defect formation, and the relaxation behavior of both frozen and unfrozen fractions. In all frozen polymer/protein—water mixtures, the α-relaxation persists even below the crystallization temperature, signifying the contribution of solute segmental motion and interfacial polarization within confined amorphous regions.

Overall, the dielectric relaxation behavior of frozen aqueous systems captures the delicate balance between structural rigidity imposed by the ice phase and dynamic flexibility maintained in uncrystallized domains. Understanding the dielectric relaxation behavior of polymer–water and frozen–unfrozen water mixtures provide a powerful framework for interpreting the complex dynamics of hydrated proteins. The cooperative α-relaxation observed in these model systems parallels the collective motions of protein backbones and hydration shells, while the local ν-process resembles the reorientation of water molecules at protein surfaces. By elucidating how hydrogen bonding and molecular flexibility influence these relaxation modes, such studies enable a deeper comprehension of protein stability, conformational transitions, and function under physiological and cryogenic conditions. Ultimately, this knowledge bridges the gap between polymer physics and biomolecular hydration dynamics, contributing to advances in cryopreservation, drug formulation, and the broader understanding of structure–dynamics–function relationships in biological systems.

References

1. Doster W (2010) The protein-solvent glass transition. Biochim Biophys Acta—Proteins Proteomics 1804:3–14. https://doi.org/10.1016/j.bbapap.2009.06.019

2. Bellissent-Funel MC, Hassanali A, Havenith M, Henchman R, Pohl P, Sterpone F, Van Der Spoel D, Xu Y, Garcia AE (2016) Water determines the structure and dynamics of proteins. Chem Rev 116:7673–7697. https://doi.org/10.1021/acs.chemrev.5b00664
3. Ngai KL, Capaccioli S, Shinyashiki N (2008) The protein "glass" transition and the role of the solvent. J Phys Chem B 112:3826–3832. https://doi.org/10.1021/jp710462e
4. Sudo S, Shinyashiki N, Kitsuki Y, Yagihara S (2002) Dielectric relaxation time and relaxation time distribution of alcohol-water mixtures. J Phys Chem A 106:458–464. https://doi.org/10.1021/jp013117y
5. Shinyashiki N, Sudo S, Abe W, Yagihara S (1998) Shape of dielectric relaxation curves of ethylene glycol oligomer–water mixtures. J Chem Phys 109:9843–9847. http://ojps.aip.org/jcpo/jcpcr.jsp
6. Shinyashiki N, Yagihara S (1999) Comparison of dielectric relaxations of water mixtures of poly(vinylpyrrolidone) and 1-vinyl-2-pyrrolidinone. J Phys Chem B 103:4481–4484. https://doi.org/10.1021/jp983800t
7. Yamamoto W, Sasaki K, Kita R, Yagihara S, Shinyashiki N (2015) Dielectric study on temperature-concentration superposition of liquid to glass in fructose-water mixtures. J Mol Liq 206:39–46. https://doi.org/10.1016/j.molliq.2015.01.057
8. Shinyashiki N, Sengwa RJ, Tsubotani S, Nakamura H, Sudo S, Yagihara S (2006) Broadband dielectric study of dynamics of poly(vinyl pyrrolidone)—ethylene glycol oligomer blends. J Phys Chem A 110:4953–4957. https://doi.org/10.1021/jp0605193
9. Shinyashiki N, Imoto D, Yagihara S (2007) Broadband dielectric study of dynamics of polymer and solvent in poly(vinyl pyrrolidone)/normal alcohol mixtures. J Phys Chem B 111:2181–2187. https://doi.org/10.1021/jp065414e
10. Shinyashiki N, Miyara M, Nakano S, Yamamoto W, Ueshima M, Imoto D, Sasaki K, Kita R, Yagihara S (2013) Dielectric relaxation strength and magnitude of dipole moment of poly(vinyl pyrrolidone)in polar solutions. J Mol Liq 181:110–114. https://doi.org/10.1016/j.molliq.2013.02.020
11. Shinyashiki N, Yagihara S, Arita I, Mashimo S (1998) Dynamics of water in a polymer matrix studied by a microwave dielectric measurement. J Phys Chem B 102:3249–3251. https://doi.org/10.1021/jp9729627
12. Shinyashiki N, Sakai T, Yamada G, Yagihara S (1999) Dielectric study on the dynamics of an aqueous solution of gellan gum. Progr Colloid Polym Sci 114:36–40. https://doi.org/10.1007/3-540-48349-7_6
13. Hayashi Y, Shinyashiki N, Yagihara S (2002) Dynamical structure of water around biopolymers investigated by microwave dielectric measurements using time domain reflectometry method. J Non Cryst Solids 305:328–332. https://doi.org/10.1016/S0022-3093(02)01113-4
14. Ryabov YE, Feldman Y, Shinyashiki N, Yagihara S (2002) The symmetric broadening of the water relaxation peak in polymer–water mixtures and its relationship to the hydrophilic and hydrophobic properties of polymers. J Chem Phys 116:8610–8615. https://doi.org/10.1063/1.1471551
15. Kaminski K, Kaminska E, Ngai KL, Paluch M, Wlodarczyk P, Kasprzycka A, Szeja W (2009) Identifying the origins of two secondary relaxations in polysaccharides. J Phys Chem B 113:10088–10096. https://doi.org/10.1021/jp809760t
16. Sudo S, Shimomura M, Kanari K, Shinyashiki N, Yagihara S (2006) Broadband dielectric study of the glass transition in poly(ethyleneglycol)- water mixture. J Chem Phys 124:044901-1–44906. https://doi.org/10.1063/1.2149860
17. Shinyashiki N, Sudo S, Yagihara S, Spanoudaki A, Kyritsis A (2007) Relaxation processes of water in the liquid to glassy states of water mixtures studied by broadband dielectric spectroscopy. J Phys Condens Matter 19:205113-1–205212. https://doi.org/10.1088/0953-8984/19/20/205113
18. Sasaki K, Matsui Y, Miyara M, Kita R, Shinyashiki N, Yagihara S (2016) Glass transition and dynamics of the polymer and water in the poly(vinylpyrrolidone)-water mixtures studied by dielectric relaxation spectroscopy. J Phys Chem B 120:6882–6889. https://doi.org/10.1021/acs.jpcb.6b05347

19. Capponi S, Arbe A, Cerveny S, Busselez R, Frick B, Embs JP, Colmenero J (2011) Quasielastic neutron scattering study of hydrogen motions in an aqueous poly(vinyl methyl ether) solution. J Chem Phys 134:204906-1–204915. https://doi.org/10.1063/1.3592560
20. Takatsuka M, Arai Y, Sasaki K, Kita R, Yagihara S, Shinyashiki N (2024) Dielectric relaxations of polymer and primary and slow water in water mixtures of poly(vinyl methyl ether) as a polymer with low glass transition temperature. Macromolecules 57:8338–8350. https://doi.org/10.1021/acs.macromol.4c00616
21. Cerveny S, Colmenero J, Alegría A (2005) Dielectric investigation of the low-temperature water dynamics in the poly(vinyl methyl ether)/H_2O system. Macromolecules 38:7056–7063. https://doi.org/10.1021/ma050811t
22. Ngai KL, Capaccioli S, Shinyashiki N, Thayyil MS (2010) Recent progress in understanding relaxation in complex systems. J Non Cryst Solids 356:535–541. https://doi.org/10.1016/j.jnoncrysol.2009.03.011
23. Capaccioli S, Ngai KL, Shinyashiki N (2007) The Johari–Goldstein β-relaxation of water. J Phys Chem B 111:8197–8209. https://doi.org/10.1021/jp071857m
24. Sudo S, Shimomura M, Shinyashiki N, Yagihara S (2002) Broadband dielectric study of α-β separation for supercooled glycerol-water mixtures. J Non Cryst Solids 307–310:356–363. https://doi.org/10.1016/S0022-3093(02)01496-5
25. Sudo S, Shinyashiki N, Yagihara S (2001) The dielectric relaxation of supercooled ethyleneglycol-water mixtures. J Mol Liq 90:113–120. https://doi.org/10.1016/S0167-7322(01)00113-1
26. Sudo S, Tobinai S, Shinyashiki N, Yagihara S (2006) Broadband dielectric study on glass transition of 1-propanol-water mixture. AIP Conf Proc 832:149–152. https://doi.org/10.1063/1.2204477
27. Sudo S, Shimomura M, Saito T, Kashiwagi T, Shinyashiki N, Yagihara S (2002) Dielectric study on α- and β-processes in supercooled diethyleneglycol- and pentaethyleneglycol—Water mixtures. J Non Cryst Solids 305:197–203. https://doi.org/10.1016/S0022-3093(02)01094-3
28. Sudo S, Tsubotani S, Shimomura M, Shinyashiki N, Yagihara S (2004) Dielectric study of the α and β processes in supercooled ethylene glycol oligomer-water mixtures. J Chem Phys 121:7332–7340. https://doi.org/10.1063/1.1796232
29. Shinyashiki N, Shinohara M, Iwata Y, Goto T, Oyama M, Suzuki S, Yamamoto W, Yagihara S, Inoue T, Oyaizu S, Yamamoto S, Ngai KL, Capaccioli S (2008) The glass transition and dielectric secondary relaxation of fructose-water mixtures. J Phys Chem B 112:15470–15477. https://doi.org/10.1021/jp807038r
30. Sasaki K, Takatsuka M, Shinyashiki N, Ngai KL (2021) Relating the dynamics of hydrated poly(vinyl pyrrolidone) to the dynamics of highly asymmetric mixtures and polymer blends. J Mol Liq 333:115907. https://doi.org/10.1016/j.molliq.2021.115907
31. Safna Hussan KP, Inoue S, Arai Y, Sasaki K, Kita R, Ito T, Yagihara S, Shinyashiki N (2025) The unique dynamics of water and polymer in polyethylene imine–water mixtures across the glass transition. Polym J 328:128447. https://doi.org/10.1016/j.polymer.2025.128447
32. Combarro Palacios I, Olsson C, Kamma-Lorger CS, Swenson J, Cerveny S (2019) Motions of water and solutes—slaving versus plasticization phenomena. J Chem Phys 150:124902-1–7. https://doi.org/10.1063/1.5030064
33. Shinyashiki N, Shimomura M, Ushiyama T, Miyagawa T, Yagihara S (2007) Dynamics of water in partially crystallized polymer/water mixtures studied by dielectric spectroscopy. J Phys Chem B 111:10079–10087. https://doi.org/10.1021/jp0730489
34. Safna Hussan KP, Inoue S, Arai Y, Sasaki K, Kita R, Takeru I, Shinyashiki N (2025) Molecular insights into an electrolytic polymer, polyethylene imine. Polymer 321:128101. https://doi.org/10.1016/j.polymer.2025.128101
35. Fujii M, Sasaki K, Matsui Y, Inoue S, Kita R, Shinyashiki N, Yagihara S (2020) Dynamics of uncrystallized water, ice, and hydrated polymer in partially crystallized poly(vinylpyrrolidone)-water mixtures. J Phys Chem B 124:1521–1530. https://doi.org/10.1021/acs.jpcb.9b11552
36. Shinyashiki N, Matsumura Y, Miura N, Yagihara S, Mashimo S (1994) Dielectric study of water structure in polymer solution. J Phys Chem 98:13612–13615. https://doi.org/10.1021/j100102a028

37. Shinyashiki N, Yamamoto W, Yokoyama A, Yoshinari T, Yagihara S, Kita R, Ngai KL, Capaccioli S (2009) Glass transitions in aqueous solutions of protein (bovine serum albumin). J Phys Chem B 113:14448–14456. https://doi.org/10.1021/jp905511w
38. Sasaki K, Kita R, Shinyashiki N, Yagihara S (2014) Glass transition of partially crystallized gelatin-water mixtures studied by broadband dielectric spectroscopy. J Chem Phys 140:124506-1–124507. https://doi.org/10.1063/1.4869346
39. Sasaki K, Panagopoulou A, Kita R, Shinyashiki N, Yagihara S, Kyritsis A, Pissis P (2017) Dynamics of uncrystallized water, ice, and hydrated protein in partially crystallized gelatin-water mixtures studied by broadband dielectric spectroscopy. J Phys Chem B 121:265–272. https://doi.org/10.1021/acs.jpcb.6b04756
40. Kawai K, Suzuki T, Oguni M (2006) Low-temperature glass transitions of quenched and annealed bovine serum albumin aqueous solutions. Biophys J 90:3732–3738. https://doi.org/10.1529/biophysj.105.075986
41. Panagopoulou A, Kyritsis A, Sabater R, Serra I, Gómez Ribelles JL, Shinyashiki N, Pissis P (2011) Glass transition and dynamics in BSA-water mixtures over wide ranges of composition studied by thermal and dielectric techniques. Biochim Biophys Acta 1814:1984–1996. https://doi.org/10.1016/j.bbapap.2011.07.014
42. Panagopoulou A, Kyritsis A, Shinyashiki N, Pissis P (2012) Protein and water dynamics in bovine serum albumin-water mixtures over wide ranges of composition. J Phys Chem B 116:4593–4602. https://doi.org/10.1021/jp2105727
43. Kyritsis A, Panagopoulou A, Pissis P, Serra RSI, Gómez Ribelles JL, Shinyashiki N (2012) Water and protein dynamics in protein-water mixtures over wide range of composition. IEEE Trans Dielectr Electr Insul 19:1239–1246. https://doi.org/10.1109/TDEI.2012.6259997
44. Johari GP, Whalley E (1981) The dielectric properties of ice lh in the range 272–133 K. J Chem Phys 75:1333–1340. https://doi.org/10.1063/1.442139
45. Auty RP, Cole RH (1952) Dielectric properties of ice and solid D2O. J Chem Phys 20:1309–1314. https://doi.org/10.1063/1.1700726
46. Sasaki K, Kita R, Shinyashiki N, Yagihara S (2016) Dielectric relaxation time of ice-Ih with different preparation. J Phys Chem B 120:3950–3953. https://doi.org/10.1021/acs.jpcb.6b01218
47. Yasuda T, Sasaki K, Kita R, Shinyashiki N, Yagihara S (2017) Dielectric relaxation of ice in gelatin-water mixtures. J Phys Chem B 121:2896–2901. https://doi.org/10.1021/acs.jpcb.7b00149
48. Popov I, Puzenko A, Khamzin A, Feldman Y (2015) The dynamic crossover in dielectric relaxation behavior of ice Ih. Phys Chem Chem Phys 17:1489–1497. https://doi.org/10.1039/c4cp04271a
49. Popov I, Lunev I, Khamzin A, Greenbaum A, Gusev Y, Feldman Y (2017) The low-temperature dynamic crossover in the dielectric relaxation of ice Ih. Phys Chem Chem Phys 19:28610–28620. https://doi.org/10.1039/c7cp05731h
50. Khamzin AA, Nasybullin AI (2018) Trap-controlled proton hopping: interpretation of low-temperature dielectric relaxation of ice Ih. Phys Chem Chem Phys 20:23142–23150. https://doi.org/10.1039/C8CP03468K
51. Khamzin AA, Nasybullin AI, Nikitin AS (2021) Theoretical description of dielectric relaxation of ice with low concentration impurities. Chem Phys 541:111040. https://doi.org/10.1016/j.chemphys.2020.111040
52. Vijayakumar B, Takatsuka M, Sasaki K, Kita R, Shinyashiki N, Yagihara S, Rathnasabapathy S (2023) Dielectric relaxation of ice in a partially crystallized poly(N-isopropylacrylamide)microgel suspension compared to other partially crystalized polymer-water mixtures. Phys Chem Chem Phys 25:22223–22231. https://doi.org/10.1039/d3cp02116e
53. Tsukahara T, Sasaki K, Kita R, Shinyashiki N (2022) Dielectric relaxations of ice and uncrystallized water in partially crystallized bovine serum albumin-water mixtures. Phys Chem Chem Phys 24:5803–5812. https://doi.org/10.1039/d1cp05679d

54. Nakanishi M, Sokolov AP (2015) Protein dynamics in a broad frequency range: dielectric spectroscopy studies. J Non Cryst Solids 407:478–485. https://doi.org/10.1016/j.jnoncrysol.2014.08.057

Protein Dynamics Slaved by Solvent Dynamics as Revealed by Dielectric Spectroscopy

Jan Swenson and Helén Jansson

Abstract Broadband dielectric spectroscopy (BDS) is a powerful experimental technique since it can probe dynamics on widely different time scales. This makes it possible to study the relation between protein and solvent dynamics, despite that they usually occur on very different time scales. The results from such investigations show that the major protein and solvent relaxation processes (i.e. the viscosity related α-relaxation in the case of the solvent) exhibit the same temperature dependence, although the protein relaxations occur on a much slower time scale. This finding verifies the correctness of Fraunfelder's slaving model, where it is proposed that protein dynamics is slaved by the motions in the surrounding solvent. Furthermore, by comparing the BDS data with data from differential scanning calorimetry (DSC) measurements it is evident that both the α-relaxation in the solvent as well as several different protein relaxations are involved in the broad calorimetric glass transition, i.e. both techniques demonstrate that glass transition related protein dynamics occur on widely different time scales. These findings underscore the critical role of solvent mobility in governing protein dynamics and biological activity, with implications for cryopreservation and the design of protein-stabilizing formulations.

Keywords Protein dynamics · Solvent dynamics · Hydration water · Glass transition · Slaving · Dielectric spectroscopy · Differential scanning calorimetry

1 Introduction

Relaxations of proteins and other biological macromolecules are intricately determined by the surrounding water, which modifies their structural and dynamic properties. Without hydration water, proteins would not fold properly and thus lack the

J. Swenson (✉)
Department of Physics and Astronomy, Chalmers University of Technology, Göteborg, Sweden
e-mail: jan.swenson@chalmers.se

H. Jansson
Department of Architecture and Civil Engineering, Chalmers University of Technology, Göteborg, Sweden

S. Cerveny and J. Swenson (eds.), *Dynamics of Biomolecules Studied by Dielectric Spectroscopy*, Advances in Dielectrics, https://doi.org/10.1007/978-3-032-30588-6_6

conformational flexibility and structural properties required for their specific functions and activities. This is because hydration shell fluctuations interact dynamically with protein motions, driving protein motions and influencing the protein energy landscape. Therefore, the function of a protein is critically dependent on its surrounding hydration water. This can, for example, be seen by the fact that the protein lysozyme, as shown in Fig. 1, exhibits no enzymatic activity at a water content below ~ 0.2 g water per gram of protein, and full activity requires roughly equal masses of water and protein [1, 2]. Furthermore, it is well established that protein function involves large-scale conformational changes and that these essential motions are affected by the solvent [3–8].

Based on this knowledge, the central question is: *how* does the solvent promote protein dynamics and associated functions [3–8]? While a definitive answer remains elusive, we demonstrate here how broadband dielectric spectroscopy (BDS) can be used to probe both solvent and protein motions, making it an ideal tool for elucidating protein-solvent coupling and highlighting the necessity of solvent mobility for protein function. Especially when combined with other techniques such as differential scanning calorimetry (DSC).

The conformational energy landscape of a protein is highly complex and cannot be mapped in full [9]. Nonetheless, as illustrated in Fig. 2, it features a hierarchical structure with a few major states (shown as "valleys") in the top level CS0, only one of which represents the native and biologically active structure (i.e. A_1 in this illustration). Each of these major states contains multiple substates CS1α, accessible only via large conformational changes. Frauenfelder and co-workers [4–8] showed that transitions between these CS1α states, which are given by the rate constant k_{α}, are "slaved" to the cooperative, viscosity-dependent α-relaxation in the surrounding solvent. Each substate also has a finer structural hierarchy CS1β, accessible through

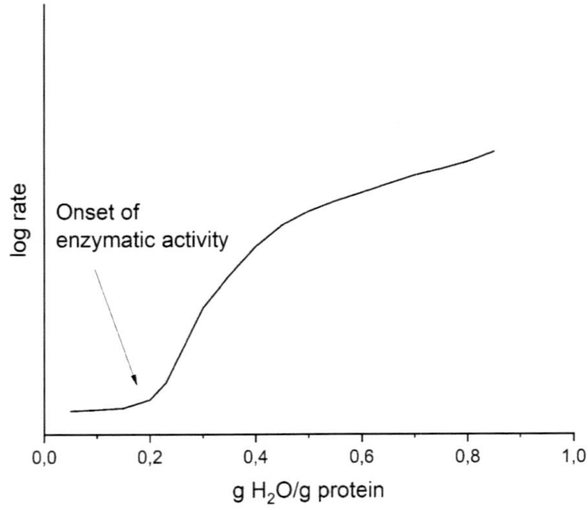

Fig. 1 Enzymatic activity of the protein lysozyme starts at a water content of ~ 0.2 g water per g of protein and for full activity, about the same amount of water as protein is required [1]

more localized motions near a main conformation. These are governed by the β-relaxation in the hydration shell and given by the rate constant k_β [4–8].

In Fig. 3, the temperature dependence of local and more large-scale relaxations of the protein myoglobin and its surrounding solvent are shown by blue and red lines, respectively. At low temperatures, the time scale and temperature dependence for the α- and β-relaxations in the solvent differ, i.e. they are well separated and easy to follow. However, at some temperature above the solvent's T_g, β- and α-relaxations merge, making it difficult to isolate β-relaxation effects. In this regime, dielectric spectra are dominated by solvent α-relaxation and the slower large-scale protein motions, which the slaving model predicts share the same non-Arrhenius temperature dependence [3, 4, 8, 10].

Further, a consequence of the slaving model is that when the α-relaxation in the solvent freezes in or disappears around T_g (T_g solvent in Fig. 3), also the more large-scale protein motions vanish. On the other hand, below T_g where the α-relaxation in the solvent is absent, and therefore also the large-scale protein motions, it is possible to observe more local (β-like) internal protein motions, which are slaved by the β-relaxation in the hydration shell. Since this β-relaxation exhibits an Arrhenius temperature dependence, it further implies that the internal local protein motions

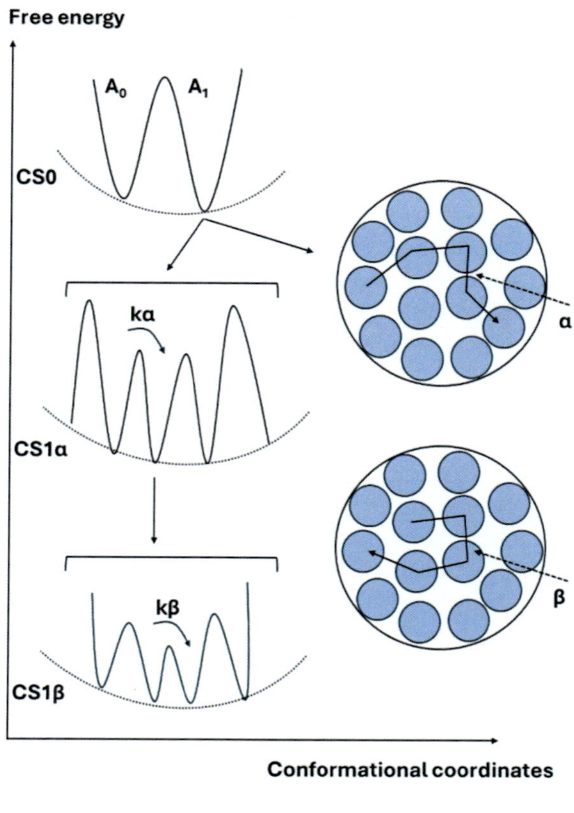

Fig. 2 Typical energy landscape of a protein. The top level CS0 is represented by A_0 and A_1, which are two completely different structures of the protein. In this illustration, the protein can only be active in one of the conformations, here represented by state A_1. Each of the conformations in CS0 can assume a large number of substates CS1, which in turn can be divided into CS1α and CS1β. Transitions between CS1α, which are given by the rate constant k_α, involve large-scale structural changes and are slaved to the α-relaxation in the surrounding solvent (see the top figure to the right). Every CS1α substate contains CS1β. The transitions, with the rate constant k_β, at this level occur due to the β-relaxation in the hydration water (see the bottom figure to the right)

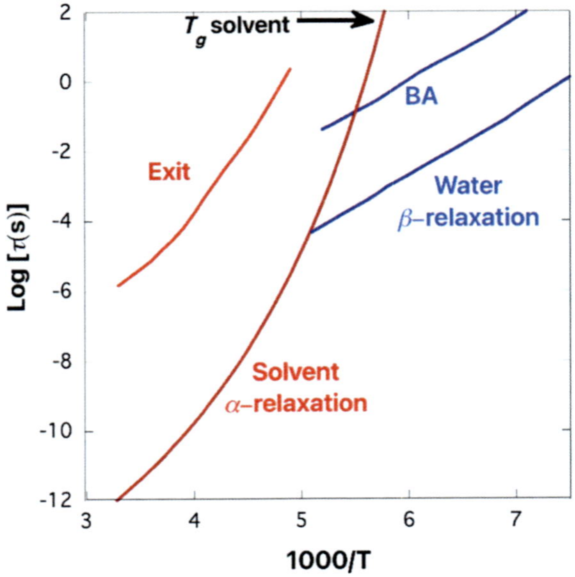

Fig. 3 Large-scale protein motions, as here exemplified by the motions responsible for the escape of CO from myoglobin (Exit), are determined by the α-relaxation in the solvent, whereas local CO migration (BA) within the protein myoglobin is determined by the β-relaxation, as evidenced by the difference in temperature dependence. T_g of the solvent is defined as the temperature where the α-relaxation in the solvent reaches a time scale of 100 s

also follow an Arrhenius temperature dependence. Another difference between the large-scale conformational changes and the more local protein motions is that the large-scale motions do not require water as the solvent [4, 8]. This is in contrast to the local motions, which are slaved by the β-relaxation in the water closest to the protein surface [4, 8]. However, the finding that large-scale protein motions do not require water does not imply that proteins can function equally well in other solvents than water. First, the alternative solvent needs to preserve the native structure of the protein, and second, it has to maintain the timescale of the biological important large-scale motions, i.e. the α-relaxation in the solvent needs to be given by the same rate constant k_α as for water. Furthermore, for full activity of the protein also the internal local motions, which are slaved by the β-relaxation in the surrounding water layer, are needed, and therefore full protein activity cannot be obtained without water.

Another important implication of the slaving model is that increased protein mobility with temperature is not driven directly by thermal energy but by the corresponding increase in solvent motions [6]. Molecular dynamics (MD) simulations by Tarek and Tobias [5] confirm that solvent translational motions are essential for substantial protein dynamics, whereas local solvent rotations and reorientations have little effect. Consistent with this, Doster and Settles [7] proposed that the key coupling parameter between a protein and its solvent is the viscosity of the hydration water. However, it should be noted that it is not only the protein which is affected by the solvent. The other way around is also true, particularly in the case of the hydration water, i.e. the layer of water molecules hydrogen bonded to the protein surface. It has been shown [11–13] that this hydration water is slowed down, compared to bulk water, by the interactions with the protein. Due to this interaction, there is also dynamics of surface groups of the protein on the same timescale as the hydration water, i.e.

they exhibit coupled dynamics [14, 15]. Such coupled protein-water dynamics has also been observed in MD simulations [16], where the authors also found that there is a pronounced anticorrelation between the energy of the protein and the energy in the surrounding water. Thus, due to the law of energy conservation the protein energy increases when the water energy decreases and vice versa [16]. This further implies that the energy (in terms of entropic contribution) provided by the solvent is needed for the protein to overcome the energy (or enthalpy, to be more precise) barriers between the different substates, as shown in Fig. 2.

The solvent driven protein fluctuations are furthermore directly related to the calorimetric glass transition, as measured by differential scanning calorimetry (DSC). The calorimetric glass transition temperature, T_g, of a protein occurs when the time scale of these fluctuations reaches approximately 100 s, in analogy to how the structural α-relaxation time is commonly used to define T_g, of ordinary glasses [6]. However, as mentioned above, since the protein motions are driven by the solvent, in contrast to ordinary glasses, this further implies that the glass transition is not an intrinsic property of proteins, i.e. proteins cannot undergo a glass transition without a dynamic surrounding. Therefore, calorimetric and also rheological methods can be used to probe protein motions and their relation to solvent dynamics [17–21]. Such studies have shown that the glass transition of proteins is exceptionally broad [18, 20, 21], thus lacking a single characteristic temperature. This broadness has been attributed either to a distribution of water cluster sizes on the protein surface [21] or to a wide range of relaxation times arising from numerous locally relaxing regions within the protein–water system [18]. Our previous work [11, 22] supports the latter explanation and shows that multiple distinct protein relaxations contribute to the glass transition region in hydrated proteins.

To systematically investigate protein dynamics, solvent dynamics and their interrelation, it is valuable to vary both the total solvent content and the solvent's dynamical properties. In this study, we achieve this by using water–glycerol mixtures. Such mixtures offer several advantages:

1. Biological function can be maintained at high glycerol contents (up to at least 90 wt% for globular proteins such as lysozyme [23]).
2. Glycerol forms hydrogen bonds with water molecules and reduces the tendency of water to crystallize in the supercooled state.
3. Like water, glycerol forms hydrogen bonds with itself and with biological surfaces.
4. The solvent's dynamical properties can be tuned by adjusting the glycerol/water ratio.

Our calorimetric and dielectric measurements link the calorimetric glass transition in the protein–solvent (water–glycerol) system to the relaxation processes identified by dielectric spectroscopy. This enables direct probing of protein–solvent dynamic coupling and reveals how the relaxation behaviour differs from that of hydrated proteins. The results show that the calorimetric glass transition arises from the freezing of both the solvent α-relaxation and multiple protein relaxations occurring on different timescales. Furthermore, protein motions are directly related to

Table 1 Investigated sample compositions, where h is grams of solvent (both water and glycerol) per grams of protein

Water 100 wt%	Glycerol 100 wt%	Water: glycerol 20:80 wt%	Water: glycerol 33:67 wt%	Water: glycerol 40:60 wt%	Water: glycerol 50:50 wt%
$h = 0.33$	$h = 1$	$h = 1$	$h = 1$	$h = 1$	$h = 1$
$h = 0.5$		$h = 2$	$h = 2$	$h = 2$	$h = 2$

solvent dynamics. In hydrated proteins, hydration water exhibits a dynamic crossover at 170 ± 10 K, where a local low-temperature water relaxation transforms into an α-like relaxation [24]. This onset of more cooperative, long-range water motions appears to drive the glass-transition-like onset of large-scale protein fluctuations.

2 Experimental Section

2.1 *Sample Preparation*

For this study, freeze-dried horse heart myoglobin powder (Sigma) was hydrated or mixed with water–glycerol mixtures of varying composition. Hydrated proteins were prepared in 100% humidity (desiccator) whereas homogeneous water-glycerol samples were obtained by gradually adding small amounts of protein powder to the solvent while continuously stirring. The investigated compositions, expressed as wt% water and glycerol in the solvent and total solvent level h (grams solvent/gram protein), are shown in Table 1. As reference samples, the corresponding pure solvent mixtures were used. A residual water content of ≈ 0.1 wt% was present but not included in the stated hydration levels.

2.2 *Calorimetric Measurements*

The DSC measurements were performed on a TA Instruments Q1000 using ~ 5–10 mg of sample in hermetic aluminium pans. An empty pan of the same type was used as reference. Each measurement was performed by cooling the sample to 95 K where it was equilibrated before heated up to 295 K at 5 K/min. The glass transition temperature (T_g) was defined as the inflection point in the heat flow curve (maximum in its derivative). The broadness of the glass transition, i.e. the glass transition range (ΔT_g), was determined from the derivative curve of the heat flow with respect to the temperature. All measurements were repeated to ensure reproducibility.

2.3 Dielectric Measurements

Dielectric spectra were recorded using a Novocontrol Concept 80 broadband dielectric spectrometer. The samples were investigated in the frequency interval 10^{-2}–10^{9} Hz, where the low-frequency measurements (10^{-2}–10^{7} Hz) were performed by employing an Alpha-S High Resolution Dielectric Analyser and the high-frequency data (10^{6}–10^{9} Hz) were obtained using an Agilent 4291B RF Impedance Analyser. For the measurements, the samples (0.1 mm thick, defined by silica spacers) were placed between gold-plated electrodes (20 and 10 mm in diameter for the low frequency and high-frequency measurements, respectively). All measurements were performed upon heating from 120 to 350 K, with isothermal (± 0.2 K) scans at every 5 K.

The imaginary permittivity (with dc conductivity subtracted) and the temperature derivative of the real permittivity (Eq. 1) [25] were analysed due to strong conductivity contribution at lower frequencies, especially at higher temperatures.

$$\varepsilon''(\omega)_{\text{deriv}} = -\frac{\pi}{2}\frac{\delta\varepsilon'(\omega)}{\delta\ln(\omega)} \approx \varepsilon''_{\text{real}} \tag{1}$$

The advantage of this derivative analysis of the real permittivity is that it approximately corresponds to the imaginary part of the dielectric function, but without any contribution from DC conductivity which can submerge relaxation processes in the low-frequency region.

Dielectric loss peaks were fitted using multiple Havriliak–Negami functions [26] (Eq. 2) and a generalized fit function [27] (Eq. 3):

$$\varepsilon''(\omega) = \sum\left(\frac{\varepsilon_{\text{s}} - \varepsilon_{\infty}}{(1 + (i\omega\tau)^{\alpha})^{\beta}}\right) \tag{2}$$

$$\varepsilon''(\omega) = \frac{\varepsilon''_p}{\frac{(1-C)}{a+b}\left[b(\omega/\omega_p)^{-a} + a(\omega/\omega_p)^{b}\right] + C} \tag{3}$$

In Eq. 2, τ is the relaxation time, ε_{s} the static dielectric constant and ε_{∞} its high-frequency limit. The parameters α and β describe the symmetric and asymmetric broadening of the peak. In Eq. 3, ω_p and ε''_p denotes the peak position and height, respectively. The parameters a and b define the slopes in a log–log plot on the low- and high-frequency sides, and C describes the peak broadening without altering asymptotic slopes.

Temperature dependences of relaxation times were fitted using the Vogel–Fulcher–Tammann (VFT) equation [28–30] (Eq. 4) for the structural α-relaxation in the solvent and protein relaxations above T_{g}:

$$\tau = \tau_0 \cdot \exp\left(\frac{DT_0}{T - T_0}\right) \tag{4}$$

Here, τ_0 is the high-temperature limit of τ, T_0 is the divergence temperature, and D quantifies deviation from Arrhenius behaviour (thus related to the fragility in supercooled liquids). More local β-relaxations in the solvent (water relaxation below the 170 ± 10 K crossover in temperature dependence) and protein relaxations below T_g were fitted by the Arrhenius law (Eq. 5):

$$\tau = \tau_0 \cdot \exp\left(\frac{E_a}{k_B T}\right) \tag{5}$$

where E_a is the activation energy.

3 Results

3.1 Calorimetric Results

The DSC scans shown in Fig. 4 reveal calorimetric glass transitions for myoglobin in mixtures containing 20, 33 and 40 wt% water at solvent levels ranging from $h = 0.33$ to $h = 2$. From this figure and Fig. 5, it can be seen that both T_g, taken from the inflection point, and ΔT_g, i.e. the broadness of the transitions, determined from the onset and end temperatures, decrease systematically with increasing water content at fixed solvent level. For example, raising the water fraction from 20 to 40 wt% lowers T_g by ≈ 15 K and slightly narrows ΔT_g, independent of solvent content. Increasing the solvent level produces a similar reduction in T_g, and ΔT_g, indicating that protein dynamics are influenced not only by solvent mobility but also by the absolute amount of solvent, consistent with quasielastic neutron scattering observations [13].

3.2 Dielectric Results

To connect the calorimetric observations with molecular-scale dynamics, dielectric loss spectra were recorded for all the sample compositions. At higher temperatures it can be difficult to resolve the relaxation processes due to a relatively strong contribution from DC conductivity, but as evident in Fig. 6 multiple relaxation processes are clearly observable in the derivative analysis of ε' (Eq. 1) or after the DC conductivity has been subtracted from ε''. Figure 7 shows the temperature dependence of ε'' with the DC conductivity subtracted for two of the investigated samples. The shift of the relaxation peaks to higher frequencies with increasing temperature illustrates the temperature induced speeding-up of the dynamics. Figure 7c also shows how the relaxation times of the observed relaxation processes were determined by curve fitting using the Havriliak–Negami function (Eq. 2) for all relaxation processes

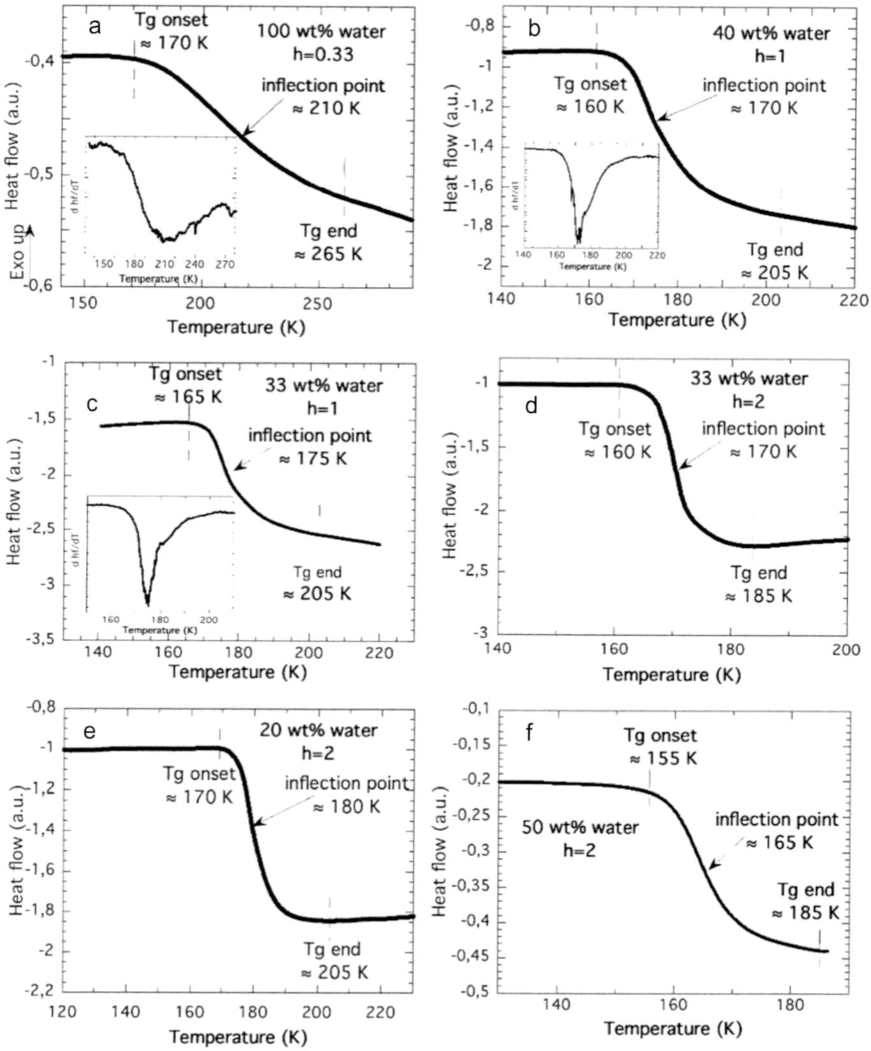

Fig. 4 DSC heating curves obtained for hydrated (33 wt%) myoglobin (**a**) and myoglobin in water-glycerol mixtures with 40 wt% water at the solvent levels $h = 1$ (**b**), 33 wt% water at $h = 1$ (**c**) and $h = 2$ (**d**), 20 wt% water at $h = 2$ (**e**) and 50 wt% water at $h = 2$ (**f**). The insets in (a)–(c) show the derivative of the heat flow (hf) with respect to the temperature (T), since the broadness of the glass transition range was difficult to determine for these samples without this derivative analysis

except for the slowest one for which Eq. 3 was used. Figure 8 provides the temperature dependent relaxation times obtained from this curve fitting for the same samples as shown in Fig. 4. Furthermore, the temperatures where these relaxation times reach 100 s are compared with the DSC-derived T_g ranges (ΔT_g) from Fig. 4, here shown

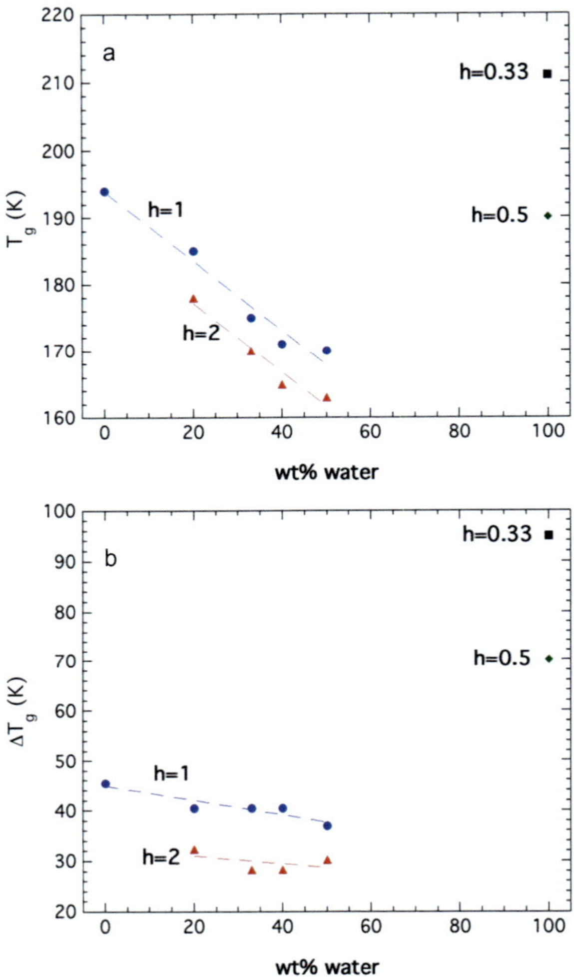

Fig. 5 In (**a**) the DSC glass transition temperature T_g (determined by the inflection point), and in (**b**) the DSC glass transition range ΔT_g (the whole transition range of the heat flow, as determined by the onset and end temperatures of the transition, with respect to the temperature) are shown as a function of wt% water in the solvent for all measured samples. The error in ΔT_g is $\pm$ 5 K

by arrows on top of each figure. In total, seven different relaxation processes were observed, they are:

- **Process I**—a local water relaxation, likely due to reorientations of individual water molecules, and it is typical of confined supercooled water [31]. There are no indications for that this process is directly linked to any protein motions and associated protein activity, and therefore it will not be further discussed.
- **Process IIa**—the primary viscosity related solvent α-relaxation, consistent with bulk solvent behaviour [32, 33]. It has also minor contributions from local protein motions [34, 35].
- **Process IIb**—Process IIa transforms to Process IIb below the crossover temperature around 170 K, where it becomes a local β-relaxation of confined water, most likely of Johari–Goldstein type [24, 31, 36, 37], also known as w- or ν-relaxation

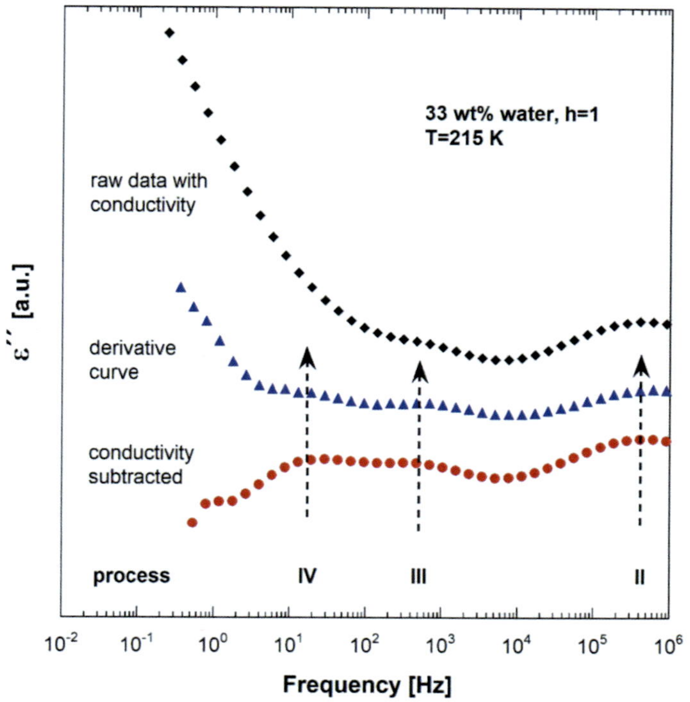

Fig. 6 Experimental data on myoglobin in a water-glycerol solvent of 33 wt% water taken at $T = 215$ K. The ordinary data, shown as the imaginary part of the permittivity is shown in the top and below the derivative data of the real part of the permittivity is shown (see Eq. 1) and in the bottom the imaginary part of the permittivity is shown, where conductivity and polarization effects have been subtracted

[37, 38], as supported by ^{2}H NMR studies [39, 40]. The crossover marks a decoupling of local water motions from the structural α-relaxation. In this regime, the α-relaxation is weak or absent for hydration water [24, 41, 42], and its temperature dependence is extrapolated via VFT fits.

- **Process III**—based on previous results from NMR [35] and time-domain dielectric spectroscopy (TDDS) [43], this process is attributed to the relaxation of polar side chains with associated water.
- **Process IV**—assigned to conformational fluctuations of the protein, in agreement with results from hole-burning spectroscopy [44].
- **Process V**—also likely due to conformational protein fluctuations since there are many types of such fluctuations occurring on somewhat different timescales.
- **Process VI**—a protein relaxation process which is only observed below the crossover temperature of the solvent for the sample containing the largest amount of amorphous water (50 wt% in total solvent content $h = 2$). The exact nature of this process is unknown, but it should be due to local protein motions since

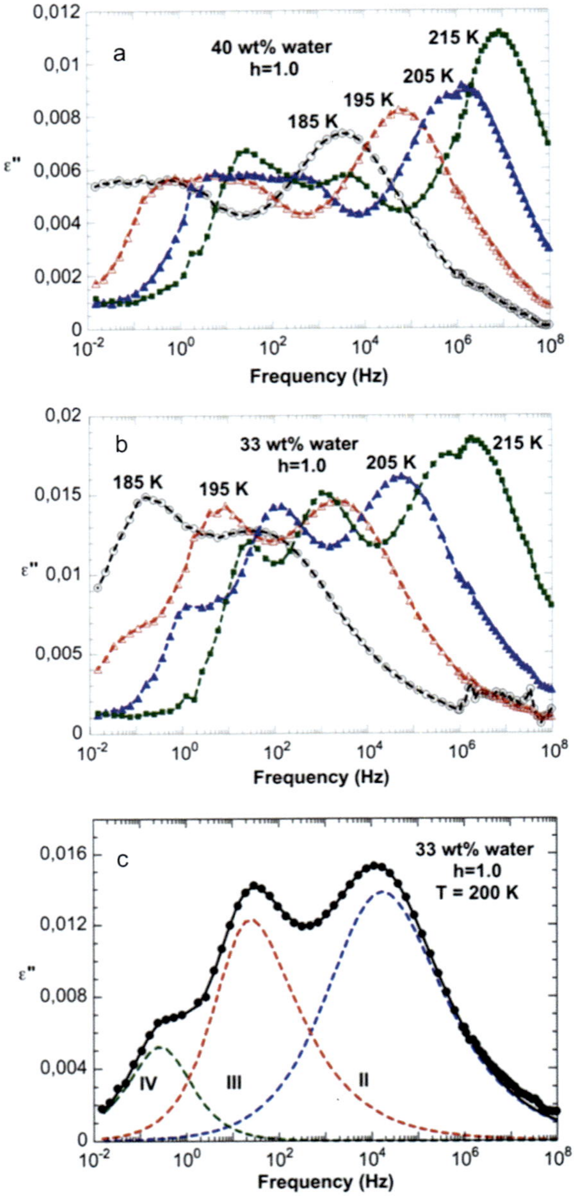

Fig. 7 Temperature evolution of the imaginary part of the dielectric permittivity with conductivity and polarization effects subtracted vs frequency for two of the measured myoglobin samples with solvent compositions and total solvent contents as given in the figures. In (c) an example of the curve fitting is shown to visualize how the relaxation times of the different relaxation processes were extracted from the measured data by use of Eqs. 2 and 3

it follows an Arrhenius temperature dependence and is only present below the crossover temperature of the solvent.

Further evidence for that processes III–V in Fig. 8 are due to protein motions comes from the finding that the T_g range determined by DSC extends well above the α-relaxation freezing temperature (i.e. where the α-relaxation reaches 100 s) of

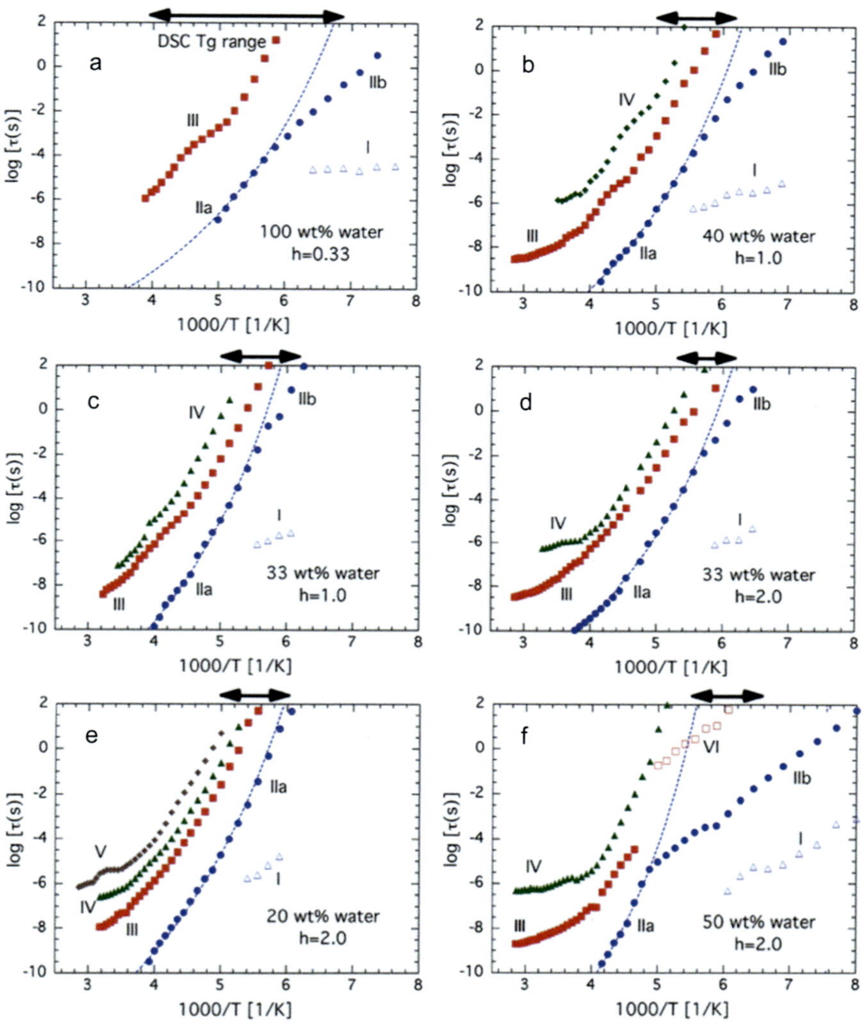

Fig. 8 Dielectric relaxation times for six myoglobin samples with total solvent contents (h) and water fractions (given in wt% of the total solvent) as given in the figure. The arrows on top of each figure shows the DSC glass transition ranges, ΔT_g, obtained as presented in Fig. 4. The dotted lines show the results of VFT (Eq. 4) curve fitting of the temperature dependence of the main solvent relaxation in the high-temperature range. The origins of the processes I–VI are provided in the main text

the solvent. Thus, the solvent motions are generally much too fast to contribute to the relaxation processes associate with the high-temperature T_g range. The only exception to this is the relatively small amount of water which is hydrogen bonded to polar side groups. These water molecules tend to be dragged along with the amino acid side chains, causing the water dynamics to be coupled to the motion of the side chains, i.e. Process III. However, in terms of energy there is an anticorrelation between the hydration water and the polar side groups, where energy flows from the water to the protein and vice versa [16]. Thus, also in this case the protein motions are dependent on the energy provided by the surrounding.

Compositional trends in the solvent and protein dynamics are summarized in Fig. 9. Process IIa accelerates significantly with higher water fractions and slightly with greater solvent content, reflecting faster structural relaxation in more hydrated environments. At low water fractions (20 wt%), process IIa in Fig. 8 retains α-relaxation character across the full temperature range. However, at higher water fractions, a pronounced crossover occurs near 170 K, shifting from the non-Arrhenius α-relaxation regime (IIa) to an Arrhenius β-relaxation regime (IIb), mirroring the behaviour in hydrated proteins [24]. The IIb activation energy and timescale vary with local hydrogen bonding and thus with water fraction [38], causing the crossover temperature to be composition-dependent.

The protein processes III and IV show less dependence on solvent composition, although they tend to follow the same composition dependence as the α-relaxation in the solvent (process IIa). Their weaker composition dependences, compared to the solvent, is most like due to preferential hydration effects [46–50], which implies that the protein molecules are preferably interacting with water, giving rise to a similar water-rich environment closest to the protein surface irrespective of the exact sample composition. To sum up, these dielectric and calorimetric results converge to show that the water content controls both the location of the glass transition and the solvent–protein relaxation relation.

4 Discussion

4.1 The Dynamics Crossover and Low-Temperature Water Relaxation

In hydrated myoglobin, the low-temperature water process (IIb in Fig. 8a) reaches a relaxation time of 100 s at ~ 125 K, which is far below the onset of the broad glass transition range (170–265 K, as shown in Fig. 4a). Since secondary processes are not associated with a liquid-to-glass transition, this supports the assignment of process IIb to a secondary β-relaxation of hydration water below the dynamic crossover temperature.

In hydrated protein, the crossover from process IIa to IIb at 170 ± 10 K (Fig. 8a and Refs. [22, 24, 39, 40]) coincides with the onset of T_g (Fig. 4a). This likely reflects the disappearance of large-scale α-like hydration water relaxation close to the crossover temperature, preventing protein conformational changes without viscosity related solvent motions. Consequently, the crossover in temperature dependence of this process appears to initiate the glass transition of the protein–water system.

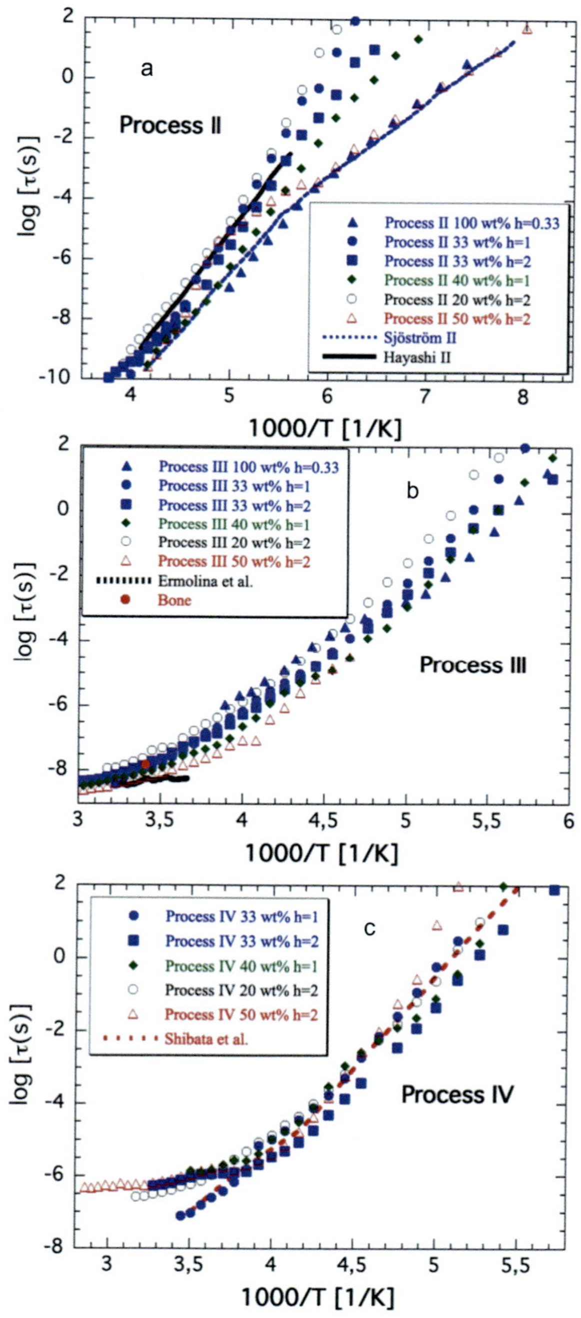

Fig. 9 Dielectric relaxation times for the three main processes in the myoglobin samples, together with data from the literature as indicated in the legends of the figure. In (**a**) the main solvent relaxation (process II) is compared with results for a glycerol:water mixture of about 31 wt% water by Hayashi et al. [33] and the main water relaxation found in MCM-41 by Sjöström et al. [45]. (**b**) Shows the fastest of the obtained protein relaxations (process III), which is attributed to the motion of polar groups on the protein surface, in accordance with the interpretation made by Bone in Ref. [43]. In addition, results from a combined NMR and TDDS study by Ermolina et al. [35] on protein motions are also included in the figure. In (**c**) process IV is shown together with data from Shibata et al. [44] on conformational changes of protein molecules as obtained by hole-burning spectroscopy

4.2 Protein Relaxation Processes

As shown in Fig. 8a, the calorimetric T_g range (ΔT_g) for the hydrated protein sample extends to substantially higher temperatures than the crossover temperature (~ 165 K) where the only clearly observed protein process (III in Fig. 8A) reaches 100 s. This obviously indicates that additional, slower protein processes contribute to the calorimetric T_g, although they are too weak to be detected dielectrically due to much stronger contributions from conductivity and polarization effects at higher temperatures. A reason for that the protein relaxations are weak for this sample is probably the low total solvent content ($h = 0.33$). Since these glass transition related protein relaxations only occur if there is an α-relaxation in the surrounding solvent, it is likely that they almost vanish when the solvent content is so low that an α-relaxation hardly exists. However, myoglobin in water–glycerol solvents (Fig. 8) and at higher hydration levels [22] displays such additional slower protein processes, in consistency with the broad calorimetric T_g range.

4.3 Glass Transition of Proteins in Water–Glycerol Mixtures

In myoglobin–water–glycerol mixtures with less than 40 wt% water, the α-relaxation in the solvent (process IIa in Fig. 8) reaches 100 s (i.e. the dielectric or dynamic glass transition) at temperatures matching the onset of the calorimetric T_g. For instance, in the case of 20 wt% water in the solvent ($h = 2$), its relaxation time is 100 s at 169 K and the onset temperature of the calorimetric T_g is at 168 K. Thus, it can be concluded that the glycerol containing solvents give a direct contribution to the calorimetric T_g. This seems to be in contrast to hydration water, where no direct contribution can be seen from the experimental results, as will be discussed below. Furthermore, by comparing the dielectric and calorimetric results, it can be concluded that the calorimetric T_g range must involve also the slower dielectric processes due to protein dynamics. When these protein fluctuations occur on widely different time scales, the T_g range becomes particularly broad, in accordance with previous studies of the protein glass transition [18, 20]. However, if the calorimetric T_g (see Fig. 4) is analysed in detail, it can be seen that its inflection point is considerably closer to its onset than to its end point. Hence, the glass transition range is asymmetric and seems to contain at least two components, one relatively sharp main component due to the solvent in the lower part of the T_g range and one weak and broad component due to several different protein relaxations at slightly higher temperatures.

4.4 Glass Transition of Hydrated Proteins

The glass transition range of hydrated myoglobin is shown by an arrow on top of Fig. 8a. As mentioned above, it has an onset temperature of about 170 K. However, process IIa reaches a relaxation time of 100 s at about 147 K. This implies that the calorimetric T_g is located at a too high temperature to involve the freezing-in of this water relaxation. The same conclusion was obtained in a dielectric study of the protein bovine serum albumin in water [51]. Thus, for hydrated myoglobin, the calorimetric T_g arises from protein motions which freezes-in on multiple time scales [22]. The small heat capacity change during T_g, compared to myoglobin in water-glycerol mixtures (see Fig. 4), further suggests that the water of hydrated proteins is not giving any direct contribution to the calorimetric T_g. On the other hand, it is well established that dry myoglobin powder lacks a calorimetric Tg. This means that there is no protein glass transition without a surrounding solvent, and in the case of hydrated proteins it is the α-like motions in the hydration water which give rise to the protein motions that freeze-in at T_g. Thus, even if the hydration water does not give an evident relaxational contribution to the observed calorimetric glass transition it is responsible for the glass transition by causing the protein motions. A further consequence of the importance of α-like water motions for global protein fluctuations is that such motions vanish at the same temperature as there is a crossover from process IIa to IIb in Fig. 8a.

4.5 Relationship Between Protein and Solvent Dynamics

In Fig. 8, it is evident that the protein relaxations III–V are only present at temperatures above the glass transition of the solvent, or the dynamic crossover temperature in the case of hydration water. Furthermore, the protein processes show similar temperature dependences as the α-relaxation in the solvent. This is even more evident in Fig. 10, which shows the protein relaxation times as a function of the solvent relaxation time (i.e. the α-relaxation in the solvent for processes III–V and the water β-relaxation in the case of process VI) in a log–log representation. In such a plot, a protein relaxation process exhibits the same temperature dependence as the solvent relaxation if their relaxation times give rise to a linear dependence, i.e. a slope of unity in a log–log plot. As shown in Fig. 10, such a linear dependence is obtained, more or less exact, for all the protein relaxations III–VI. This is an interesting finding since such a behaviour is predicted by the slaving model, proposed by Fenimore et al. [3, 4, 8]. According to their model the motions of a protein are slaved or governed by the surrounding solvent. The more local protein motions are governed by the β-relaxation in the hydration water, whereas the viscosity related α-relaxation in the solvent is responsible for large-scale protein fluctuations. However, as shown in Figs. 8 and 10, the solvent relaxation and its associated protein motions do not occur on the same time scale (i.e. having the same relaxation rate), in consistency with the slaving model. The reason that the protein motions are generally slower than their related solvent motions is that a conformational change of a protein requires that much larger energy barriers (see Fig. 2) are overcame compared to the energy barriers in the solvent, which, in turn, results in that a large number of elementary steps in the solvent are needed before an energy barrier in the protein can be passed [16]. This can be conceptually compared to Brownian motion of macromolecules in a solution, where the slow macromolecular motions are also governed by faster solvent motions.

4.6 Is This Slaving Concept Unique for Proteins?

The results presented above raise the following two questions:

- **Can proteins be slaved by all types of solvents?**
- **Can other types of biomolecules also be slaved by motions in their surrounding solvent?**

The answer to the first question seems to be yes, provided that the solvent does not damage the native structure of the protein. In fact, the protein does not necessarily need to be surrounded by a solvent since all molecules which exhibit a viscosity related α-relaxation and are able to maintain the native protein structure appear to work. This implies that many types of organic materials can be used, but not all. Monohydroxy alcohols had been ideal molecules for high protein activity due to their low viscosity (i.e. fast α-relaxation), but they cause denaturation of the protein and therefore destroy the protein activity [52, 53]. However, for many other types of liquids and melts proteins remain in their native state and global conformational protein fluctuations are slaved by the viscosity related α-relaxation in the surrounding. One such example is myoglobin surrounded by a poly(ethylene oxide)-based polymer surfactant [54]. Despite that only 0.01 g water per gram myoglobin remained after the sample preparation, the protein was able to perform similar dynamics and activity as in an aqueous solution [54]. The authors suggested that the similarity to water was due to that the polymer surfactant was able to provide a similar flexibility in the molecular matrix surrounding each protein molecule as the water hydration layer [54]. A drawback to replace the water is, however, that the viscosity of the protein surrounding tends to increase, which thereby slows down the protein dynamics and associated activity. Furthermore, full activity of a protein requires also more local protein motions, which are slaved by the β-relaxation in the water and therefore absent in a non-aqueous environment. An example of this is seen in Fig. 3, where it is shown that the local CO migration (process BA) within myoglobin cannot occur without the β-relaxation in the surrounding water.

The answer to the second question is also yes since it has been found that the slaving concept is not unique for proteins. As discussed in more detail in Chap. 3 of this book, a dielectric relaxation study of aqueous solutions of monomers and oligomers of the amino acid L-lysine shows [55] that their dynamics are also slaved

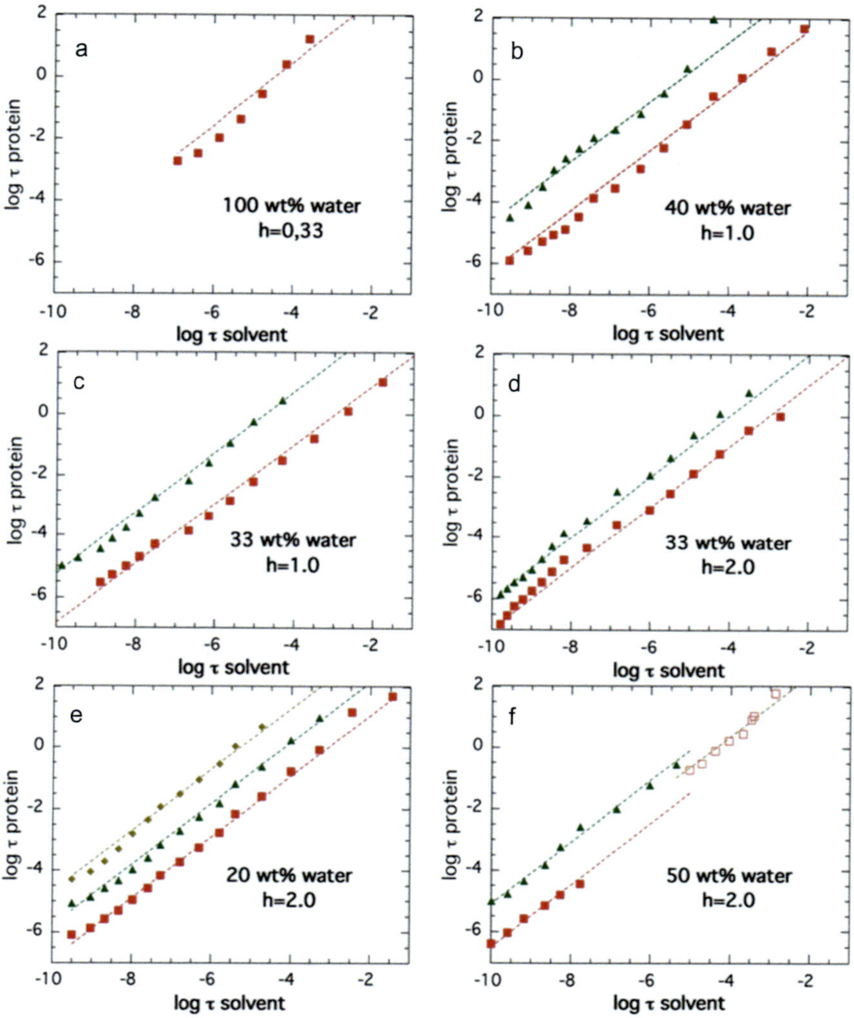

Fig. 10 Relaxation times of the protein processes are shown as a function of the relaxation time for the α-process in the solvent, except in the case of process VI (open squares) in (*F*), which is plotted as a function of the β-relaxation in the water below the crossover temperature (i.e. below the T_g of the protein). The protein processes are the same (same symbols) as shown in Fig. 8. Note the almost perfect linear dependences for all protein processes and samples, except for the hydrated sample (a) where the protein process III is very weak and therefore can only be compared with the water relaxation IIa over a limited temperature range

by the surrounding water. In fact, the results indicate that solvent-slaved motions are not unique for proteins but are present in all kinds of solute molecules (even nonbiological) which are dependent on the dynamics in the surrounding solvent to reach different conformational substates. Using thermodynamics and the concept of energy landscape, the criteria for solvent-slaved conformational motions can be defined as motions which cannot occur without a substantial entropic contribution from the solvent to overcome enthalpy barriers between different substates in the solute molecules [55]. This further implies that the slaving phenomenon is not directly related to biological activities, but when the motions between different substates are linked to biological activities, which is the case for both proteins and peptides, there is a close relation and the biological activities are completely dependent on the solvent-slaved motions. However, there are differences also between peptides and proteins, since the peptide molecules are smaller they cannot occupy as many conformational substates as proteins, and thereby there are less types of conformational fluctuations, leading to a considerably narrower T_g range. Another difference is that the peptide relaxations probably involve both conformational intrachain fluctuations as well as collective interchain motions, and for the monomer of lysine only an intermolecular relaxation is clearly observable [55]. Thus, although the slaving phenomenon is similar as for proteins, the physical nature of the peptide and amino acid dynamics is slightly different compared to the completely intramolecular dynamics of proteins, and this difference is due to the large size difference.

Both the observation that proteins do not need water for exhibiting up to about 90% of its full activity [23] and the finding that also other types of molecules, including nonbiological molecules, can be slaved by the surrounding solvent may have important implications for our general understanding of the role of water and hydrogen bonding for life as well as how proteins are functioning. It also opens up further questions, such as what the difference is between biologically active molecules and nonbiological molecules?

4.7 *Biological Implications of the Slaving Mechanism*

It is well documented that large-scale protein fluctuations are essential for biological activity [56]. Therefore, the present findings imply that in water-rich solvents at low temperatures, below the crossover temperature where the α-relaxation is absent, proteins become nearly inactive. Thus, in the case of, for instance, cryopreservation of protein-rich materials, lowering the temperature below the crossover point not only slows but can completely pause biologically significant protein motions. However, it should here be noted that since some local protein motions are slaved by the local β-relaxation in the hydration water, which remain far below the crossover temperature, also some biological activity of proteins remain below T_g of the protein, and this low-temperature activity is directly related to the timescale of the β-relaxation [57]. The behaviour of only local motions below T_g is in close analogy with ordinary glass-forming liquids, where the cooperative α-relaxation time basically goes to infinity at T_g, leaving only local β-relaxations below that temperature. However, proteins differ in that they cannot undergo a glass transition without its surrounding solvent. Hence, in contrast to ordinary glasses, the glass transition is not an intrinsic property of proteins. Instead, solvent motions drive the protein glass transition and associated biological functions, though protein conformational changes typically occur on timescales 10^3–10^6 times slower than the solvent α-relaxation, as shown in Fig. 10. This substantial slowing down of global protein fluctuations, compared to the solvent dynamics, reflecting the complex nature of the protein-solvent energy landscape, where the energy barriers in the protein are fewer but much larger than in the surrounding.

Since different protein motions, and related functions, are determined by different relaxation processes in the protein surrounding it implies that protein functions and activity can be controlled by the temperature, viscosity and water content in the environment. We have already mentioned that the large-scale protein fluctuations, which are most important for protein functions, can be paused by simply lowering the temperature below T_g of the solvent (or the dynamic crossover temperature in the case of hydration water). This is the reason for why preservation of protein and protein-rich materials is so effective at cryogenic temperatures. However, if cooling is not a possibility an alternative approach is to slow down the same biologically relevant protein motions by replacing most of the surrounding water by disaccharides or other organic materials with a high T_g. This is a common approach in, e.g. pharmaceutical industry [58] for stabilizing proteins that must be stored in room temperature. Thus, with knowledge about how the viscosity and water content in the environment of proteins affect protein motions and related biological activity it is possible to tailor the components and their contents in pharmaceuticals.

5 Conclusion

The present results support the slaving model by Frauenfelder et al. [3, 4, 8] and show that the protein relaxation processes observed above T_g exhibit the same non-Arrhenius temperature dependence as the viscosity related α-relaxation in the solvent. For the hydrated protein sample (i.e. without glycerol in the solvent), we also observed a local protein relaxation having the same Arrhenius temperature dependence as the β-relaxation of water below T_g. However, the solvent dynamics is also somewhat affected by interactions with the protein surface, giving rise to slower dynamics compared to the corresponding bulk solvent. Furthermore, some protein and solvent dynamics are coupled in the sense that they occur on the same time scale, since parts of the hydration water moves together with polar side chains of the protein, due to hydrogen bonding between them. Nevertheless, the biologically most important protein fluctuations occur on a timescale 10^3–10^6 times slower than the solvent α-relaxation.

The nature of the calorimetric glass transition of hydrated proteins and proteins in a water-glycerol mixture is also interesting, since there is a clear difference between the two types of systems. In the case of myoglobin in water-glycerol the main contribution to the calorimetric T_g comes from the solvent, with only a weak and broad contribution from several protein relaxations in the high-temperature range of T_g. On the other hand, for hydrated myoglobin there is no visible contribution from the hydration water and only the weak and broad contribution from several protein relaxations is giving rise to T_g. However, despite that the hydration water does not give a direct contribution the calorimetric T_g-step, it is still responsible for the protein motions that freeze-in at T_g. The reason for this difference between the glycerol containing samples and the hydrated proteins may be that for the glycerol rich samples both protein and solvent dynamics is of global character in the temperature range of the calorimetric glass transition (i.e. when their relaxation times reach 100 s), while in the case of hydrated proteins the hydration water has undergone a dynamic crossover to a local β-relaxation, which does not contribute to the calorimetric T_g. When the high temperature α–like relaxation in the hydration water transforms to this local process also the large-scale protein motions disappear, in accordance with the slaving model by Frauenfelder and co-workers [3, 4, 8]. An important implication of the slaving model is also that cryopreservation of protein-rich materials should be very effective if the temperature is lowered below this crossover temperature around 170 K or the T_g of the solvent in the case when the solvent contains a cryoprotectant such as glycerol or sugar.

Finally, it should be noted that the slaving phenomenon is much broader than being restricted to proteins in aqueous solutions. Studies have shown that protein dynamics and activity can even be slaved by the dynamics in a surrounding polymeric melt, and that also other types of solute molecules can be slaved by the dynamics in their surroundings. It seems as solvent-slaved conformational motions occur when a substantial entropic contribution from the solvent (or surroundings) is needed to overcome enthalpy barriers between different substates in the solute molecules.

Acknowledgements This work was financially supported by the Swedish Research Council.

References

1. Rupley JA, Gratton E, Careri G (1983) Trends Biochem Sci 8:18
2. Frauenfelder H, Gratton E (1986) Methods Enzymol 127:207
3. Fenimore PW, Frauenfelder H, McMahon BH, Parak FG (2002) Proc Natl Acad Sci USA 99:16047
4. Fenimore PW, Frauenfelder H, McMahon BH, Young RD (2004) Proc Natl Acad Sci USA 101:14408
5. Tarek M, Tobias DJ (2002) Phys Rev Lett 88:138101
6. Vitkup D, Ringe D, Petsko GA, Karplus M (2000) Nat Struct Biol 7:34
7. Doster W, Settles M (2005) Biochim Biophys Acta Proteins Proteomics 1749:173
8. Frauenfelder H et al (2009) Proc Natl Acad Sci USA 106:5129
9. Frauenfelder H, Sligar SG, Wolynes PG (1991) Science 254:1598
10. Jansson H, Bergman R, Swenson J (2011) J Phys Chem B 115:4099
11. Hong L, Smolin N, Lindner B, Sokolov AP, Smith JC (2011) Phys Rev Lett 107:148102

12. Sterpone F, Stirnemann G, Laage D (2012) J Am Chem Soc 134:4116
13. Jansson H, Kargl F, Fernandez-Alonso F, Swenson J (2009) J Chem Phys 130:205101
14. Khodadadi S, Sokolov AP (2015) Soft Matter 11:4984
15. Khodadadi S, Pawlus S, Sokolov AP (2008) J Phys Chem B 112:14273
16. Mukherjee S, Mondal S, Bagchi B (2019) Phys Rev Lett 122:058101
17. Sartor G, Hallbrucker A, Hofer K, Mayer E (1992) J Phys Chem 96:5133
18. Sartor G, Mayer E, Johari GP (1994) Biophys J 66:249
19. Brownsey GJ, Noel TR, Parker R, Ring SG (2003) Biophys J 85:3943
20. Miyazaki Y, Matsuo T, Suga H (2000) J Phys Chem B 104:8044
21. Doster W, Bachleitner A, Dunau R, Hiebl M, Luscher E (1986) Biophys J 50:213
22. Jansson H, Swenson J (2010) Biochim Biophys Acta Proteins Proteomics 1804:20
23. Rariy RV, Klibanov AM (1997) Proc Natl Acad Sci USA 94:13520
24. Swenson J, Jansson H, Bergman R (2006) Phys Rev Lett 96:247802
25. Wübbenhorst M, van Turnhout J, Non-Cryst J (2002) Solids 305:40
26. Havriliak S, Negami S (1967) Polymer 8:161
27. Bergman R (2000) J Appl Phys 88:1356
28. Vogel H (1921) Phys Z 22:645
29. Fulcher GS (1925) J Am Ceram Soc 8:789
30. Tammann G, Hesse W, Anorg Z (1926) Allg Chem 156:245
31. Swenson J, Cerveny S (2015) J Phys: Condens Matter 27:033102
32. Puzenko A et al (2005) J Phys Chem B 109:6031
33. Hayashi Y et al (2005) J Phys Chem B 109:9174
34. Khodadadi S et al (2008) J Chem Phys 128:195106
35. Ermolina I, Fedotov V, Feldman Y, Ivoylov I, Non-Cryst J (1994) Solids 172:1103
36. Cerveny S, Schwartz GA, Bergman R, Swenson J (2004) Phys Rev Lett 93:245702
37. Capaccioli S, Ngai KL, Shinyashiki N (2007) J Phys Chem B 111:8197
38. Sjöström J et al (2010) Phys Chem Chem Phys 12:10452
39. Vogel M (2008) Phys Rev Lett 101:225701
40. Lusceac SA, Vogel M (2010) J Phys Chem B 114:10209
41. Swenson J, Teixeira J (2010) J Chem Phys 132:14508
42. Swenson J (2018) Phys Chem Chem Phys 20:30095
43. Bone S (1987) Biochim Biophys Acta 916:128
44. Shibata Y, Kurita A, Kushida T (1998) Biophys J 75:521
45. Sjöström J, Swenson J, Bergman R, Kittaka S (2008) J Chem Phys 128:154503
46. Arakawa T, Timasheff SN (1982) Biochem 21:6536
47. Arakawa T, Timasheff SN (1982) Biochem 24:6756
48. Sinibaldi R et al (2007) J Chem Phys 126:235101
49. Olsson C, Genheden S, Garcia Sakai V, Swenson J (2019) J Phys Chem B 123:3679
50. Ahlgren K, Olsson C, Ermilova I, Swenson J (2023) Phys Chem Chem Phys 25:21215
51. Shinyashiki N et al (2009) J Phys Chem B 113:14448
52. Hirota-Nakaoka N, Goto Y (1999) Bioorg Med Chem 7:67
53. Yoshikawa H, Hirano A, Arakawa T, Shiraki K (2012) Int J Biol Macromol 50:1286
54. Gallat F-X et al (2012) J Am Chem Soc 134:13168
55. Cerveny S, Combarro-Palacios I, Swenson J (2016) J Phys Chem Lett 7:409
56. Parak FG (2003) Curr Opin Struct Biol 13:552
57. Cicerone MT, Douglas JF (2012) Soft Matter 8:2983
58. Wang W, Ohtake S (2019) Int J Pharm 568:118505

Combined Broadband Dielectric Spectroscopy and Nuclear Magnetic Resonance Studies on the Dynamics of Cooled Protein–Solvent Mixtures

Michael Vogel

Abstract Broadband dielectric spectroscopy is combined with quasielastic neutron scattering and, in particular, nuclear magnetic resonance for component-selective studies on the dynamics of nonfreezing concentrated protein–solvent mixtures in wide frequency and temperature ranges. Results for globular and fibrous proteins are reviewed, where, in addition to water, various alcohols are used as solvents. The solvents show quasi-isotropic reorientation coupled to translational diffusion, which are slower than in the bulk. In the considered mixtures, they are governed by broad but not bimodal distributions of correlation times. For water at reduced temperatures, this faster β relaxation may be accompanied by a slower α relaxation, depending on the structure of the protein. If this is the case, a change in the temperature dependence of the β relaxation occurs in response to the vitrification of the α relaxation near 170 K. Both globular and fibrous proteins exhibit fast localized diffusion within single energy minima, which involves all backbone segments. The amplitude of this highly restricted motion is somewhat larger for higher solvation levels but is reduced upon cooling. In addition, slower conformational transitions between different energy minima are observed, which qualify as β relaxation of the protein. On the other hand, there is no evidence that noticeable fractions of the protein backbone show quasi-isotropic reorientation typical of α relaxation. Insights into protein–solvent dynamical couplings available from such component-selective studies are discussed.

Keywords Broadband dielectric spectroscopy · Nuclear magnetic resonance · Protein dynamics · Water dynamics · Glass transition

1 Introduction

The structure and stability of proteins critically depend on the presence of an appropriate solvent. Moreover, their functions typically rely on a complex interplay of protein and solvent dynamics. To gain a basic understanding, many studies analyzed

M. Vogel (✉)
Institute for Condensed Matter Physics, Technical University of Darmstadt, Darmstadt, Germany
e-mail: michael.vogel@pkm.tu-darmstadt.de

S. Cerveny and J. Swenson (eds.), *Dynamics of Biomolecules Studied by Dielectric Spectroscopy*, Advances in Dielectrics, https://doi.org/10.1007/978-3-032-30588-6_7

the temperature-dependent dynamics of protein–solvent mixtures [1, 2], including their dynamical transition upon cooling [3–5], which is commonly accompanied by a loss of biological functionality [6]. It was proposed that protein dynamics are "slaved" or "plasticized" by solvent motions [4, 7], but this topic remains controversial and it was alternatively conjectured that there is a mutual interplay of protein and solvent dynamics [8–11].

The current knowledge gap relating to protein–solvent dynamical couplings is largely attributable to the intricacy of the situation. The structures and dynamics of proteins are highly complex and the properties of their solvents depend on the distance to the protein surface. For example, small-scale and large-scale protein dynamics were proposed to be coupled to the motions of bound and free solvent fractions, respectively [7]. However, simulation studies found that protein and solvent dynamics may even differ along the interface, depending on the local polarity and curvature of the protein surface [12, 13]. In addition, the interplay of protein and solvent dynamics was affected by macromolecular crowding in biological cells [14].

In order to gain fundamental insights, it is therefore useful to study well-defined protein–solvent mixtures and to combine various experimental and computational methods for a detailed characterization of their dynamical behaviors. Accordingly, results for proteins with different structures were compared [1, 9, 15, 16] and the type and concentration of the solvent were varied [17–20]. Moreover, proteins were replaced by peptides with desired amino-acid sequences [10, 21–23]. Many studies employed broadband dielectric spectroscopy (BDS), which provided access to motions in wide frequency ranges [24–26]. Other approaches obtained component-selective information from nuclear magnetic resonance (NMR) [15, 27–29], neutron scattering (NS) [11, 30–34], and molecular dynamics (MD) simulations [12, 13, 35, 36].

It also proved to be advantageous to cover broad temperature ranges and to compare the temperature dependence of protein and solvent dynamics so as to identify dynamical couplings, which enable biological functionality [37]. Such approaches often utilized solvation levels s, which are sufficiently high to activate functional protein motions and, at the same time, sufficiently low to avoid solvent freezing, in particular, for aqueous systems. Depending on the protein and solvent, these requirements are usually met for solvation levels of $s \sim 0.3$–0.4 g/g, as specified by the mass ratio, $s = m_{\mathrm{solvent}}/m_{\mathrm{protein}}$, throughout this contribution. For such moderate values of s, all solvent molecules are typically confined to narrow spaces between densely packed proteins, which to some extent mimics in vivo macromolecular crowding. A few studies varied s in a considerable range, e.g., to ascertain effects of ice formation [20, 38, 39].

Here, main results of such approaches to dynamical couplings between proteins or peptides and their solvents will be summarized. In doing so, the author does not strive for a full coverage of this wide field. Rather, the discussed topics reflect to some extent the author's interests. Specifically, the focus will be on the globular proteins lysozyme (LYS) and myoglobin (MYO), the fibrous proteins collagen (COL) and elastin (ELA), the light-harvesting protein complex C-phycocyanin (CPC), and the peptide ε-polylysine (ε-PLL). Not only differ the globular and fibrous proteins with respect to their molecular structures, but the former also show glass transition like calorimetric features at much lower temperatures than the latter [40–43], which may lead to different protein–solvent dynamical couplings. In addition to water (H_2O/D_2O), ethylene glycol (EG) and glycerol (GL) will be considered as solvents. In most cases, there will be moderate solvation levels of $s \sim 0.3$–$0.4\,\text{g/g}$. First, typical BDS results for protein–solvent mixtures will be presented. Afterward, selective NS and, in particular, NMR studies on solvent and protein dynamics will be discussed. Finally, BDS and NMR findings for peptide–water mixtures will be compared, followed by conclusions.

2 Broadband Dielectric Spectroscopy

BDS was used to investigate the temperature-dependent dynamics of a variety of protein–solvent mixtures. This large body of work includes studies of hydrated LYS [8, 16, 26, 32, 44], MYO [8, 26, 45, 46], COL [16, 38, 47], ELA [24, 38, 43], and other hydrated proteins [16, 48–50]. Moreover, besides water, various solvents were utilized, e.g., the alcohols EG [51] and GL [17, 18, 52], disaccharides [53], and aqueous mixtures [41, 54, 55].

Figure 1 shows exemplary BDS results for proteins with moderate solvation levels s. Typically, the complex dielectric permittivity $\varepsilon(\omega)$ shows several overlapping relaxation processes and a strong conductivity contribution. Therefore, several BDS studies exploited the fact that a conductivity-free representation of the imaginary part $\varepsilon''(\omega)$ is available from the frequency derivative of the real part $\varepsilon'(\omega)$, explicitly, from [56]

$$\varepsilon''_{\text{der}}(\omega) \equiv -\frac{\pi}{2}\frac{\partial \varepsilon'(\omega)}{\partial \ln(\omega)}. \tag{1}$$

In Fig. 1a, this conductivity-free representation of ε'' is displayed for LYS–H_2O [26] and LYS–EG [51] mixtures. For both solvents, we see three relaxation processes, which are denoted as P1, P2, and P3. In Fig. 1b, ε'' is presented for ELA–H_2O [38] and ELA–EG [51]. In these examples, the fastest process P1, which is commonly attributed to solvent reorientation, is supplemented by at most one slower process. Thus, the dielectric spectra of globular LYS appear to be more complex than those of fibrous ELA, which may be related to the different calorimetric properties of these proteins [40, 42]. For a quantitative analysis, most BDS studies of protein–solvent mixtures considered the significant broadening of the relaxation processes by fitting the experimental results with a superposition of Cole-Cole (CC) processes:

$$\varepsilon_{\text{CC}}(\omega) = \frac{\Delta\varepsilon}{1 + (i\omega\tau_{\text{CC}})^{\alpha_{\text{CC}}}} \tag{2}$$

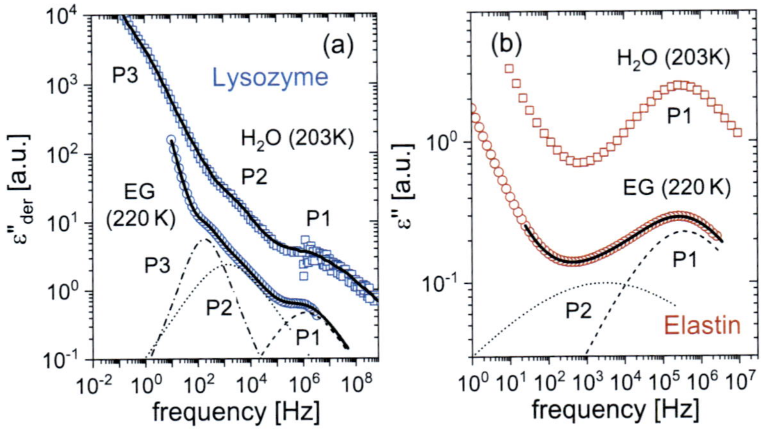

Fig. 1 **a** Conductivity-free representation of the dielectric loss, ε''_{der}, for LYS–H_2O (203 K, 0.36 g/g [26]) and LYS–EG (220 K, 0.59 g/g [51]). **b** Dielectric loss ε'' for ELA–H_2O (203 K, 0.46 g/g [38]) and ELA–EG (220 K, 0.59 g/g [51]). In both panels, the lines are fits with superpositions of CC processes supplemented by a low-frequency power law

Here, $\Delta\varepsilon$ denotes the relaxation strength, while τ_{CC} and α_{CC} are the time constant and width parameter, respectively. In Fig. 1, it is evident that this fit approach usually enables a satisfactory description of the dynamical heterogeneity.

Correlation Times of Protein–Water Mixtures

Figure 2 shows the peak correlation times $\tau \equiv \tau_{CC}$ obtained from such BDS analyses of hydrated LYS [8, 26], MYO [26, 41, 57], ELA [24, 38], and COL [38]. We see that P1, which was related to water dynamics, showed similar but not identical correlation times and, despite noticeable deviations in some cases, an Arrhenius-like temperature dependence. More precisely, below ~170 K, P1 followed a common Arrhenius law in all protein–water mixtures, whereas, above this temperature, P1 was somewhat faster than expected from an extrapolation of the Arrhenius law for some of the systems but not for others. These findings will be discussed in more detail in the context of NS and NMR results in Sec. 3.

The slower processes P2 and P3 differed more strongly for various hydrated proteins. For COL, these relaxations were not observed [38] and, in the case of ELA, they were absent in one study [38] but present in another [24]. For LYS and MYO, both slower processes were observed, but the correlation times depended on the type of the protein and the concentration of water or varied between different studies. Moreover, Arrhenius and non-Arrhenius temperature dependencies were reported. These differences regarding P2 and P3 may be related to diverse findings for fibrous and globular proteins in differential scanning calorimetry (DSC) studies. While hydrated COL and ELA are in glass-like states, hydrated LYS and MYO showed broad calorimetric signatures below 300 K, which were usually considered to be glass transitions [40–43].

Fundamentally different microscopic origins of the slow BDS processes were proposed. P2 was assigned to large-scale protein motions [26, 32] or localized non-cooperative motions [24]. Other studies attributed it to water. They argued that P1 and P2 result from weakly and strongly bound water [25] or from less cooperative β and more cooperative α relaxations of water [57], respectively. The latter conjecture receives support from a comparison with results for β and α relaxations observed for nonfreezable water confined in MCM-41 silica pores [58, 59]. Specifically, the correlation times of the β relaxation resemble those of P1, as will be discussed in Sec. 3, and those of the α relaxation of confined water are similar to those of P2, at least for hydrated MYO, as is shown in Fig. 2.

For P3, which had an "anomalously high intensity" [26] and involved cooperative motions [24], the temperature dependence was described by a Vogel-Fulcher-Tammann (VFT) relation [26]. Based on the VFT behavior and correlation times of $\tau = 10^2$ s at temperatures inside a broad glass transition like step of hydrated globular proteins, P3 was attributed to the α relaxation of the protein [9, 26]. However, a combined BDS and DSC study on hydrated MYO provided evidence that not one but several protein relaxations were involved in the broad step, which extended from 170–255 K [41]. A comparison of the real and imaginary parts of the complex specific heat for hydrated LYS led to a completely different interpretation of the broad calorimentric feature [60]. Specifically, it was concluded that, even for a moderate

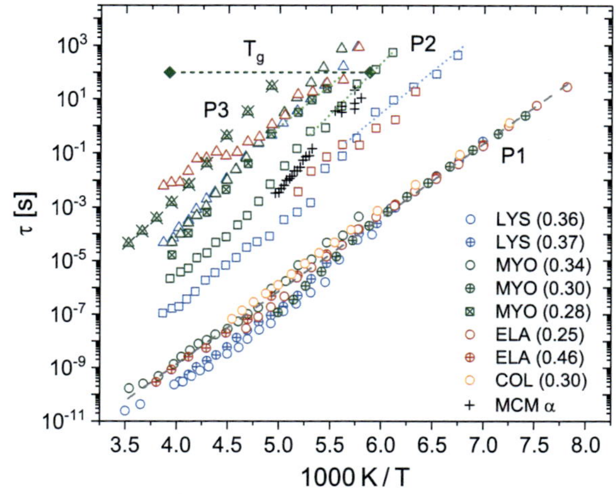

Fig. 2 BDS peak correlation times $\tau \equiv \tau_{CC}$ of protein–water mixtures: LYS (0.36 g/g [26], 0.37 g/g [8]), MYO (0.34 g/g [26], 0.30 g/g [41], 0.28 g/g [57]), ELA (0.25 g/g [24], 0.46 g/g [38]), and COL (0.30 g/g, [38]). For MYO, not all observed processes P1–P3 are shown for every study to avoid overcrowding. The gray dashed line and the green and blue dotted lines are guides to the eye. The solid green diamonds mark the broad glass transition zone in a calometric study of MYO (0.30 g/g [41]). In addition, correlation times for the α process of water confined to MCM-41 silica pores from dielectric and calorimetric studies are included [58]

solvation level of $s = 0.35\,\text{g/g}$, the effect did not result from a glass transition but rather from a formation of ice-like aggregates. Thus, the microscopic origin of the broad low-temperature calorimetric signature of globular proteins remains elusive. A more detailed discussion of the assignments of the slower BDS processes will be provided in Sec. 4.

Correlation Times of Protein–Alcohol Mixtures

In Fig. 3, BDS correlation times of LYS–EG and ELA–EG mixtures [51] are compared with those of EG confined to MCM-41 silica pores [61]. As was the case for the hydrated proteins, the properties of P1 weakly depended on the protein, while the slower processes differed significantly in the studied mixtures. For EG in MCM-41, two BDS processes were observed. The faster one, which showed bulk-like correlation times, was assigned to EG near the pore center, whereas the slower one was attributed to EG near the pore wall [61]. Thus, the microscopic interpretation of faster and slower BDS processes differed for various liquids in identical silica pores. While they were explained by faster and slower molecules in different pore regions for EG, they were rationalized by faster β and slower α relaxations of all molecules for H_2O. The P1 process of EG in LYS and ELA matrices and the slower process of EG in MCM-41 pores had very similar correlation times. If the assignment of the BDS processes of EG in MCM-41 is true, this agreement implies that EG dynamics

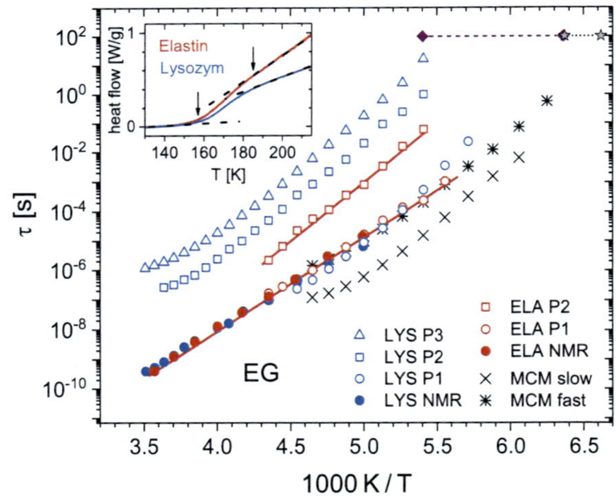

Fig. 3 BDS and NMR correlation times for mixtures of selectively deuterated EG-d_4 with LYS or ELA (0.59 g/g) [51]. The NMR data are from ^{2}H spin-lattice relaxation (SLR) analysis and reflect EG reorientation. The red lines are Arrhenius fits of the ELA–EG data. For comparison, BDS data for fast and slow EG reorientations in MCM-41 silica pores are included [61]. The solid purple diamonds and gray stars indicate the temperature range of the glass transition step in DSC studies of both mixtures (inset) and bulk EG-d_4 [61], respectively

near protein and silica surfaces resemble each other, while they are about an order of magnitude slower than in the bulk at ~200 K. P2 of ELA–EG was substantially weaker, see Fig. 1, and faster than both slower processes of LYS–EG. However, the temperature dependence of all the slow processes differed little.

Extrapolation of the P2 and P3 correlation times leads to $\tau = 10^2$ s at temperatures in the broad glass transition range, which extended from ~157–185 K for both LYS–EG and ELA–EG, see Fig. 3 [61]. This finding may suggest that P2 or P3 is the α process of the protein. However, if this assignment is valid, the glass transition of ELA will occur at much lower temperatures in mixtures with EG than in mixtures with H_2O, which show $T_g > 300$ K [42, 43]. Therefore, one may alternatively argue that there was a significant solvent contribution to the glass transitions of LYS–EG and ELA–EG. This conjecture is consistent with the observation that the glass transition took place at only slightly higher temperatures in these mixtures than in bulk EG [61]. In such a situation, P1, which is too fast to contribute to the glass transition step, should be considered as the β relaxation of EG, while the α relaxation of EG should manifest itself as P2. We will come back to the microscopic origins of the BDS processes of LYS–EG and ELA–EG when discussing the NMR results for these systems.

BDS studies on protein–glycerol mixtures revealed a different dynamical scenario. Unlike EG dynamics, GL dynamics was bulk-like in mixtures with LYS [52], COL [18], and hemoglobin [31]. Specifically, the same VFT temperature dependence was observed for GL reorientation in mixtures with solvation levels of $s = 0.2$–0.8 g/g and in the bulk. However, the presence of the proteins caused a considerable broadening of the corresponding loss peak. Like for the hydrated proteins, additional slower BDS relaxations were observed for hemoglobin [31] but not for COL [18].

3 Selective Studies of Solvent Dynamics

The isotope selectivity of NMR and NS enabled component-resolved studies of protein–solvent mixtures. Such studies clearly confirmed that BDS P1 can be assigned to solvent dynamics, as will be discussed for water and alcohols in the following.

Correlation Times of Solvent Dynamics

Figure 4 compares the above discussed BDS P1 correlation times for hydrated proteins with NS and NMR results, which selectively probed water dynamics in similar samples. The NS correlation times were obtained for LYS [32] and CPC [33] mixtures from quasielastic scattering (QENS). The NMR data were extracted for D_2O in MYO [15], ELA [62], and COL [15] from ^{2}H spin-lattice relaxation (SLR) analysis using a CC spectral density. We see that the BDS, QENS, and SLR results agree for all studied hydrated proteins, clearly confirming that P1 can be attributed to water

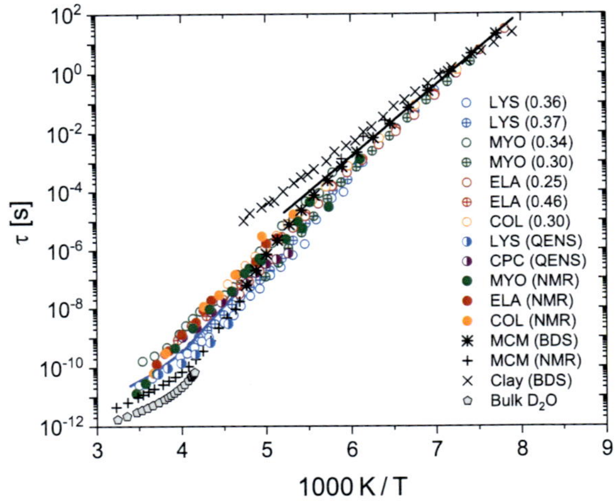

Fig. 4 Correlation times of water dynamics in protein and solid matrices obtained from BDS, NMR, and QENS. The BDS data for the hydrated proteins are the same as in Fig. 2. ^{2}H NMR (SLR) data for D_2O in MYO (0.35 g/g [15]), ELA (0.25 g/g [62]), and COL (0.25 g/g [15]) are shown. QENS results were obtained from H_2O in perdeuterated CPC (0.30 g/g [33]) and from contrasting LYS–H_2O and LYS–D_2O data (0.30 g/g [30]). For comparison, NMR data for bulk D_2O [63], BDS data for H_2O in vermiculite clay [64], and BDS and NMR results for D_2O in MCM-41 silica pores [59] are included. The black line is an Arrhenius fit of the latter data at $T \leq 160$ K, yielding an activation energy of $E_a = 0.48$ eV [59]. Additional QENS data of LYS–D_2O (0.37 g/g [32]), which characterize protein dynamics, are displayed as blue line

reorientation. For the fibrous proteins COL and ELA, a scaling analysis showed that thermally activated motion governed by a Gaussian distribution of activation energies $g(E_a)$ with a mean value of $\overline{E}_a = 0.54$ eV and a standard deviation of $\sigma_E = 0.04$ eV consistently described the temperature and the frequency dependence of the BDS and SLR data in the whole accessible ranges [47]. This implies that water reorientation was considerably affected by a disordered energy landscape imposed by largely static protein molecules. Considering the strong effects of the protein matrices and the, if at all, weak deviations from Arrhenius behavior, the probed water motion was regarded as secondary β relaxation rather than structural α relaxation. For globular MYO [26, 41, 45] and LYS [8, 26], BDS analyses revealed that water reorientation followed an Arrhenius law with an activation energy of $\sim$0.5 eV below $\sim$170 K, while VFT behaviors were observed at higher temperatures. Another study on hydrated MYO reported even two dynamical crossovers in water dynamics at $\sim$175 K and $\sim$200 K [57]. For LYS, QENS correlation times of water dynamics [30] agreed with the P1 results at higher temperatures, but deviated at lower ones. In Fig. 4, the low-temperature data are omitted because of a controversial discussion relating to their validity and interpretation [33]. A comparison with D_2O bulk data [63] shows that water reorientation was significantly slower in all studied protein mixtures than in the neat liquid near ambient temperatures. In addition, for LYS and ELA, QENS studies [32, 34] reported protein dynamics that occurred on the same time scale as water dynamics, providing evidence for a coupled protein–water process, consistent with results for other hydrated proteins [4, 65]. The nature of the involved protein motion will be discussed in Sec. 4.

Further insights are available when we compare the reorientation of nonfreezable water in protein mixtures and in solid confinements, like vermiculite clay [64] and MCM-41 silica pores [59]. In Fig. 4, we see that, below ~170 K, the correlation times of water are similar in all these matrices and obey an Arrhenius law with an universal activation energy of $E_a \sim 0.5\,\mathrm{eV}$. Like in COL and ELA, the Arrhenius law continued to higher temperatures in the vermiculite clay [64], where water was severely confined to two layers between the clay platelets. By contrast, for water in wider (2 nm) silica pores, there was a dynamical crossover to a VFT behavior [59], which was somewhat more prominent than in MYO and LYS.

These results imply that the existence of the crossover in water dynamics depends on the degree and softness of the confinement. The dynamical crossover is suppressed when water is severely confined to one or two layers between solid surfaces. Contrarily, it occurs in confinements, which are sufficiently wide and/or become flexible when the temperature is increased. Specifically, for confinements, which are sufficiently narrow to prevent crystallization but at the same time sufficiently wide so that water–water contacts outweigh water–matrix ones, slower cooperative α water relaxation can set in near the crossover and affect the probed faster water relaxation, which is β type. Additionally, for protein–water mixtures, an onset of protein dynamics can soften the water confinements and lead to an altered temperature dependence of water dynamics.

Both effects are known from studies of various glass-forming liquids [66]. Specifically, even for neat bulk liquids, the arrest of the slower strongly cooperative α process at the glass transition temperature T_g is usually accompanied by a kink in the temperature dependence of the faster weakly cooperative β process [66]. Moreover, for water in solid MCM-41 silica pores, the glass transition, see Fig. 2, and the dynamical crossover, see Fig. 4, occurred at similar temperatures [58, 59]. In addition, for dynamically asymmetric glass-forming mixtures, the glassy arrest of the slower component is commonly accompanied by a kink in the temperature-dependent dynamics of the fast component [67, 68]. Based on these analogies, it was argued for hydrated MYO [57] that P1, which was attributed to the water β relaxation, actually showed two dynamical crossovers when the α relaxations of the water and the protein, which were associated with P2 and P3, vitrify at ~175 K and ~200 K, respectively, see Fig. 2. While a coupling of the probed water β relaxation to slower dynamics is highly plausible, the proposed assignments of the slower BDS processes will be reconsidered in Sec. 4.

Either way, we may conclude that differences between various proteins relating to the existence of a dynamical crossover of water result from different water distributions, which do or do not allow for cooperative α water motion, and different protein dynamics, which can manifest themselves in diverse dielectric and calorimetric properties. These aspects differ for globular and fibrous proteins but may even depend on sample preparation.

Revisiting Fig. 3, we see that BDS and NMR also yield consistent results for the LYS–EG and ELA–EG mixtures. Specifically, the P1 correlation times agreed with ^{2}H SLR data, which were obtained for selectively deuterated EG-d$_4$ and exclusively monitored solvent reorientation. Moreover, a ^{2}H NMR study of selectively deuterated GL-d$_5$ in mixtures with COL and ELA arrived at the same result [69]. Thus, P1 can undoubtedly be attributed to solvent reorientation in all studied protein–water and protein–alcohol mixtures. For the latter systems, the available data do not allow us to decide whether or not a glass transition concurs with a crossover in alcohol dynamics.

Mechanisms for Solvent Dynamics

NMR and QENS inform about not only the rates but also the mechanisms for molecular dynamics. NMR experiments in homogeneous magnetic fields yield insights into rotational motion while an application of magnetic field gradients provides access to diffusion. Likewise, QENS studies probe both aspects of molecular dynamics when the magnitude of the scattering vector q is appropriately varied.

NMR line-shape analysis (LSA) exploits the fact that broad rigid-lattice spectra denoted as Pake spectra collapse to motionally narrowed lines when the correlation time of molecular dynamics crosses the experimental time scale, which amounts to $\tau_{\mathrm{LSA}} \sim 1\,\mu$s, depending, among other things, on the probe nucleus. Thereat, the geometry of the probed motion determines the shape of the motionally averaged line. Isotropic reorientation results in a narrow Lorentzian line, whereas restricted reorientation merely causes a mild narrowing of the broad Pake spectrum. When slow and fast molecules coexist at a given temperature due to prominent dynamical heterogeneity, the NMR spectrum is a superposition of static and narrowed lines.

Figure 5 shows ^{2}H NMR spectra of deuterated solvents in mixtures with ELA or LYS [47, 51, 62, 69]. Lorentzian lines and Pake spectra were found at high and low temperatures, indicating quasi-isotropic solvent reorientation, which was fast and slow on the time scale $\tau_{\mathrm{LSA}} = 1\,\mu$s, respectively. At intermediate temperatures, the spectral shape $S(\nu)$ could be described by a weighted superposition of normalized Lorentzian and Pake contributions:

$$S(\nu; T) = W_{\mathrm{fast}}(T) S_{\mathrm{Lorentz}}(\nu) + [1 - W_{\mathrm{fast}}(T)] S_{\mathrm{Pake}}(\nu) \tag{3}$$

Monitoring the ^{2}H NMR line-shape transition of viscous bulk liquids, e.g. GL [70], such a superposition was not observed, see Fig. 5b. This indicated that solvent reorientation showed a considerably broader distribution of correlation times $G(\log \tau)$ in protein matrices than in the bulk liquid. In the protein mixtures, the fast ($\tau < \tau_{\mathrm{LSA}}$) and slow ($\tau > \tau_{\mathrm{LSA}}$) solvent molecules from the distribution contributed to the Lorentzian and Pake spectra, respectively, and the ratio of both line-shape components varied when $G(\log \tau)$ shifted over the experimental time scale upon cooling, explicitly

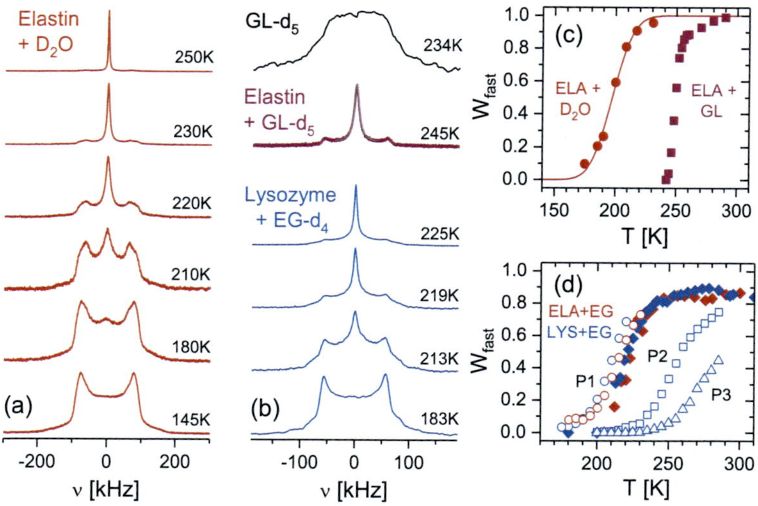

Fig. 5 ^{2}H NMR spectra of deuterated solvents in mixtures with proteins at the indicated temperatures: **a** D_2O in ELA (0.43 g/g [47, 62]) and **b** EG-d_4 in LYS (0.59 g/g [51]), GL-d_5 in ELA (0.60 g/g [69]) with a fit using Eq. (3), and neat GL-d_5 [70]. Resulting weighting factor $W_{\text{fast}}(T)$ for **c** D_2O and GL-d_5 in ELA and **d** EG-d_4 in ELA and LYS. In panel (**c**), the line is a prediction calculated from the distribution of activation energies $g(E_a)$ obtained from BDS and SLR analyses using Eq. (4) [47]. In panel (**d**), the open symbols are predictions calculated from the fits of the observed BDS processes [51]

$$W_{\text{fast}}(T) = \int_{-\infty}^{\log \tau_{\text{LSA}}} G(\log \tau; T) \mathrm{d} \log \tau \tag{4}$$

Using the temperature-dependent distributions $G(\log \tau)$ obtained from BDS and SLR analyses in Eq. (4), the weighting factors $W_{\text{fast}}(T)$ determined from the ^{2}H NMR spectra were successfully reproduced. In Fig. 5c, we observe for ELA–D_2O that the increase of $W_{\text{fast}}(T)$ occurred in a broad temperature range of 180–240 K and was well described by $G(\log \tau)$ distributions derived from the above discussed temperature-independent Gaussian distribution of activation energies $g(E_a)$ from BDS [47]. For ELA–GL [69], the increase of $W_{\text{fast}}(T)$ was shifted to higher temperatures, reflecting the slower GL than D_2O reorientation. In Fig. 5d, the NMR weighting factors for EG in mixtures with LYS and ELA are compared with BDS predictions. It can be seen that $W_{\text{fast}}(T)$ agreed with the expectation from P1, whereas the NMR line-shape transition would occur at higher temperatures if it was related to P2 or P3. Therefore, these LSA approaches showed that BDS and NMR yielded consistent results for not only the position but also the width of broad $G(\log \tau)$ distributions for solvent dynamics. Moreover, the observation of a Lorentzian component at low temperatures indicated that a quasi-isotropic mechanism was retained for solvent reorientation upon cooling. In particular, in ELA mixtures, D_2O, EG, and GL showed quasi-isotropic reorientation well below the glass transition of the protein matrix at $T_g > 300\,K$ [42, 43].

For hydrated proteins, NMR and QENS studies revealed that the water reorientation observed in the above discussed approaches is coupled to water diffusion. For hydrated LYS [71], NMR static field gradient (SFG) experiments probed the self-diffusion of water on the micrometer scale down to 250 K. In Fig. 6a, it can be seen that the measured NMR self-diffusion coefficients D of water agree with values calculated from the BDS correlation times [26] using $a^2/6\tau$ with $a = 0.45\,\text{Å}$. For hydrated ELA [34], the temperature-dependent inelastic (quasielastic) fixed window intensity for various scattering vectors q was consistently described by diffusive motion subject to a Gaussian distribution of activation energies $g(E_a)$. In Fig. 6b, we observe that the obtained QENS self-diffusion coefficients D agreed with the BDS and SLR data in Fig. 4, specifically, with $a^2/6\tau$ using $a = 0.34\,\text{Å}$. In particular, all methods yielded very similar mean activation energies of $\overline{E}_a \sim 0.55\,\text{eV}$. Hence, the water molecules continue to diffuse down to ~200 K, at least on the nanometer scale of the QENS experiment ($q = 0.60$–$1.76\,\text{Å}^{-1}$). The relatively small jump lengths a resulting from these analyses probably arise from the fact that the large protein molecules obstruct water diffusion. Consistently, MD simulations of hydrated COL and ELA ($s = 0.3\,\text{g/g}$) showed that water diffusion occurred down to temperatures below 200 K but was slowed down by the geometrical obstruction due to the presence of the protein molecules [72]. It should be noted that although these NMR, QENS and MD approaches do not yield evidence that reorientation decouples from diffusion upon cooling, the time and length scales of these studies do not allow one to rule out that water dynamics becomes localized at temperatures much lower than 200 K.

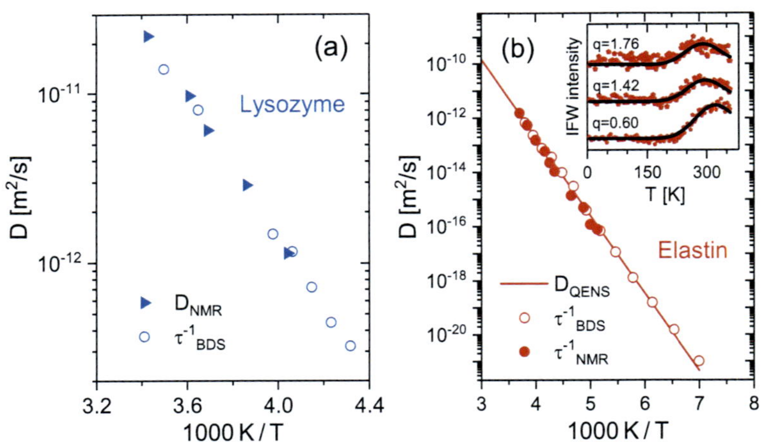

Fig. 6 Self-diffusion coefficients D of water in **a** LYS and **b** ELA. For LYS, D values from SFG NMR [71] are compared with scaled inverse correlation times from BDS [26], explicitly, $a^2/6\tau$ with $a = 0.45\,\text{Å}$. For ELA, D from QENS [34] is shown together with $a^2/6\tau$ ($a = 0.35\,\text{Å}$) from BDS and NMR [47]. The inset shows the inelastic fixed window intensity for various scattering vectors q (in Å^{-1}) together with fits (lines) for D determination [34]

4 Selective Studies of Protein Dynamics

Neutron Scattering Studies of Protein–Water Mixtures

QENS was also successfully employed to selectively investigate protein motions. The above discussed water-coupled protein motion of hydrated LYS, which occurs on a time scale of $\sim 10^{-11}$ s at room temperature, see Fig. 4, was first regarded as the α relaxation of the protein [32], but later reassigned to a secondary relaxation in combined scattering and simulation approaches [1, 35]. Specifically, it was identified as localized ($\sim$1–3 Å) diffusion of essentially all residues in a single energy minimum. In addition, non-methyl conformational transitions between different energy minima on the nanoseconds time scale were reported for some residues [1, 35].

QENS results for hydrated ELA revealed that the above discussed water diffusion, see Fig. 6, was accompanied by a highly localized motion of the protein backbone [34]. Figure 7a shows the obtained temperature-dependent mean squared displacement (MSD) $u^2(T)$ of dry and hydrated ELA [34]. At lower temperatures, the MSD was independent of the water content with a first crossover at $\sim$125 K due to the onset of methyl group reorientation on the experimental time scale $\tau_{NS} = 2$ ns. At higher temperatures, water-coupled protein dynamics resulted in larger $u^2(T)$ for hydrated ELA than for dry ELA. In particular, the former sample showed two further crossovers at $\sim$195 K and $\sim$320 K, which were absent for the latter sample. The water-coupled ELA dynamics was attributed to caged dynamics related to the nearly constant loss (NCL) of disordered systems in the QENS study [34] but was

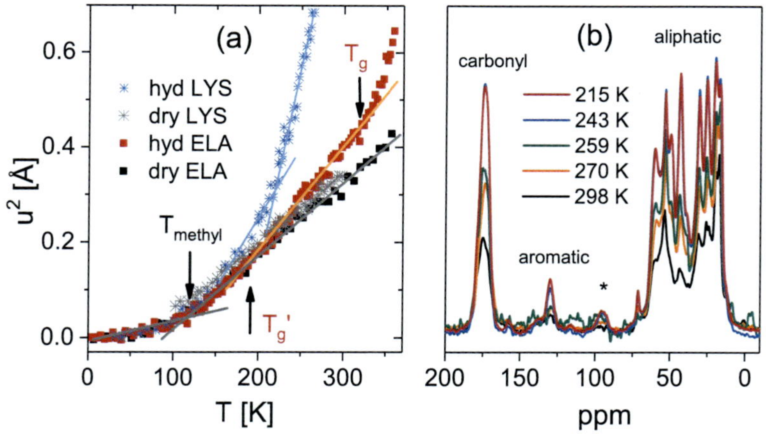

Fig. 7 **a** Temperature-dependent MSD $u^2(T)$ of dry ELA and ELA–D_2O (0.20 g/g) [34] and of dry LYS and LYS–D_2O (0.45 g/g) [8]. The lines indicate approximately linear behaviors in various temperature ranges. The arrows mark crossovers for ELA–D_2O due to an onset of methyl group reorientation at $T_{methyl} \sim 125$ K, a secondary glass transition at $T_g' \sim 195$ K, and the regular glass transition at $T_g \sim 320$ K. **b** ^{13}C CP MAS NMR spectra of ELA–H_2O (0.60 g/g [15]) at various temperatures. Resonances from aliphatic, aromatic and carbonyl carbons are indicated

found to also involve cooperative forward-backward jumps of neighboring backbone segments in MD simulations [36]. It was argued that this ELA motion responded to not only the regular glass transition at $T_g \sim 320\,\text{K}$ [42, 43] but also to a secondary glass transition at $T_g' \sim 195\,\text{K}$. The latter low-temperature kink of $u^2(T)$ at $\sim 195\,\text{K}$ should not be mistaken with the much discussed protein dynamical transition [3–5]. In Fig. 7a, we see that the secondary glass transition also manifested itself in an onset of MSD differences between dry and hydrated LYS [73] when heating toward 200 K, whereas the dynamical transition of this globular protein was observed as another $u^2(T)$ enhancement at a higher temperature of $\sim 215\,\text{K}$. As was previously discussed in more detail [34, 57], the protein dynamical transition, which occurs when the aforementioned coupled protein–water motion enters the experimental time window, was not observed for hydrated ELA.

Nuclear Magnetic Resonance Studies of Protein–Water Mixtures

Likewise, a variety of NMR methods provided valuable information about the mechanisms for protein dynamics. The dependence of ELA dynamics on the temperature and the concentration and composition of the solvent was obtained from ^{13}C NMR spectra, which were measured using ^{1}H–^{13}C cross polarization (CP) under magic angle spinning (MAS) conditions [15, 19, 74]. In these experiments, ^{13}C signals from different chemical environments are resolved but the signal intensities are reduced when protein dynamics in the microseconds regime lead to decreased CP and MAS efficiencies. The ^{13}C CP MAS line shapes of ELA–H_2O mixtures indicated higher protein flexibility for higher water concentration at room temperature [74]. This solvation-enhanced protein flexibility was continuously reduced when the solvent composition was changed, explicitly, when water was replaced by glycerol, to increase the solvent viscosity [19]. Figure 7b shows ^{13}C CP MAS spectra of ELA–H_2O at various temperatures. We see that the intensities of all resolved ^{13}C signals were reduced due to restricted ELA dynamics at room temperature but were recovered upon cooling to $\sim 215\,\text{K}$, where ELA became static on the experimental time scale [15]. Thus, these results imply that, below the glass transition at $T_g \sim 320\,\text{K}$, solvent-coupled restricted protein dynamics existed, which became less pronounced when reducing the solvent concentration and increasing the solvent viscosity and which vanished upon cooling. However, ^{13}C LSA did not allow for a discrimination between changes in the rates and the amplitudes of ELA dynamics. For hydrated COL, such information was obtained by measuring ^{1}H–^{13}C NMR order parameters [75]. The analysis revealed fast backbone motions with hydration-dependent amplitudes of $\sim 10°$, while fast puckering motions of Pro and Hyp side chains were somewhat less restricted (10–30°).

Various ^{2}H NMR methods were applied to perdeuterated CPC in dry and hydrated states [76]. Figure 8a shows the ^{2}H NMR spectrum of CPC–H_2O ($s = 0.30\,\text{g/g}$) at 300 K. Apart from a very small ($\sim 1\%$) Lorentzian intensity, it features a superposition of narrow and broad Pake spectra, which resulted from the methyl and non-methyl C–D bonds of the protein. Thus, the vast majority of the residues did not show

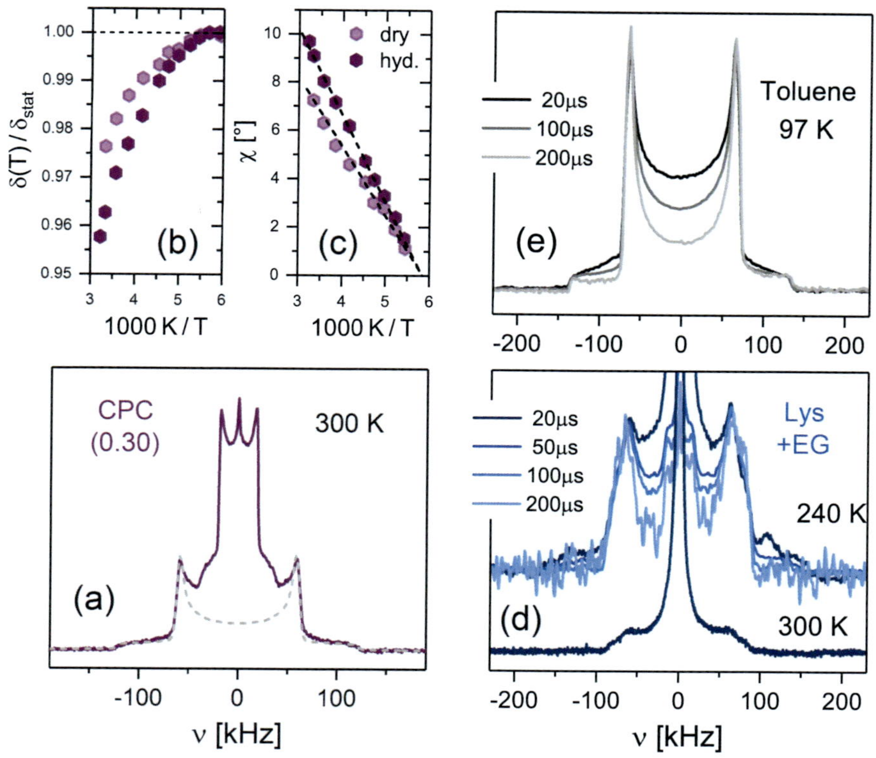

Fig. 8 ^{2}H NMR spectra of proteins: **a** Deuterated CPC–H_2O (300 K, 0.3 g/g) and Pake fit of the non-methyl contribution [76]. LSA yielded **b** the spectral width parameter δ relative to the static value δ_{stat} for dry and hydrated CPC and **c** the corresponding amplitude of backbone reorientation χ. ^{2}H NMR spectra for various solid-echo delays between 20–200 μs: **d** LYS–EG-d_2/EG-d_4 (240 and 300 K) and **e** toluene-d_5 (97 K) [70] for comparison

a Lorentzian line, ruling out α-like isotropic reorientation with $\tau < \tau_{LSA} = 1\,\mu$s. Rather, temperature-dependent ^{2}H LSA revealed that, apart from methyl group rotation, there was only highly restricted protein reorientation on such time scales [76]. In Fig. 8b, it can be seen that this motion resulted in a hydration-dependent reduction of the line width by a few percent at 310 K, which diminished upon cooling until the protein backbone was static below ~175 K. Based on these and further ^{2}H NMR findings, it was concluded that the protein backbone showed water-coupled highly restricted motion, which lost amplitude upon cooling but did not show a glassy slowdown [76]. Specifically, the reorientation had an amplitude of a little less than $\chi = 10^\circ$ at room temperature and vanished at ~175 K, see Fig. 8c. It should be noted that all reorientation amplitudes discussed here were obtained based on the assumption that energetically similar orientations are involved. The real amplitude will be larger than the specified ones if jumps occur between orientations with largely different energies such that a forward jump to the less favored orientation is almost immediately followed by a backward jump.

Hence, ^{2}H and ^{13}C NMR approaches to hydrated ELA, COL, and CPC consistently revealed fast and restricted protein reorientation, the amplitude of which increased upon hydration and decreased upon cooling, but they did not yield evidence for quasi-isotropic or at least large-amplitude backbone reorientation typical of an α process. These observations are consistent with the static (glassy) states of the studied fibrous proteins below room temperature. However, α-like large-amplitude reorientation may be observable in NMR investigations of hydrated MYO and LYS, for which broad glass transition like DSC signatures extended to temperatures well below 200 K [40, 41]. This is particularly true if P2 or P3 monitor the α relaxation of these globular proteins because these BDS processes occur well inside the time windows of various NMR experiments at room temperature. Explicitly, we expect correlation times of $\sim 10^{-9}$–10^{-8} s for P2 and of $\sim 10^{-7}$–10^{-6} s for P3 based on an extrapolation of the LYS–H_2O data in Fig. 2.

Several ^{13}C and ^{15}N NMR studies informed about the dynamics of hydrated LYS. The observation of a broad ^{13}C NMR room-temperature spectrum clearly ruled out the existence of large-amplitude motion on the experimental time scale [77], although the P2 and P3 correlation times $\tau < \tau_{LSA}$ suggest that line narrowing could be expected. ^{13}C SLR analyses showed that, apart from methyl group reorientation, there were two types of LYS motions [78, 79]. A faster ($\tau \sim 10^{-9}$ s) reorientation within a given energy minimum was denoted as γ relaxation. In the present context, it corresponds to NCL-related caged dynamics. A slower ($\tau \sim 10^{-5}$–10^{-4} s) reorientation, referred to as β relaxation, was attributed to conformational transitions, which involved different energy minima but did not break secondary structure elements. The amplitudes of the β and γ relaxations were very small for both the protein backbone and side chains and mildly increased upon hydration, while their rates were broadly distributed but remained unchanged [78, 79]. Hence, despite some differences in the reported correlation times, the above discussed QENS and MD analyses [32, 35] and various ^{13}C NMR studies [77–79] on hydrated LYS revealed similar mechanisms for two small-amplitude secondary protein motions, while neither of these approaches yielded evidence for α-like large-amplitude dynamics.

In order to further explore the latter aspect, ^{15}N NMR correlation functions $F_{NMR}(t)$, which are sensitive to slower dynamics with $\tau \sim 10^{-3}$–10^{0} s, are displayed for hydrated LYS [79] in Fig. 9. In panel (a), $F_{NMR}(t)$ decays for $s = 0.32$ g/g are independent of temperature, indicating that they did not result from molecular dynamics but rather from spin diffusion. In panel (b), $F_{NMR}(t)$ decays for $s = 0.67$ g/g shift to longer times upon cooling, implying a contribution from protein dynamics in the milliseconds regime. Based on these and further observations, it was concluded that the protein as a whole showed restricted reorientation, which became possible when water acted as a lubricant at high solvation levels [79]. Therefore, the combined ^{13}C and ^{15}N NMR studies of hydrated LYS exclude large-amplitude backbone reorientation in wide dynamic ranges and, thus, rule out that P2 or P3 of hydrated LYS can be identified with the α relaxation of the protein. Rather, P2 may be related to the α relaxation of the solvent, as was discussed in Sec. 3, and P3 may be linked to the β relaxation of the protein observed in ^{13}C SLR analyses [78, 79].

Protein–Alcohol Mixtures

^{2}H LSA enabled valuable insights into the protein dynamics of the LYS–EG mixture. Figure 8d presents ^{2}H NMR spectra, which, in addition to the solvent contribution, feature a noticeable protein contribution due to proton-deuteron exchange between the components, particularly for EG-d_2 [51]. We see a Pake shape of the LYS contribution at 300 K although P2 and P3 were sufficiently fast to cause line narrowing at this temperature, see Fig. 5. Hence, neither of these BDS processes was the α process of the protein. This, in turn, supports the above argument that it was not the arrest of the protein's α process, which caused the calorimetric signature of LYS–EG at 157–185 K, see Fig. 3. In view of this result, the sensitivity of the ^{2}H NMR line shape to restricted reorientation was enhanced by increasing the solid-echo delay of the used pulse sequence. In Fig. 8e, we see for neat toluene at $T < T_g$ that β-related small-amplitude reorientation ($\chi \sim 6°$) manifested itself in a characteristic decrease of the intensity in the center of the ^{2}H NMR spectrum when the solid-echo delay was extended [70, 80, 81]. In Fig. 8d, we observe similar line-shape changes for LYS–EG at 240 K [51], where P2 and P3 occurred on the experimental time scale τ_{LSA}. These findings indicated that P2 and P3 of LYS–EG did not involve quasi-isotropic reorientation, indicating that neither of these processes can be identified with the α relaxation of the protein. However, at least one of the slow BDS processes involved β-like highly restricted protein dynamics.

^{2}H NMR correlation functions $F_{NMR}(t)$ of LYS–EG provided further support for these conclusions [51]. Figure 9c shows $F_{NMR}(t)$ at 235 K. While EG dynamics was too fast to produce an observable loss of correlation, P2 and P3 were located well inside the time window of this NMR experiment, $\sim 10^{-4}$–10^0 s, at the studied temperature. Evidently, decays calculated from the BDS results for P2 and P3

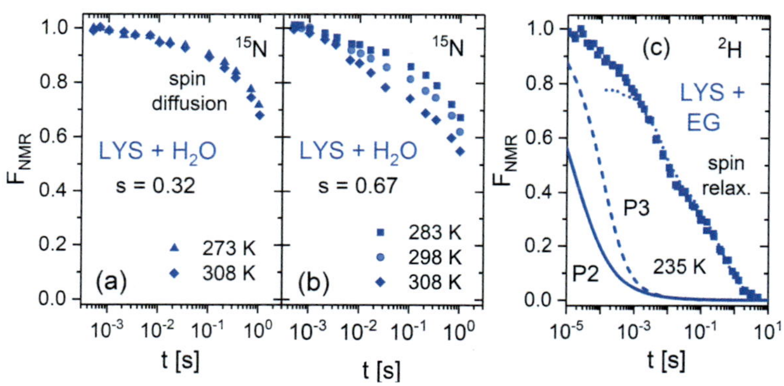

Fig. 9 NMR rotational correlation functions of LYS dynamics: ^{15}N $F_{NMR}(t)$ of LYS–H_2O [79] at various temperatures for **a** $s = 0.32$ g/g and **b** $s = 0.67$ g/g and **c** ^{2}H $F_{NMR}(t)$ of LYS–EG-d_2 (235 K, 0.30 g/g [51]) together with the ^{2}H SLR damping function (dotted line) and expectations calculated from P2 (solid line) and P3 (dashed line) of the BDS study, see Figs. 1 and 3, assuming isotropic reorientation

assuming isotropic reorientation were incompatible with the experimental findings. Specifically, the observed $F_{NMR}(t)$ decay largely resulted from spin relaxation rather than from molecular dynamics. These $F_{NMR}(t)$ findings confirm that the slower BDS processes of LYS–EG did not involve α-like large-amplitude protein dynamics. Considering that the P2 and P3 relaxations extrapolated to the broad glass transition zone of the mixture, these findings imply that the latter did not indicate the glassy arrest of an α relaxation of the protein. Rather, one may expect that the broad DSC signature of LYS–EG received contributions from the vitrification of the solvent's α process, as was discussed in Sec. 3, and from the arrest of the protein's β relaxation, consistent with the above conclusions for hydrated proteins.

5 Peptide–Water Mixtures

Peptide–water mixtures are suitable models to investigate dynamical couplings between biomolecules and their solvents. BDS proved to be very useful to determine the dynamical behaviors of such mixtures over broad frequency and temperature ranges [10, 22, 23, 82]. Figure 10 presents results for hydrated ε-PLL [23, 82]. Four BDS relaxation processes were observed. The fastest process showed an Arrhenius temperature dependence but was too weak for a detailed analysis and assignment.

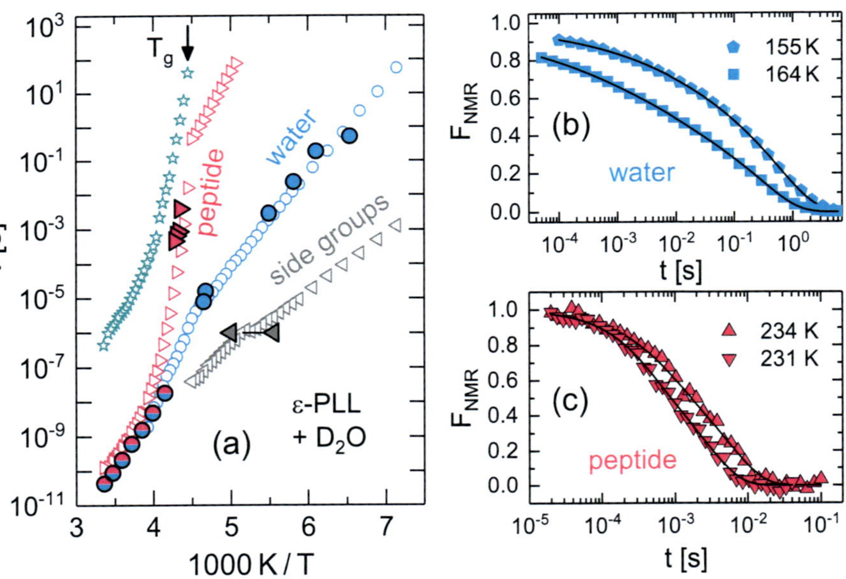

Fig. 10 Results for mixtures of ε-PLL with H_2O or D_2O (0.52 g/g) [82]: (a) Correlation times from BDS (open symbols) and ^1H and ^2H NMR (solid symbols) as well as ^2H NMR correlation functions $F_{NMR}(t)$ for (b) water and (c) peptide

The slowest process exhibited a VFT temperature dependence, which was consistent with the DSC glass transition temperature of $T_g = 223$ K, and, hence, it was assigned to the α relaxation of the peptide. Both intermediate processes showed a crossover in the temperature dependence at T_g. Based on their concentration dependence, they were assigned to water and denoted as "fast" and "slow" water processes [10, 23].

^{1}H and ^{2}H NMR studies on mixtures of ε-PLL with H_2O or D_2O were performed to assist the assignment of the BDS processes [82]. LSA indicated that the fastest BDS process resulted from the reorientation of primary amine or ammonium groups about their C–N bond axes. ^{2}H NMR correlation functions $F_{\text{NMR}}(t)$ selectively characterized the reorientation of the water molecules and the peptide backbone. These measurements used the proton-deuteron exchange in the ε-PLL–D_2O mixture and exploited the different SLR behaviors of the water and peptide deuterons for component selectivity [82]. For the water component, the rate and the shape of the correlation loss indicated that the second fastest BDS process is due to quasi-isotropic water reorientation, in which, despite a broad distribution of correlation times, all water molecules participate even below T_g, see Fig. 10b. For the peptide component, the NMR findings suggested that the second slowest BDS process can be attributed to a reorientation of polar moieties in the peptide backbone, see Fig. 10c. Altogether, the ^{1}H and ^{2}H NMR studies on hydrated ε-PLL did not provide evidence for an existence of two water fractions, which interacted differently with the peptide and showed distinguishable dynamics probed as fast and slow water BDS processes.

MD simulations shed light on the origin of this discrepancy between the interpretations of BDS and NMR findings for hydrated ε-PLL [83]. Explicitly, they showed that the difference resulted from the existence of peptide caused long-lived water–water cross correlations, which contributed to BDS multi-particle correlation functions (MPCFs) but not to NMR single-particle correlation functions (SPCFs). Considering a single molecular component, the BDS MPCF of a molecule i surrounded by $N - 1$ other molecules (j) of the same species is given by

$$F_{\text{BDS}}(t) = \sum_{j=1}^{N} \langle \mathbf{u}_i(0) \cdot \mathbf{u}_j(t) \rangle \tag{5}$$

Here, the brackets $\langle \ldots \rangle$ denote the ensemble average and $\mathbf{u}$ specifies the normalized molecular dipole moment. The sum involves the self-term $i = j$ and the cross terms $i \neq j$. It is often tacitly assumed that the cross terms are negligible so that the self-term is probed and, hence, $F_{\text{BDS}}(t)$ is identical to the SPCF $F_1(t)$. Figure 11a shows that this assumption was not valid for the simulated ε-PLL–H_2O model [83]. Specifically, focusing on $F_{\text{BDS}}(t)$ of the water subsystem, we see that the cross-term contribution was almost twice as high as the self-term contribution. $F_{\text{BDS}}(t)$ showed a stronger fast decay followed by a weaker slow decay. When mistakenly considering the MPCF $F_{\text{BDS}}(t)$ to be the SPCF $F_1(t)$, one may be misled to conclude that this two-step decay is indicative of a coexistence of fast and slow water molecules.

The actual origin of the two-step decay was revealed by a closer look at the self and cross terms of $F_{\text{BDS}}(t)$ [83]. In Fig. 11a, we find from the self-term contribution,

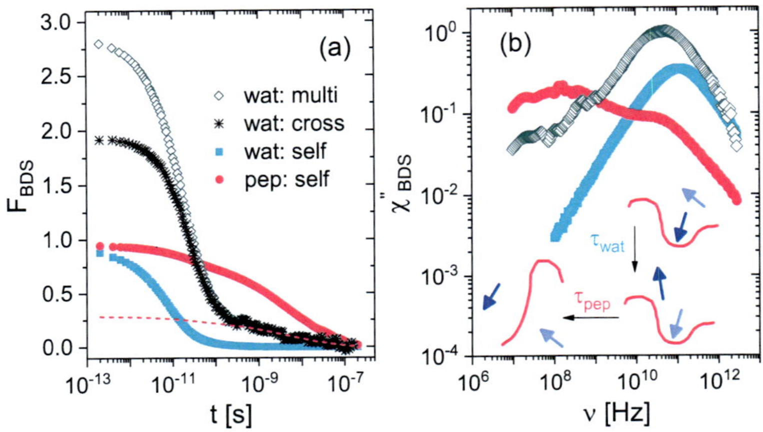

Fig. 11 MD simulations for ε-PLL–H_2O (0.30 g/g, 300 K [83]): **a** Calculated BDS MPCF $F_{\mathrm{BDS}}(t)$ of the water subsystem together with its cross-term and self-term contributions. For comparison, the self-term contribution of the peptide subsystem is included, which was calculated from the dipole moments of the individual lysine segments and corresponds to their BDS SPCF $F_1(t)$. A scaled variant is shown as dashed line. **b** Corresponding dynamical susceptibilities together with a sketch illustrating the origins of the self-term and cross-term decays on the time scales τ_{wat} and τ_{pep}, respectively. The pink line depicts a section of the peptide backbone, while the light blue and dark blue arrows delineate the dipole moments of two water molecules

i.e., from the true SPCF $F_1(t)$, that all water molecules completely lost their orientational correlation during the fast $F_{\mathrm{BDS}}(t)$ decay. Hence, there was no slow water fraction. Rather, the slow $F_{\mathrm{BDS}}(t)$ decay exclusively resulted from the cross-term contribution. Further insights were available from a comparison with $F_{\mathrm{BDS}}(t)$ of the peptide subsystem, which was calculated from the dipole moments of the individual lysine monomers. The comparison showed that the self-term contribution, i.e., the SPCF $F_1(t)$ of the lysine segments, decayed on the same time scale as the cross terms between the water molecules. This agreement implied that the long-time decay of $F_{\mathrm{BDS}}(t)$ for the water subsystem and the associated low-frequency shoulder in the corresponding dynamical susceptibility $\chi''_{\mathrm{BDS}}(\nu)$, see Fig. 11b, did not inform about water dynamics but about peptide dynamics.

To clarify why water–water cross correlations surprisingly probe peptide rather than water dynamics and what consequences result for the interpretation of BDS data for dynamically asymmetric mixtures, the dynamical scenario obtained from further MD analyses [83] is sketched in Fig. 11b. It turned out that the peptide molecules imposed preferable positions and orientations to neighboring water molecules, e.g., to allow for hydrogen bonds. Because the peptide was less mobile than the water, these "sites" were long-lived and occupied successively by different water molecules. As a result, the orientations of various water molecules occupying a given site at different times were correlated, leading to prominent cross terms as long as the sites were preserved. Owing to their microscopic origin, the cross terms did not decay due to water dynamics but only when peptide motion caused a renewal of the preferred sites for the water molecules. In other words, $F_{BDS}(t)$ showed a slow decay, which was associated with water–water cross terms, but did not report on water reorientation but on those peptide dynamics that led to fluctuations of preferable water sites near the peptide. These effects are by no means restricted to ε-PLL–H_2O but they occur for all mixtures of appreciably interacting components with substantially different dynamics [83]. In particular, one may speculate for protein–solvent mixtures that slow BDS processes may arise from a decay of solvent–solvent cross terms, which results when protein dynamics leads to a redistribution of preferable solvent sites. Because of their inherent link to the multi-particle nature of the BDS correlation functions, such slow processes will not be probed in NMR and NS single-particle studies.

6 Conclusions

Combining BDS studies with component-selective NMR and QENS analyses and using additional information from DSC experiments and MD simulations, it is possible to obtain detailed insights into the complex molecular dynamics of protein–solvent and peptide–solvent mixtures. However, despite valuable progress from such comprehensive approaches, several aspects of the intricate dynamical behaviors are still elusive.

Manifold experimental and computational studies yield a coherent picture of solvent dynamics. Specifically, they show that various proteins have similar but not identical effects on solvent dynamics. Thereat, the effects may differ for various solvents. Specifically, H_2O/D_2O and EG dynamics are significantly slower in protein matrices than in the bulk liquids, whereas there is no change in the time scale but only an increase of the dynamical heterogeneity for GL. Moreover, for the former solvents, deviations from Arrhenius behavior are, if at all, weak in protein mixtures, while GL exhibits VFT behavior. For all considered solvents, the reorientation remains quasi-isotropic upon cooling and it is linked to diffusion, as explicitly observed for water down to $\sim 200\,\text{K}$. This solvent dynamics shows itself as a clear BDS process denoted as P1 throughout this contribution. Because the solvent dynamics probed by P1 is isotropic and diffusive, but significantly affected by the much less mobile protein matrix, it should be considered as a delocalized β relaxation, maybe with the exception of GL. It shares many similarities with the dynamics of the fast component in other dynamically asymmetric glass-forming mixtures, such as polymer-plasticizer systems.

Water dynamics is very similar in various protein matrices below $\sim 170\,\text{K}$, where it follows an Arrhenius law with an universal activation energy of $E_a \sim 0.5\,\text{eV}$. At $\sim 170\,\text{K}$, a crossover in water dynamics occurs for some proteins but not for others. The Arrhenius behavior continues to higher temperatures for hydrated COL and ELA, where a temperature-independent Gaussian distribution of activation energies describes the effects of the protein matrices on water dynamics in broad frequency and temperature ranges. A crossover to weak non-Arrhenius behavior is found for hydrated LYS and MYO. As will be discussed in more detail below, such a dynamical crossover can occur when appropriate protein motion starts to soften the local water environments and/or when cooperative α water relaxation sets in for suitable water distributions in the mixtures.

For various protein–solvent mixtures, a highly restricted protein motion, which involves essentially all residues and is strongly heterogeneous, is coupled to the solvent dynamics associated with P1. Thereat, higher translational displacements and larger rotational amplitudes in hydrated than dry states are indicative of a solvent coupling. This highly restricted protein motion occurs for globular and fibrous proteins alike. Moreover, it involves both backbone segments and side chains although the amplitude may differ for various moieties. It can be attributed to localized diffusion within a single energy minimum, i.e., caged dynamics. The amplitude of the highly restricted motion continuously decreases upon cooling until it becomes undetectable somewhere near 200 K. One may speculate that the vanishing of the highly restricted motion is related to the glassy arrest of a slower β relaxation of the protein. Similar relations between the caged dynamics, i.e., the NCL, and the β relaxation exist for many disordered systems.

Despite valuable progress, the microscopic origins of slow BDS processes, denoted as P2 and P3, and broad DSC signatures of, in particular, globular proteins are not yet completely understood. This knowledge gap results from the fact that the findings not only strongly depend on the protein type and solvation level, but may also differ across various studies. Moreover, NMR data are often missing for, e.g., a detailed comparison with BDS correlation times. Nonetheless, a combined view of such results enables various insights. In particular, it shows that neither P2 nor P3 should be considered as a structural α relaxation of the proteins discussed in this contribution. Specifically, NMR studies rule out that, on the P2 and P3 time scales, a noticeable fraction of the protein backbone shows large-amplitude, quasi-isotropic reorientation, which is typical of the α relaxation, or, equivalently, segmental motion, of various glass-forming polymers. Considering the necessity that specific structure elements be preserved, the absence of an unrestricted motion of the protein backbone is actually no surprise for the studied cases. Because the situation may be different for disordered proteins, a closer analysis of their dynamics may be rewarding.

There is evidence that, in BDS studies on protein–solvent mixtures, not only P1 but also P2 is related to solvent dynamics. However, this does not mean that, beside a possible small amount of structural solvent molecules, fast and slow solvent fractions coexist. Rather, one may expect that P1 and P2 are the less cooperative β relaxation and more cooperative α relaxation of essentially all solvent molecules, respectively. For water, the latter assignments are supported by comparison of results for protein and silica confinements. Explicitly, the P1 and P2 correlation times of the protein mixtures resemble those of the α and β relaxations of nonfreezable water confined in silica pores, respectively. If the assignments are correct, the dynamical crossover observed for water at ~170 K can be explained based on common knowledge for glass-forming liquids. Namely, the temperature dependence of the β relaxation (P1) changes upon vitrification of the α relaxation (P2) at the glass transition temperature. In such a situation, the findings relating to the dynamical crossover indicate that cooperative α water motion is suppressed in fibrous proteins but not in globular ones because, e.g., water clusters are too small for cooperative motion in the former but not in the latter cases due to somewhat different water distributions in the mixtures. In addition, diverse dynamics of structurally different proteins may or may not lead to significant fluctuations of the local water environments and, in this way, contribute to various crossover scenarios, as aforementioned.

Protein conformational transitions, which involve different energy minima but do not break secondary structure elements, may be related to P3. Based on various NMR findings for LYS, this protein motion occurs on time scales of $\tau \sim 10^{-5}$–10^{-4} s at room temperature and has hydration-dependent small amplitudes. Accordingly, it should be regarded as a slow β relaxation of the protein. It is currently unknown to what extent the microscopic origins of such secondary relaxations depend on specific structural motifs of proteins. However, one may speculate that the onsets of such secondary relaxations of proteins somewhere near 200 K cause the softening of the local environments in protein–water mixtures, which are related to (additional) crossovers in the β relaxation of water and to the enhancement of caged protein dynamics in this temperature range.

Although further studies are required to scrutinize the proposed P2 and P3 assignments, the discussed findings show that a broad calorimetric signature observed for, in particular, hydrated globular proteins at subzero temperatures may not be considered as a regular α-related glass transition of the protein. Rather, combined knowledge from BDS and NMR studies suggests that the feature has several possible other origins, including an arrest of the β relaxation of the protein, a vitrification of the α relaxation of the solvent and, in the case of hydrated proteins, even a formation of ice-like aggregates.

MD simulations are very suitable to assist the interpretation of BDS and NMR data for protein–solvent mixtures. For a peptide–water mixture, such combined analyses show that prominent water–water cross correlations strongly affect BDS but not NMR results. Due to their microscopic origin, decays of such cross correlations require a renewal of preferable sites for water near the peptide and, hence, associated dielectric relaxations do not inform about water dynamics but peptide dynamics. Because similar effects are expected for various protein–solvent mixtures, the role of cross correlations in BDS approaches to these systems should be investigated in more detail. For example, it may be worthwhile to determine whether or not solvent–solvent cross correlations strongly magnify the BDS signatures of β-like protein motions and cause anomalously high intensities, which are found for, e.g., P3.

Competing Interests The authors have no conflicts of interest to declare that are relevant to the content of this chapter.

References

1. Khodadadi S, Sokolov AP (2015) Soft Matter 11:4984
2. Bellissent-Funel M-C, Hassanali A, Havenith M, Henchman R, Pohl P, Sterpone F, van der Spoel D, Xu Y, Garcia AE (2016) Chem Rev 116:7673
3. Doster W, Cusack S, Petry W (1989) Nature 337:754
4. Doster W (2008) Eur Phys J 37:591
5. Schiro G, Fichou Y, Gallat F-X, Wood K, Gabel F, Mouilin M, Härtlein M, Heyden M, Colletier J-P, Orecchini A, Paciaroni A, Wuttke J, Tobias DJ, Weik M (2015) Nat Commun 6:6490
6. Rasmussen BF, Stock AM, Ringe D, Petsko GA (1992) Nature 357:423
7. Frauenfelder H, Chen G, Berendzen J, Fenimore PW, Jansson H, McMahon BH, Stroe IR, Swenson J, Young RD (2009) Proc Natl Acad Sci 106:5129
8. Khodadadi S, Roh JH, Kisliuk A, Mamontov E, Tyagi M, Woodson SA, Briber RM, Sokolov AP (2010) Biophys J 98:1321
9. Ngai KL, Capaccioli S, Paciaroni A (2017) Biochim Biophys Acta - Gen Subj 1861:3553
10. Combarro-Placios I, Olsson C, Kamma-Lorger CS, Swenson J, Cerveny S (2019) J Chem Phys 150:124902
11. Yamamoto N, Kofu M, Nakajima K, Nakagawa H, Shibayama N (2021) J Phys Chem Lett 12:2172
12. Laage D, Elsaesser T, Hynes JT (2017) Chem Rev 117:10694
13. Dahanayake JN, Mitchell-Koch KR (2018) Phys Chem Chem Phys 20:14765
14. Ellis RJ, Minton AP (2003) Nature 425:27
15. Lusceac SA, Vogel MR, Herbers CR (2010) Biochim Biopys Acta - Proteins Proteom 1804:41
16. Sasaki K, Popov I, Feldman Y (2019) J Chem Phys 150:204504
17. Jansson H, Bergman R, Swenson J (2005) J Phys Chem B 109:24134
18. Fillmer A, Gainaru C, Böhmer R, Non-Cryst J (2010) Solids 356:743
19. Demuth D, Vogel M (2015) Biochim Biophys Acta - Proteins Proteom 1854:995
20. Sasaki K, Panagopoulou A, Kita R, Shinyashiki N, Yagihara S, Kyritsis A, Pissis P (2017) J Phys Chem B 121:265
21. Russo D, Murarka RK, Copley JRD, Head-Gordon T (2005) J Phys Chem B 109:12966
22. Pagnotta SE, Cerveny S, Alegria A, Colmenero J (2010) Phys Chem Chem Phys 12:10512
23. Cerveny S, Combarro-Placios I, Swenson J (2016) J Phys Chem Lett 7:4093
24. Samouillan V, Tintar D, Lacabanne C (2011) Chem Phys 385:19
25. Wolf M, Gulich R, Lunkenheimer P, Loidl A (2012) Biochim Biophys Acta - Proteins Proteom 1824:723
26. Nakanishi M, Sokolov AP, Non-Cryst J (2015) Sol 407:478
27. Lewandowski JR, Halse ME, Blackledge M, Emsley L (2015) Science 348:578
28. Roos M, Ott M, Hofmann M, Link S, Rössler E, Balbach J, Krushelnitsky A, Saalwächter K (2016) J Am Chem Soc 138:10365
29. Weissheit S, Kuttich B, Vogel M, Thiele CM (2024) ChemPhysChem 25:e202400117
30. Chen S-H, Liu L, Fratini E, Baglioni P, Faraone A, Mamontov E (2006) Proc Natl Acad Sci USA 103:9012
31. Jansson H, Bergman R, Swenson J, Non-Cryst J (2006) Solids 352:4410
32. Khodadadi S, Pawlus S, Sokolov AP (2008) J Phys Chem B 112:14273
33. Doster W, Busch S, Gaspar AM, Appavou MS, Wuttke J, Scheer H (2010) Phys Rev Lett 104:098101
34. Kämpf K, Demuth D, Zamponi M, Wuttke J, Vogel M (2020) J Chem Phys 152:245101
35. Hong L, Smolin N, Lindner B, Sokolov AP, Smith JC (2011) Phys Rev Lett 107:148102
36. Kämpf K, Klameth F, Vogel M (2012) J Chem Phys 137:205105
37. Fenimore PW, Frauenfelder H, McMahon BH, Young RD (2004) Proc Natl Acad Sci USA 101:14408
38. Gainaru C, Fillmer A, Böhmer R (2009) J Phys Chem B 113:12628
39. Tsukahara T, Sasaki K, Kita R, Shinyashiki N (2022) Phys Chem Chem Phys 24:5803

40. Sartor G, Mayer E, Johari GP (1994) Biophys J 66:249
41. Jansson H, Swenson J (2010) Biochim Biophys Acta - Proteins Proteom 1804:20
42. Samouillan V, Cédric A, Dandurand J, Lacabanne C (2004) Biomacromol 5:958
43. Panagopoulou A, Kyritsis A, Vodina M, Pissis P (2013) Biochim Biophys Acta - Proteins Proteom 1834:977
44. Panagopoulou A, Kyritsis A, Aravantinou A-M, Nanopoulos D, i Serra S, Gomez Ribelles JL, Shinyashiki N, Pissis P(2011) Food Biophys 6:199
45. Swenson J, Jansson H, Bergman R (2006) Phys Rev Lett 96:247802
46. Fomina M, Schiro G, Cupane A (2014) Biophys Chem 185:25
47. Lusceac SA, Rosenstihl M, Vogel M, Gainaru C, Fillmer A, Böhmer R, Non-Cryst J (2011) Solids 357:655
48. Panagopoulou A, Kyritsis A, i Serra RS, Gomez Ribelles JL, Shinyashiki N, Pissis P (2011) Biochim Biophys Acta - Proteins Proteom 1814:1984
49. Capaccioli S, Ngai KL, Ancherbak S, Rolla PA, Shinyashiki N, Non-Cryst J (2011) Solids 357:641
50. Panagopoulou A, Kyritsis A, Shinyashiki N, Pissis P (2012) J Phys Chem B 116:4593
51. Reuhl M, Vogel M (2022) J Chem Phys 157:074705
52. Pawlus S, Khodadadi S, Sokolov AP (2008) Phys Rev Lett 100:108103
53. Kaminski K, Kaminska E, Wlodarczyk P, Pawlus S, Kimla D, Kasprzycka A, Paluch M, Ziolo J, Szeja W, Ngai KL (2008) J Phys Chem B 112:12816
54. Roy S, Richert R (2014) Biochim Biophys Acta - Proteins Proteom 1844:323
55. Demuth D, Reuhl M, Hopfenmüller M, Karabas N, Schoner S, Vogel M (2020) Molecules 25:4127
56. Wübbenhorst M, van Turnhout J, Non-Cryst J (2002) Solids 305:40
57. Capaccioli S, Zheng L, Kyritsis A, Paciaroni A, Vogel M, Ngai KL (2021) ACS Omega 6:340
58. Melillo JH, Cangialosi D, Di Lisio V, Steinrücken E, Vogel M, Cerveny S (2024) Proc Natl Acad Sci USA 121:e2407030121
59. Steinrücken E, Weigler M, Schiller V, Vogel M (2023) J Phys Chem Lett 14:4104
60. Tombari E, Johari GP (2013) J Chem Phys 139:105102
61. Reuhl M, Weigler M, Brodrecht M, Buntkowsky G, Vogel M (2020) J Phys Chem C 124:20998
62. Vogel M (2008) Phys Rev Lett 101:225701
63. Qvist J, Mattea C, Sunde EP, Halle B (2012) J Chem Phys 136:204505
64. Bergman R, Swenson J (2000) Nature 403:283
65. Wood K, Frölich A, Paciaroni A, Moulin M, Härtlein M, Zaccai G, Tobias DJ, Weik M (2008) J Am Chem Soc 130:4586
66. Ngai KL (2023) Prog Mater Sci 139:101130
67. Capaccioli S, Ngai KL, Ancherbak S, Bertoldo M, Ciampalini G, Thayyil MS, Wang L-M (2019) J Chem Phys 151:034504
68. Swenson J, Cerveny S (2015) J Phys: Condens Matter 27:033102
69. Herbers CR, Sauer D, Vogel M (2012) J Chem Phys 136:124511
70. Vogel M, Rössler E (2000) J Phys Chem B 104:4285
71. Rosenstihl M, Vogel M (2011) J Chem Phys 135:164503
72. Vogel M (2009) J Phys Chem B 113:9386
73. Caliskan G, Briber RM, Thirumalai D, Garcia-Sakai V, Woodson SA, Sokolov AP (2006) J Am Chem Soc 128:32
74. Perry A, Stypa MP, Tenn BK, Kumashiro KK (2002) Biophys J 82:1086
75. Reichert D, Pascui O, deAzevedo ER, Bonagamba TJ, Arnold K, Huster D (2004) Magn Reson Chem 42:276
76. Kämpf K, Kremmling B, Vogel M (2014) Phys Rev E 89:032710
77. Kennedy SD, Bryant RG (1990) Biopolymers 29:1801
78. Krushelnitsky A, Reichert D (2004) Appl Magn Reson 27:501
79. Krushelnitsky A, Zinkevich T, Mukhametshina N, Tarasova N, Gogolev Y, Gnezdilov O, Fedotov V, Belton P, Reichert D (2009) J Phys Chem B 113:10022
80. Vogel M, Rössler E (2001) J Chem Phys 114:5802

81. Vogel M, Rössler E (2001) J Chem Phys 115:10883
82. Weigler M, Combarro-Palacios I, Cerveny S, Vogel M (2020) J Chem Phys 152:234503
83. Böhmer T, Horstmann R, Gabriel JP, Pabst F, Vogel M, Blochowicz T (2021) Macromolecules 54:10340

Terahertz Spectroscopy of Water and Proteins: Methodology and Insights

Fabio Novelli and Ellen Adams

Abstract Terahertz spectroscopy probes electromagnetic radiation in the ~0.1–10 THz range, a spectral region highly sensitive to collective motions in condensed matter. Liquid water, with its transient hydrogen-bond network and strong terahertz absorption, is a prime example: its dielectric response can be minimally described by two Debye relaxations and two Lorentz oscillators associated with intermolecular translations and librations. Temperature-dependent measurements reveal that weakening the hydrogen-bond network accelerates collective dipolar reorientation and increases both refraction and absorption. Extending to aqueous solutions, effective medium theories relate changes in the terahertz absorption to volume exclusion and hydration shells. Deviations from "water plus void" models provide qualitative access to hydration water structure, distinguishing "structure-making" from "structure-breaking" solutes. For proteins, terahertz absorption as a function of concentration reveals long-range hydration shells extending up to several layers and thousands of water molecules. Moreover, the sign and magnitude of the terahertz response correlate with surface electrostatic potential, offering a route to infer surface charge patterns. Analysis of Debye relaxation times links protein hydration dynamics to a complex interplay of surface charge, hydrophobicity, and local topology, emphasizing hydration as an emergent, collective property.

Keywords Terahertz · Infrared · Spectroscopy · Dielectric · Function · Liquid · Water · Aqueous · Solution · Effective · Medium · Protein · Colloid · Dispersion · Surface · Electrostatic · Debye · Relaxation

F. Novelli (✉)
School of Physics and Astronomy, Faculty of Engineering and Physical Sciences, University of Southampton, Southampton, UK
e-mail: F.Novelli@soton.ac.uk

E. Adams
Cluster of Excellence Physics of Life, TU Dresden, Dresden, Germany

Institute of Resource Ecology, Helmholtz Zentrum Dresden Rossendorf, Dresden, Germany

S. Cerveny and J. Swenson (eds.), *Dynamics of Biomolecules Studied by Dielectric Spectroscopy*, Advances in Dielectrics, https://doi.org/10.1007/978-3-032-30588-6_8

1 Introduction to Terahertz Spectroscopy

Electromagnetic radiation with frequencies between about 0.1 terahertz (THz) and 10 THz is typically referred to as terahertz radiation. Different units are used by different communities, so this spectroscopic range corresponds to ca. 3–333 cm^{-1}, 3000–30 μm, or 0.4–41 meV.

Terahertz radiation overlaps in part with the traditional far-infrared region. The different terminology originates from technical details: while Fourier-transform infrared (FTIR) spectrometers make use of interferometric techniques to retrieve spectral intensities, terahertz spectroscopy can detect both phase and amplitude of phase-stable pulses that can propagate in free space [1]. FTIR spectroscopy can be performed with incoherent sources of radiation, but terahertz spectroscopy typically entails the use of a coherent source, such a laser. In most cases, FTIR spectroscopy allows the absorption coefficient to be estimated over a broad frequency range spanning from about 2–30 THz. Terahertz time-domain spectroscopy often works best at frequencies between ca. 0.1 and 3 THz, can detect both refraction and absorption changes, and has at least an order of magnitude better signal-to-noise ratio than a standard FTIR spectrometer. For these reasons, FTIR and THz-TDS could be currently considered complementary techniques.

Terahertz radiation triggers a variety of significant transitions in matter and is sensitive to different phases: rotational lines in gases, internal vibrations in molecules, dipolar relaxations and intermolecular modes in liquids, phonons and superconducting gaps in solids, and charge carriers' dynamics in both soft and hard condensed matter systems, to name a few [2]. In the following, we focus on the terahertz response of liquid water.

1.1 Liquid Water

Water in the liquid phase can be thought of as a dynamic network of molecules rapidly forming and breaking hydrogen bonds (HBs). At any instant of time, these bonds are approximately arranged in tetrahedral structures with oxygen at the vertices and one water molecule embedded in the middle. Each molecule forms about four HBs—two donated and two accepted—with the surrounding waters. The average lifetime of each HB is about 1 ps [3]. Thus, we expect picosecond-long fields to be able to probe the dynamic HB network. As electromagnetic radiation oscillating on picosecond timescales have corresponding spectral components covering the terahertz frequency range, terahertz spectroscopy is a powerful tool to study liquid water. Indeed, liquid water interacts strongly with THz radiation. For example, the absorption coefficient of water at 1 THz and 20 °C is ~230 cm^{-1} [4], corresponding to a penetration depth of only ~43 μm. That is, if we were to illuminate with terahertz radiation a water layer with a thickness of a few tens of microns, most of the input radiation would be absorbed. If our eyes could perceive only THz radiation, water would appear black.

The dielectric function of liquid water at terahertz frequencies has been reported multiple times, each confirming the same qualitative trend of increasing absorption with frequency and the presence of three broad spectroscopic features centered at ~0.02 THz, ~5 THz, and ~18 THz. However, quantitative variations up to about 5% can be present when comparing the literature values. These discrepancies likely arise from systematic uncertainties related to the different techniques used, i.e., FTIR or THz-TDS in reflection or transmission geometries. The dashed black line in Fig. 1 is the average of the results reported in refs. [5–7]. Reference [4] summarizes in a table the optical properties of water between 0.1 THz and 1 THz, which were averaged over multiple THz-TDS results reported in the literature.

Key theory results [8–10] clarified the intermolecular, microscopic motions triggered by the interaction of water with electromagnetic radiation between about 2 and 30 THz. The broad band around 5 THz is associated with restricted translations of H-bonded waters, whereby the oxygen are mostly displaced along the HB directions. The feature spanning from ca. 10–30 THz is linked to restricted rotations of H-bonded waters, also called librations. Two types of librations involve rocking, where the water molecule rotates mostly in the plane of incidence, and wagging motions, where the hydrogens tilt out of the plane. Cartoons of the IR-active modes are highlighted in Fig. 1 in light and dark blue for collective stretching (T) and librations (L), respectively. The most common approach to describe the room temperature dielectric function of water below about 2 THz is by summing up two Debye relaxations with

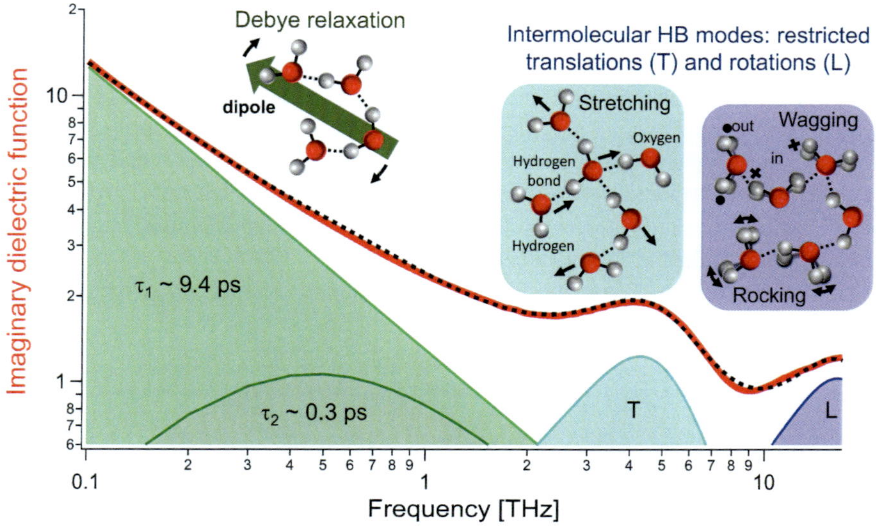

Fig. 1 Imaginary part of the dielectric function of pure liquid water at 20 °C (dashed black). A minimal fit (solid red) requires two harmonic oscillators centered at ~5 THz (light blue) and ~17 THz (dark blue), associated with intermolecular modes involving restricted translations (T) or librations (L), respectively. Two Debye features are present at lower frequencies, with relaxation time constants τ_1~9.4 ps (light green) and τ_2~0.3 ps (dark green). Cartoons of the corresponding molecular motions are displayed on the top

a slower time constant between 7–10 ps (τ_1) and a faster time constant of 0.1–0.5 ps (τ_2) [11]. Typically, τ_1 is associated with the collective dipolar reorientation of groups of H-bonded waters, while τ_2 is interpreted as rotations of molecules with fewer bonds [11]. While we will keep in line with this mainstream interpretation for the remainder of the chapter, we note that alternative explanations, including vibrational motions [7, 12], have been proposed to describe the dielectric function of water below ~2 THz [13].

In general, the interaction of electromagnetic radiation with matter results in the radiation being damped and delayed. These two fundamental processes, absorption and refraction, are typically described in a compact way with a complex-valued and frequency-dependent function such as the dielectric function, $\tilde{\varepsilon}(\nu) = \varepsilon'(\nu) + i\varepsilon''(\nu)$, with i imaginary unit. There are several related optical quantities able to describe absorption and refraction and we refer to the excellent table on page 395 of ref. [14] for an overview. Here, we only mention that the real part, $n(\nu)$, of the complex index of refraction, $\tilde{n}(\nu) = n(\nu) + ik(\nu)$, is often used to describe the refractive properties of a material. The imaginary part of the complex index, $k(\nu)$, is called extinction coefficient and is related to the absorption coefficient, $\alpha(\nu) = 2\omega k(\nu)/c$ with c speed of light.

The minimal description of the dielectric response of pure water requires two Debye and two Lorentz functions:

$$\tilde{\varepsilon}(\nu) = \varepsilon_\infty + \frac{\varepsilon_S - \varepsilon_1}{1 + i\omega\tau_1} + \frac{\varepsilon_1 - \varepsilon_\infty}{1 + i\omega\tau_2} + \frac{a^2}{\omega_a^2 - \omega^2 - i\omega\Gamma_a} + \frac{b^2}{\omega_b^2 - \omega^2 - i\omega\Gamma_b} \tag{1}$$

where $\omega = 2\pi\nu$ is the angular frequency, ν the frequency, $\varepsilon_S = 80.1$ is fixed to the value of the static dielectric constant of water at 20 °C, ε_1 is the amplitude parameter of the Debye excitations, and ε_∞ the value of the dielectric constant at frequencies much higher than the probed spectroscopic range. For the Lorentz parameters, a and b are the plasma frequencies of the translation and libration modes, respectively. The linewidths are indicated with the capital greek letter Γ. The center angular frequencies are ω_a and ω_b, the center frequencies ν_a and ν_b.

The real part of Eq. 1 describes the refractive propreties and is equal to:

$$\begin{aligned}\varepsilon'(\nu) = \mathrm{Re}[\tilde{\varepsilon}(\nu)] &= \varepsilon_\infty + \frac{\varepsilon_S - \varepsilon_1}{1 + (\omega\tau_1)^2} + \frac{\varepsilon_1 - \varepsilon_\infty}{1 + (\omega\tau_2)^2} \\ &+ \frac{a^2 \cdot (\nu_a^2 - \nu^2)}{(2\pi)^2(\nu_a^2 - \nu^2)^2 + \nu^2\Gamma_a^2} + \frac{b^2 \cdot (\nu_b^2 - \nu^2)}{(2\pi)^2(\nu_b^2 - \nu^2)^2 + \nu^2\Gamma_b^2}\end{aligned} \tag{2}$$

The solid red line in Fig. 1 is the result of a fit function to the imaginary part of Eq. 1, which accounts for the damping of electromagnetic radiation:

$$\varepsilon''(\nu) = \mathrm{Im}[\tilde{\varepsilon}(\nu)] = \frac{(\varepsilon_S - \varepsilon_1)(2\pi \cdot \nu \cdot \tau_1)}{1 + (\omega\tau_1)^2} + \frac{(\varepsilon_1 - \varepsilon_\infty)(2\pi \cdot \nu \cdot \tau_2)}{1 + (\omega\tau_2)^2}$$

$$+\frac{a^2\cdot\nu\cdot\Gamma_a/(2\pi)}{(2\pi)^2\left(\nu_a^2-\nu^2\right)^2+\nu^2\Gamma_a^2}+\frac{b^2\cdot\nu\cdot\Gamma_b/(2\pi)}{(2\pi)^2\left(\nu_b^2-\nu^2\right)^2+\nu^2\Gamma_b^2} \tag{3}$$

The obtained best-fit parameters are: $\varepsilon_1 = 3.94 \pm 0.04$, $\varepsilon_\infty = 1.81 \pm 0.01$, $\tau_1 = 9.36 \pm 0.04$ ps, $\tau_2 = 0.34 \pm 0.01$ ps, $a = 33.39 \pm 0.35$ THz, $\nu_a = 4.93 \pm 0.02$ THz, $\Gamma_a = 31.21 \pm 0.54$ THz, $b = 102.61 \pm 1.72$ THz, $\nu_b = 18.00 \pm 0.14$ THz, $\Gamma_b = 94.42 \pm 2.49$ THz, where the error bars correspond to the numerical errors provided by the fit procedure. At higher frequencies, the parameters describing the librations are less reliable because the spectrum extends only to 17 THz. Please note that we performed a global fit, accounting simultaneously for the real and imaginary parts of the dielectric function of water via Eqs. 2 and 3.

To first approximation, the response of water to low-frequency THz fields is dominated by the first Debye term, see the light green shaded area in Fig. 1. If τ_1 would be the largest term, Eq. 1 would be scaling roughly as $\tilde{\varepsilon}(\nu) \approx 1/\tau_1$. That is, the value of the dielectric function should be inversely proportional to the main reorientation lifetime. As this lifetime accounts for the reorientation of groups of H-bonded water molecules, one expects this process to be faster if the average number of HB decreases. On the contrary, the stronger the HB network, the more hindered and the slower the collective molecular reorientations should be. If this qualitative model holds, one expects τ_1 to decrease, and the magnitude of the optical functions to increase, whenever the HB is weaker. If we "weaken" the HBs by increasing the thermal energy, liquid water becomes more absorbing and refractive. Figure 2 reports the frequency-dependent changes of refractive index (left) and absorption coefficient (right) of pure water upon heating from 19.7 °C up to 43.1 °C. This was measured with THz-TDS on a 200-micron thick sample in a laboratory of the Saraceno group [15, 16]. In keeping with this observation, seminal works by the Keiding group [17, 18] reported that τ_1 (sometimes labeled with τ_D) decreases from ~10 ps at 20 °C to ~6 ps around 40 °C. Such a simple model suggests that "more structured" water displays smaller optical quantities, while "more disordered" water is associated to optical responses with larger magnitudes. This is a good starting point to rationalize the results of terahertz spectroscopy on aqueous solutions and colloids.

1.2 Aqueous Materials

To our knowledge, and except for metals or other highly conductive materials, liquid water is among the strongest absorbers of terahertz radiation. This implies that most solutes that can be dissolved, suspended, or dispersed in water have lower dielectric functions at THz frequencies. As a result, most aqueous solutions show a decreased absorption for increasing solute concentrations. A common approach to address the terahertz response of a solution is to measure a series of samples with increasing concentrations and plot the value of the optical quantities, often averaged

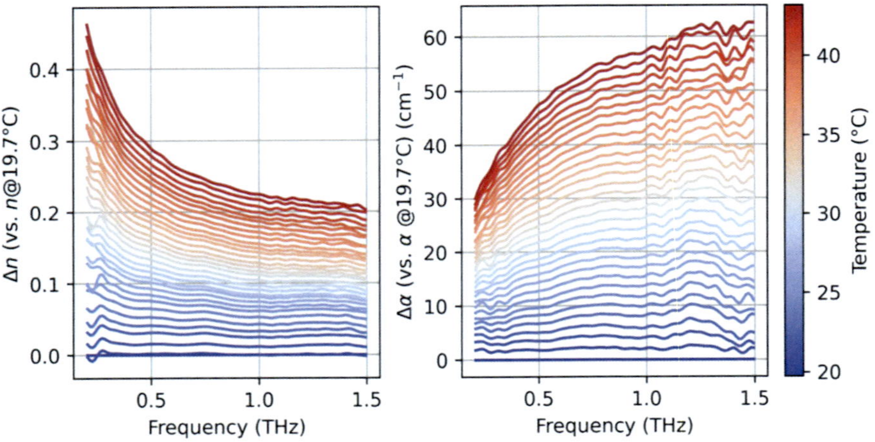

Fig. 2 Variation of the index of refraction (left) and of the absorption coefficient (right) of liquid water as a function of temperature. The temperature increased from 19.7 °C (darker blue) to 43.1 °C (darker red). Each separate trace corresponds to a temperature increase of about 0.75 °C. Both refraction and absorption components increase for increasing temperature. Small oscillations in the amount of water vapor in the spectrometer are responsible for the bumps at ~1.1 and ~1.4 THz

over the available THz frequency range, versus concentration. A cartoon of such an experiment is reported with the red curve in Fig. 3.

We can idealize the process of adding a solute to a liquid with two steps. First, the liquid needs to "make room" for the solute: that is, create a cavity of suitable size to accommodate it. The creation of this cavity has an enthalpic cost because liquid–liquid bonds are broken, and an entropic cost as the number of liquid molecules in the unit volume decreases along with their number of permutations that defines the entropy. Second, the solute fills the cavity, resulting in the formation of new solute–solvent bonds (enthalpy gain) and a generally favorable entropy of mixing. These variations of enthalpy and entropy define the solvation energy, which is an important parameter affecting chemical reactions [19]. In other words, adding a solute to water will "remove" some water molecules in any chosen unit of volume. This results in a volume-exclusion effect that is equivalent to the formation of a cavity filled by the solute. In addition, the interaction of the solute with the surrounding water molecules affects their properties, forming a so-called hydration layer or shell. This term can be used to describe both the chemisorbed water molecules—the ones forming more stable bonds directly at the solute interface—and physisorbed waters characterized by more dynamical interactions, up to several nanometers from the solute.

Effective medium theories (EMT) are powerful approaches that can highlight the spectroscopic features of the hydration water. While several EMT are available as detailed in more specialized publications [20, 21], here we aim at presenting only the simplest one: the absorption coefficient of a mixture (α_{TOT}) is equated to the sum of the absorption of each of its j components (α_j) weighted by their volume fractions (ϕ_j):

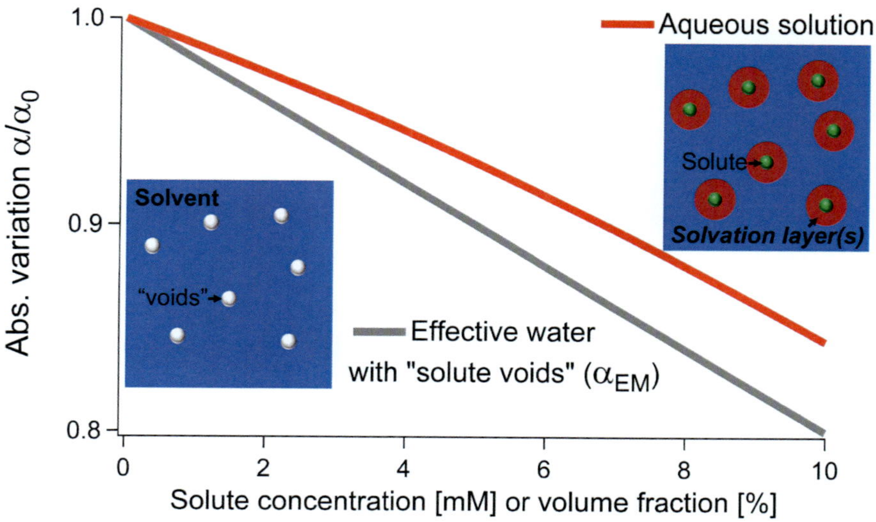

Fig. 3 Cartoon of the relative variation of the terahertz absorption coefficient of a generic aqueous solution as a function of the solute concentration (red). Details about the solvation process can be estimated by comparing the results (red curve) with an effective model (gray line) describing the response of water and empty solute volume (white spheres). The solutes are sketched with green spheres, the hydration water is in red. Please note that the absorption on the y-axis has a value of 1 at zero solute concentration, i.e., it is divided by the absorption coefficient of the bare solvent (α_0). Concentration and volume fraction are not equal but either could be used as x-axis

$$\alpha_{\mathrm{TOT}}(\nu) = \sum_j \phi_j \alpha_j(\nu) \tag{4}$$

Please note that, in general, absorption is not an additive property. For example, if we add a thin layer of an appropriate material on top of a glass slab, we end up increasing the transmission of the glass, apparently *decreasing* its absorption even if both the glass and the top layer have positive absorption coefficients. This is how anti-reflection coatings work: the multiple reflections form the thin layer destructively interfere, imposing a higher transmission due to the conservation of energy. Nonetheless, Eq. 4 is a valid approximation whenever the contributions of the index of refraction are neglected. As this holds roughly true when comparing diluted solutions to the bare solvent, this approximation can become useful and is often found in the literature.

Following Eq. 4 and dropping the dependence on frequency for simplicity, the absorption coefficient of an aqueous solution could be written as

$$\alpha_{\mathrm{TOT}} = \phi_b \alpha_b + \phi_h \alpha_h + \phi_s \alpha_s \tag{5}$$

where s indicates the solute, h the hydration water, and b the bulk water—the water that is not perturbed by the presence of the solute. The gray line in Fig. 3 is obtained from this additive-absorption EMT under the assumption that the solute is transparent ($\alpha_s = 0$) and the hydration water is identical to the water ($\alpha_h = \alpha_b$):

$$\alpha_{\mathrm{EM}} = (\phi_b + \phi_h)\alpha_b \tag{6}$$

It is straightforward to calculate this quantity from the known absorption coefficient of pure bulk water, α_b, and the concentration or volume fraction of the dissolved solute, i.e., from the identity $\phi_b + \phi_h + \phi_s = 1$ we compute $1 - \phi_s$, which is equal to $\phi_b + \phi_h$.

Equation 5 can be used to describe the red curve in Fig. 3. Equation 6 is the gray line in Fig. 3. The difference between Eqs. 5 and 6 is.

$$\alpha_{\mathrm{TOT}} - \alpha_{\mathrm{EM}} = \phi_h(\alpha_h - \alpha_b) + \phi_s\alpha_s. \tag{7}$$

As the absorption coefficient of the solute is often much smaller than the one of water [22, 23], we might simplistically assume that $\alpha_s \approx 0$ and recast Eq. 7 as

$$\alpha_{\mathrm{TOT}} - \alpha_{\mathrm{EM}} \approx \phi_h(\alpha_h - \alpha_b) \tag{8}$$

If all the approximations used above are appropriate, the result of Eq. 8 is powerful: if the experimental results (red in Fig. 3) are above the void-plus-water effective medium simulation (gray in Fig. 3), then $\alpha_{\mathrm{TOT}} - \alpha_{\mathrm{EM}} > 0$ and we can infer from Eq. 8 that there is a detectable hydration layer ($\phi_h > 0$) absorbing more than bulk water, i.e., $\alpha_h - \alpha_b > 0$ or $\alpha_h > \alpha_b$. In turn, if $\tilde{\varepsilon}(\nu) \approx 1/\tau_1$ as detailed previously, $\alpha_h > \alpha_b$ implies $\tau_{1,h} < \tau_{1,b}$. In other words, if the detected absorption is higher than the prediction of the effective medium model, the solute has a detectable hydration shell whose molecules have a reorientation timescale that is faster than bulk water, indicating a "more disordered" or weaker HB network. On the contrary, if the red experimental curve lies below the gray model in Fig. 3, $\alpha_{\mathrm{TOT}} - \alpha_{\mathrm{EM}} < 0$, $\alpha_h < \alpha_b$, and $\tau_{1,h} > \tau_{1,b}$, indicating a "structure maker" solute. Please note that this is just the simplest take on a complex problem: we do not expect it to hold when any of the many approximations under which it was derived are not valid. More sophisticated interpretations will be provided in Paragraph 3 and the other Chapters.

2 Methodology of Terahertz Time-Domain Spectroscopy

Terahertz time-domain spectroscopy requires the detection of two terahertz fields, one resulting from the interaction with a reference and known material, and one from the interaction with the sample. Often the optical properties of the sample are apparent by inspecting these two time-dependent pulses. At simplest, absorption results in a reduced amplitude of the THz pulse, while refraction adds a temporal delay. Some important quantities can be estimated directly from the measurements without further data manipulation: the maximum amplitudes of the THz fields that interacted with

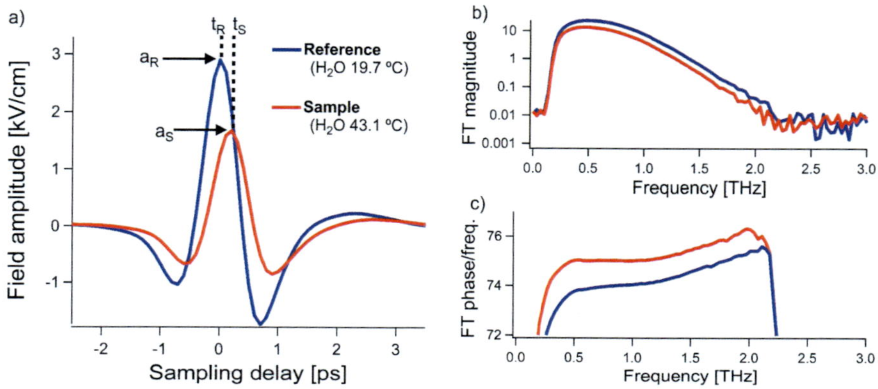

Fig. 4 THz fields recorded after the interaction with a reference material (blue) or a sample (red). In this case, these correspond to the transmissions of a 200-micron thick water layer kept at two different temperatures, 19.7 °C in blue and 43.1 °C in red, respectively. **b** Magnitude of the FT of the THz fields. **c** Phase of the FT divided by the THz frequency value, indicating a detectable signal between about 0.2 THz and 2.1 THz

sample or reference—marked with a_S and a_R in Fig. 4a, respectively—and the arrival time of these maxima, indicated by t_s and t_R. As the frequency components of each pulse are often close to being in phase in the proximity of the THz peak, the order of magnitude of the absorption coefficient and the index of refraction of the sample compared to the reference—effectively averaged over the THz spectrum—can be estimated from the following simple equations:

$$\Delta n = n_S - n_R \approx \frac{c}{d}(t_S - t_R) \tag{9}$$

$$\Delta\alpha = \alpha_S - \alpha_R \approx -\frac{2}{d}\ln\left(\frac{a_S}{a_R}\right) \tag{10}$$

where d is the (same) thickness of sample and reference materials.

The THz fields reported in Fig. 4a are mostly characterized by a main, single cycle oscillation. This is possible only because the water vapor is reduced to minimal level in the spectrometer chamber, typically few percent. If there were absorbing gas species, oscillations would appear after the main THz peak—see, for example, Fig. 1 in ref. [4]. This is because the rotational lines of a gas span well-defined and narrow frequency ranges, implying a larger temporal duration due to the time-energy uncertainty principle. Another important technical detail to consider is that, as shown in Fig. 2, the dielectric function of liquid water strongly depends on temperature. Thus, temperature is often stabilized when performing THz-TDS measurements on aqueous samples. This can be done, for example, by mounting the liquid cell into a large metallic plate whose temperature is held by an external recirculating chiller. The experiments reported in Figs. 2 and 4 were performed in this way.

While the intuitive relations described by Eqs. (9, 10) can be useful for rough estimates, we can take full advantage of the phase-sensitivity of THz-TDS and assess both the magnitude and phase of the spectral components. These quantities can be obtained by Fourier transform (FT) and are reported in Fig. 4. It is straightforward to obtain the magnitude shown in Fig. 4b, which pinpoints the reduced THz transmission by liquid water upon heating. Handling of the phase information requires a bit more care [24, 25]. The phase directly obtained by the FT will look like a sawtooth with jumps around y-values close to $+\pi$ and $-\pi$. This is because the phase is undefined by multiples of 2π. In fact, any complex number $z = r(\cos\theta + i\sin\theta)$ with magnitude r and phase θ is identical if we replace θ with $\theta + l \cdot 2\pi$, where l is integer. To correct these jumps the phase needs to be unwrapped, that is, you want to add $+2\pi$ to each y-value that follows a frequency at which the phase jumped down. Most analysis programs have built-in unwrap functions to undo modulus operations. The unwrapped phase looks like a line: you can further add or subtract integer multiples of 2π to have the linear phase extrapolate to zero. This corresponds to arbitrarily set a zero phase at zero frequency. Dividing this linear phase by the frequency axis gives a graph that is easier to read, see Fig. 4c. The positive offset separating the red and blue curves signifies an increased refraction of the liquid upon heating, as expected. If you have issues in extracting the phase information, consider recording the reference and sample THz traces over the same temporal delay, with the same temporal step size, and have the main THz pulse in the first half of the temporal range of acquisition. This ensures that the FT phases of the THz fields resulting from the interactions with reference and sample are directly comparable, even without unwrapping.

2.1 Geometrical Models

To estimate the frequency-dependent optical functions we need the two THz fields that interacted with sample and reference (see, e.g., blue and red traces in Fig. 4a), and a model that appropriately describes the experimental geometry. First of all, experiments can be performed in (total) reflection or transmission, thus affecting the spatial extent of the THz-matter interaction. In particular, the thickness is a set, external and controllable parameter in transmission experiments. On the contrary, the thickness can only be estimated indirectly in reflection experiments and cannot be independently measured. The angle of incidence of the THz pulses with respect to surface of sample and reference is another important parameter to consider, together with the polarization of the incoming electromagnetic radiation. In addition, the structure or morphology of the sample and reference materials are critical parameters.

The simplest analytical solutions are found when linearly polarized THz pulses are normally incident and transmitted by uniform, isotropic, flat, and "thick-enough" samples and references having the same thicknesses. In this case, thick-enough means that multiple reflections—also referred to as echelon or Fabry–Perot—can be neglected, typically because they are delayed beyond the detected temporal range. If the materials have also relatively high refraction and low absorption, the difference between the refractive index and absorption coefficient of sample (S) and reference (R) are:

$$\Delta n(\nu) = n_S(\nu) - n_R(\nu) = \frac{c}{d}\frac{1}{2\pi}\left(\frac{\phi_S}{\nu} - \frac{\phi_R}{\nu}\right) \tag{11}$$

where ϕ_S is the FT phase of the THz field transmitted by the sample and ϕ_R by the reference: for example, $\frac{\phi_S}{\nu}$ is the red curve in Fig. 4c and $\frac{\phi_R}{\nu}$ is the blue one;

$$\Delta\alpha(\nu) = \alpha_S(\nu) - \alpha_R(\nu) = -\frac{2}{d}\ln\left(\frac{M_S}{M_R}\right) - \frac{2}{d}\ln\left(\frac{n_R}{n_S}\frac{(1+n_S)^2}{(1+n_R)^2}\right) \tag{12}$$

where M_S and M_R are the magnitudes (not the intensities) of the FT of the THz fields transmitted by sample and reference, respectively. In the example of Fig. 4, M_S is the red trace in Fig. 4b while M_R is in blue. If the reference is vacuum or air with low water vapor content, the values $\alpha_R \approx 0$ and $n_R \approx 1$ can be substituted in Eqs. (11, 12). In summary, one records the THz fields that interacted with sample and reference materials, obtains the frequency-dependent phase and magnitude of their FT, and then uses a pair of equations like Eqs. (11, 12) to estimate the refractive and absorptive coefficients of the sample with respect to the reference ones. The equations that can be used for a liquid cell are detailed in ref. [4] together with their limits of validity.

Another important case is when sample and reference are so thin that all multiple reflections "arrive simultaneously", that is, within the temporal resolution of the performed spectroscopic experiments. In this case, the complex-valued and frequency-dependent optical conductivity of the thin sample ($\tilde{\sigma}_S$) can be estimated from

$$\tilde{\sigma}_S(\nu) = \frac{2}{Zd}\left(\frac{M_R e^{i\phi_R}}{M_S e^{i\phi_S}} - 1 + \frac{M_R e^{i\phi_R}}{M_S e^{i\phi_S}}\frac{Zd}{2}\tilde{\sigma}_R(\nu)\right) \tag{13}$$

where $Z \approx 376.7\Omega$ is the impedance of free space and $\tilde{\sigma}_R(\nu)$ is the conductivity of the reference material. If we take purged air as reference, $\tilde{\sigma}_R \approx 0$ and the last term inside the parenthesis is null. In this case, Eq. 13 is equal to the celebrated Tinkham equation [26]. Please note that the optical conductivity is related to the dielectric function and complex refraction through the relations $\tilde{n}^2(\nu) = \tilde{\varepsilon}(\nu) = 1 + i\frac{\tilde{\sigma}(\nu)}{2\pi\nu\epsilon_0}$, with ϵ_0 vacuum permittivity constant. Additional cases considering thin films on substrates in either reflection or transmission geometries are discussed in ref. [27].

The possible sample morphology varies greatly and it is not always possible to solve analytically the set of equations describing the THz propagation. Some examples are discussed in ref. [26]. The extraction of the optical parameters complicates further when dealing with non-uniform materials like pellets with air inclusions, for which EMT can be used [28].

2.2 Generation and Detection

The propagation of electromagnetic radiation is determined by the Maxwell equations. It is possible to combine these into a wave equation describing the electromagnetic field propagating in a material that is electrically neutral and nonmagnetic (see, e.g., Chap. 6 in *Introduction to Modern Optics* by Fowles):

$$\nabla \times (\nabla \times \boldsymbol{E}) + \frac{1}{c^2}\frac{\partial^2 \boldsymbol{E}}{\partial t^2} = -\mu_0 \frac{\partial^2 \boldsymbol{P}}{\partial t^2} - \mu_0 \frac{\partial \boldsymbol{J}}{\partial t} \tag{14}$$

with μ_0 magnetic permittivity of vacuum, $\boldsymbol{J}$ current density, and $\boldsymbol{P}$ polarization density [29]. This $\boldsymbol{P}$ is the number of dipoles induced in the unit volume—it should not be confused with the polarization of a wave, which describes instead the directions along which the field components oscillate. The terms on the right side of the equality in Eq. 14 are sometimes called "source terms". To obtain the electric field $\boldsymbol{E}$ of the propagating electromagnetic radiation, Eq. 14 needs to be solved for the specific material under investigation. However, one could heuristically think that whenever the polarization $\boldsymbol{P}$ or the current $\boldsymbol{J}$ vary in time over the sub-picosecond timescale, then the corresponding electric fields should have spectral components in the terahertz frequency range. That is, THz radiation could be generated or detected every time the time-derivatives of $\boldsymbol{P}$ or $\boldsymbol{J}$ are not zero.

There are indeed two main ways to generate THz radiation. One involves the second-order nonlinear polarization, the other stems from fast, sub-picosecond current effects. An insightful and accessible introduction to nonlinear optics can be found in Chap. 1 of *Nonlinear Optics* by Robert W. Boyd. In short, here we recall that the second-order polarization induced in a lossless and non-conductive (no free charges) medium is approximately

$$\boldsymbol{P}_2 = \epsilon_0\, \chi_2 \boldsymbol{E}^2 \tag{15}$$

with χ_2 material- and frequency-dependent second-order nonlinear susceptibility. As χ_2 is typically small, nonlinearities become relevant only when the driving field $\boldsymbol{E}$ is strong enough. The idea is that the input field induces a quadratic polarization change which, in turn, is the source of the emitted field. For example, if the driving field is approximately monochromatic in frequency or varies in time as a continuous wave (CW), $\boldsymbol{E} \approx \cos(\omega t)$ and Eq. 15 can be rewritten as

$$\boldsymbol{P}_2 \approx \epsilon_0\, \chi_2 \cos^2(\omega t) = \frac{1}{2}\,\epsilon_0\, \chi_2 + \frac{1}{2}\,\epsilon_0\, \chi_2 \cos(2\omega t) \tag{16}$$

Here, the last term describes second-harmonic generation, whereby an input field oscillating at angular frequency ω induces a polarization oscillating at twice the frequency (2ω). The term relevant for THz generation is $\frac{1}{2}\,\epsilon_0\, \chi_2$, which is called optical rectification. Here, an input field ω results in a zero frequency or bias effect ($\cos 0 = 1$). Terahertz radiation can be generated when the driving, input field contains frequency components that are non-zero, which can be achieved with pulses. In other words, if $\boldsymbol{E} \approx e^{-t^2/g^2}\cos(\omega t)$ describes a pulse centered at time zero with an envelope or duration defined by the width g, Eq. 15 becomes equal to:

$$\boldsymbol{P}_2 \approx \frac{1}{2} \in_0 \chi_2 e^{-2t^2/g^2} + \frac{1}{2} \in_0 \chi_2 e^{-2t^2/g^2} \cos(2\omega t) \tag{17}$$

where the spectral components of e^{-2t^2/g^2} (accessible via FT) extend to the THz spectral range if the temporal duration of the diving pulse is shorter than about one picosecond.

Another category of processes able to generate THz radiation are related to time-varying currents, i.e., stem from the $\partial\boldsymbol{J}$ term in Eq. 14. Auston switches or photoconductive antennas [30–33] are typically made of electrical contacts that are deposited on a semiconducting substrate. The contacts are separated by a small spatial gap, typically on the micrometers range, forming a capacitor. A laser pulse of sufficient energy impulsively generates electron–hole pairs by exciting the semiconductor substrate within the electrodes gap. The external electrical bias applied to the electrodes accelerates the carriers, resulting in the emission of THz radiation. Another mechanism that depends on current contributions is plasma generation, whereby intense and short laser pulses are focused in a gas, ionize it, and create a plasma. The electric field component of the driving pulse further accelerates the plasma carriers, resulting in THz emission [34]. Finally, we mention that spin polarized currents can be deflected in stacks of thin magnetic samples, resulting in THz "spintronic" generation [35, 36].

Current or polarization mechanisms can also be used to coherently detect the terahertz radiation. Coherent here refers to the ability to record the phase of the THz pulses rather than only their intensities. Antennas are routinely used for detection in combination with oscillator-based laser sources [37, 38], which are characterized by moderate energy per pulse (typically tens of nanojoules) and high repetition rates (10–100 megahertz). Air-biased coherent detection (ABCD) is some sort of "inverse" process of plasma-based THz generation. It typically requires amplified (on the order of one mJ/pulse energy) and short (<100 fs) laser pulses and can achieve broadband detection extending beyond 20 THz [37, 38]. A prominent detection technique based on polarization effects is electro-optical sampling (EOS). The idea is that the amplitude of the THz field can be encoded into the phase retardation (light polarization) of a gate or sampling laser pulse. In short, whenever an electric field is applied to a material lacking center of inversion (i.e., for which χ_2 is not zero), the material becomes (more) birefringent: the THz field induces birefringence on an EOS crystal via Pockels' effect. The polarization of a (typically near-infrared) laser pulse that co-propagates with the THz beam in the detection crystal is affected accordingly. By scanning the temporal delay between THz and near-infrared pulses, the THz field can be recorded in time as shown for example in Fig. 4a. A more detailed explanation of EOS can be found in ref. [39].

2.3 Current Developments

The improvement of THz-TDS has been ongoing for a few decades. Features under constant optimization include the signal-to-noise ratio (SNR) and broader spectral ranges. The SNR can be improved by reducing the noise and/or increasing the signal. As long as instrumental drifts or quantum limits [40] can be neglected, noise can be reduced by averaging more. Faster acquisition techniques include single-shot [41, 42], multiplexing [43], and continuously moving mechanical delays [44]. The signal can be enhanced with brighter sources if care is taken to avoid nonlinear effects [4]. This latter approach, dubbed "intense terahertz time-domain spectroscopy" (inTHz-TDS), demonstrated an order of magnitude improvement in the detectability of substances dispersed in water [4]. It was also used as an ecotoxicology probe assessing the survival rates of diatom cells in seawater [45].

Broader spectral ranges can be detected with the use of piezoelectric crystals whereby the application of an electric field—including a THz pulse—deforms the shape of a near-infrared beam that is co-propagating with the THz. By using a quadrant detector, the beam shape deformation is proportional to the amplitude of an oscillating THz field. This approach is used for example in the TeraSys-ULTRA spectrometer by Rainbow Photonics [46]. One result from this spectrometer is reported in Fig. 5. We used an empty liquid cell with 0.5 mm thick polycrystalline diamond windows (Diamond Materials) and a 13 micron thick plastic spacer. The magnitude of the FT of the THz field transmitted by the empty cell is shown in gray in Fig. 5a. When the cell is filled with ultra-pure, milli-Q water, the corresponding transmitted spectrum is reported in blue in Fig. 5a. To estimate the dielectric function of water, analytical solutions are not available, and the proper set of geometry equations should be solved, see, e.g., ref. [26]. An intuitive yet qualitative indicator of the absorption is called optical density (OD) or absorbance, which is defined as

$$\mathrm{OD} = -2 \cdot \log_{10}\left(\frac{M_S}{M_R}\right) \tag{18}$$

where M_S is the frequency-dependent FT magnitude, which is reported with the blue trace in Fig. 5a. M_R is the gray trace in Fig. 5b. The factor 2 in Eq. 18 stems from the fact that the OD is often calculated from the spectral intensities, which scale with the FT magnitudes squared. The OD we obtained for pure water is reported in red in Fig. 5b: it indicates a bump around 6 THz, as expected for the collective, intermolecular stretching band of HB networking water molecules (light blue in Fig. 1).

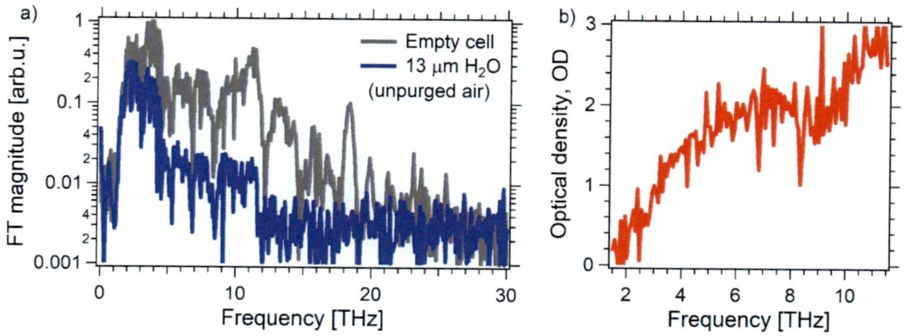

Fig. 5 **a** FT magnitudes of the THz fields transmitted by an empty cell with a 13-micron thick spacer and diamond windows (gray), and after filling it with pure water (blue). **b** The optical density of liquid water can be estimated directly from the FT magnitudes, see text for details

3 Protein Solvation Dynamics

In the current research landscape, water is recognized to play an active role in regulating the structure and function of proteins. Protein folding, ligand binding, and the efficiency of enzymatic reactions are governed by water molecules in the protein hydration shell [47–49]. As mentioned earlier, solute-water interactions fundamentally alter water's properties, creating an interfacial region where water molecules have a preferred orientation, hydrogen-bonding capacity, and range of dynamical motion. Predicting the extent to which hydration water is altered by proteins has been particularly challenging due to the complex nature of the protein surface. Their three dimensional structures are uneven and riddled with ridges and grooves. Furthermore, polar and nonpolar amino acid residues are heterogeneously dispersed over the surface. While polar groups strongly orient water and form efficient hydrogen bonds in which water is considered bound to the protein surface, nonpolar groups repel water, causing increased hydrogen bonding between neighboring water molecules above the protein surface [50], as shown in Fig. 6. As such, a residue that strongly polarizes can be close to a hydrophobic group which strongly repels water, which brings into question how these competing effects impact the relative strength of hydrogen bonds, either to the protein surface or with neighboring waters. Due to these factors, how the protein surface spatially arranges and orients water seems to be highly dependent on its surface topology and amino acid sequence.

Hydration water has distinct physico-chemical properties compared to bulk water [51]. In general, hydration water forms fewer hydrogen bonds and has slower dynamics due to restricted rotational and vibrational motions arising from solute-water interactions. Such effects can propagate several water layers from the protein surface until bulk-like properties are recovered. THz radiation excites the low frequency collective motions of the hydrogen-bonding network and is sensitive to changes in the hydrogen bond network arising from hydration water. In the last two decades, THz spectroscopy has emerged as a powerful tool for studying protein

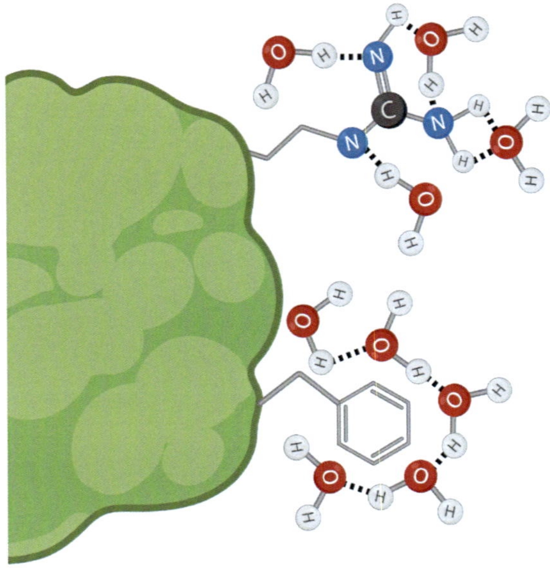

Fig. 6 Schematic illustration of polar and nonpolar hydration of protein surface residues. A positively charged arginine residue (top) polarizes surface water molecules and can form hydrogen bonds. Water molecuels near a hydrophobic phenylalanine residue (bottom) form a cavity with strengthened hydrogen bonds parallel to the benzene plane

hydration. Here, we will cover the fundamentals of how THz spectroscopy can be used to learn qualitative and quantitative information on protein solvation dynamics.

3.1 Long-Range Hydration Shell of Proteins

The range of molecular water layers influenced by the solute surface is a long standing question in the water community. As the various techniques employed to study hydration water probe various time and length scales, the size of the hydration shell has been reported to range from just the first water layer in contact with the solute surface to a long-range shell incorporating hundreds or thousands of water molecules. Utilizing the p-Ge THz laser, the Havenith group was the first to demonstrate that THz absorption spectroscopy is sensitive to the long-range hydration shell of biomolecules [52]. This has since been adopted to estimate the hydration shell size of many proteins and macromolecules.

As discussed in Sect. 1.2, the total absorption of the aqueous solvent as a function of solute concentration can be estimated assuming an ideal solution in which the solute has a negligible absorption. Equation 5 can be rewritten as Eq. 19 for a protein solution [53]:

$$\alpha_{\mathrm{TOT}} = \alpha_{\mathrm{buffer}}\left(1 - \phi_{\mathrm{protein}} - \phi_h\right) + \alpha_h \phi_h + \alpha_{\mathrm{protein}} \phi_{\mathrm{protein}} \tag{19}$$

where $\phi_{\text{buffer}} = 1 - \phi_{\text{protein}} - \phi_h$. The volume fraction of protein in solution can be estimated assuming a hard sphere model, where a single protein molecule occupies a spherical void equal to its radius of gyration, R_g. These radii values can be obtained from methods such as dynamic light scattering or X-ray scattering and can often be found in literature for numerous globular proteins. While, realistically, proteins structures are not spherical, and are further complicated by non-uniform surface topologies, the hard sphere model provides a simple method of estimating how the absorption of an aqueous solution should change with increasing protein concentration, as demonstrated by the gray line in Fig. 7.

As pointed out earlier, this gray line represents the EM model, i.e., the removal of bulk-like water equal to the volume of protein in solution, but is often denoted as "Bulk Water" or "Water Exclusion" throughout THz literature. Assuming the absorption of the protein solute is negligible compared to water, the estimation of the water exclusion can be represented mathematically by Eq. 20, which is a special case of Eq. 6

$$\alpha_{\text{EM}} = \alpha_{\text{buffer}}\left(1 - \frac{\frac{4}{3}\pi (R_g)^3 c_{\text{protein}}}{V_{\text{TOT}}}\right) \tag{20}$$

where c_{protein} is the protein concentration and V_{TOT} is the total volume of the solution. Deviations from this estimate (gray in Fig. 7) indicate nonideal solutions, in which

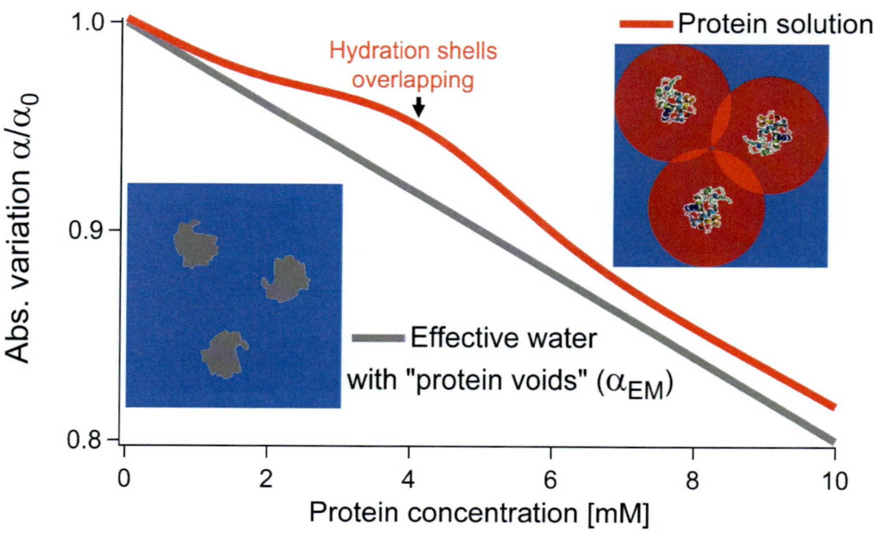

Fig. 7 Cartoon of the relative variation of the terahertz absorption coefficient of an aqueous protein solution as a function of concentration (red). A bump or relative increase in the THz response can be present at some specific concentrations, which was assigned to overlapping hydration shells [60]. The data can be compared with the simplest effective model (gray line), simulating the response of a fictitious liquid made of water plus an empty volume equal to the protein volume

protein-water interactions contribute to the measured absorption, as demonstrated by the red line in Fig. 7. Numerous THz absorption studies of globular proteins have demonstrated an initial increase in the absorption coefficient with increasing protein concentration, followed by a decline that tracks the removal of bulk water [54–59]. This change in behavior, the crossover from an increase to a decrease in the absorption coefficient, represents the point at which proteins in solution begin to have overlapping hydration shells. In the dilute regime, proteins molecules are spatially distributed and have little interaction. As the number of proteins in solution increases, the amount of hydration water increases, and hence an increase in the absorption occurs. At a threshold concentration, however, there is not enough bulk water available to hydrate more protein molecules, and proteins instead have solvent-shared hydration shells. This can cause a decrease and/or saturation in the THz absorption, resulting in a bump in the measured signal, as shown by the arrow in Fig. 7.

The consequence of this observation is significant: the concentration at which the THz absorption reaches its maximum directly reveals the size of the hydration shell. Table 1 summarizes the hydration shell sizes determined for various biomolecules from THz absorption spectroscopy. This approach was first applied to the investigation of lactose hydration, where it revealed that the solvation shell extends well beyond the first hydration layer [61]. Similar studies show that peptides and small proteins generally possess hydration shells spanning only a few water layers, while larger proteins encompass shells extending over ten or more water layers from the protein surface. Such findings indicate that the protein has the capacity to exert structural and dynamical changes on thousands of water molecules. For instance, BSA is estimated to influence 22,000 water molecules per protein molecule, while monoclonal antibodies, a larger macromolecule, perturb up to 70,000 water molecules [62]. The long-range hydration shells revealed by THz spectroscopy contribute directly to the stability and conformation of proteins in solution.

Table 1 Size and number of water layers in the hydration shell of biomolecules and proteins measured with THz absorption spectroscopy

Biomolecule	Hydration shell (Å)	Water layers	Reference
Lactose	5	2	[61]
Lamda repressor	20	6	[52]
Ubiquitin	18	6	[55]
Anti freeze glycoprotein	20–35	6–12	[63]
AK17	10.8	3	[58]
AK10G	17.3	5	[58]
AK9P	11.2	3	[58]
Bovine serum albumin	27	9	[57]
Beta lactoglobulin	15	5	[64]
Lysozyme	42	14	[59]
Matrix metalloproteinase	39	13	[65]

4 Surface Electrostatic Potential

As recognition of hydration water's integral role in shaping the solution-based chemistry of biomolecules, a central unresolved question is how the biomolecular surface dictates the orientation and organization of these neighboring water molecules. The polar nature of water, where it can be a hydrogen bond acceptor and donor, allows it to flexibly adapt to the local chemical environment, ensuring that a hydrogen-bonding network, albeit perturbed, remains intact. Surface selective methods, such as heterodyne phase sensitive sum frequency generation spectroscopy, have provided valuable information on water orientation at biomolecular interfaces [66]. The net water dipole reflects the mirror image of the surface electrostatics, where a null orientation is observed for a homogeneoulsy mixed planar surface composed of equal negative and positive charges [67].

Extrapolating such findings to understand the organization of water molecules at the heterogeneous protein surface has proven challenging. With a known protein data bank (PDB) identifier, mapping the surface electrostatic potential of a protein at specific solvent conditions (pH, salt concentration, etc) has become standard with tools such as the Adaptive Poisson-Boltzmann Solver (APBS, https://server.poissonboltzmann.org/). Such tools provide insight into the local chemistry of the protein surface, allowing one to infer how water might orient or interact with irregular topographies interspersed with polar, charged, and hydrophobic residues. Yet, it remains difficult to determine whether hydration water adopts a net dipole orientation near these highly heterogeneous surfaces, and to rationalize how local chemical patches collectively influence the structure and dynamics of the hydration layer.

While the discussion of protein hydration in the previous section focused on positive deviations from the EM model, negative deviations also occur. A positive deviation indicates that water within the hydration shell is less-hydrogen bond (or more disordered) than bulk water, while a negative deviation shows that hydration water is more hydrogen-bonded (more structured). Reminiscent of the Hofmeister series, in which the ions can "structure" or "break" the water network in the presence of proteins [68], this observation provided the first hint that the sign of the THz response may well be linked to the surface electrostatic potential of the solute.

Efforts to correlate the THz response with surface properties started with a series of amino acids. Hydrophilic amino acids, such as glycine and serine, were found to have positive THz deviations. In contrast, a negative THz deviation was observed for hydrophobic amino acids, where the THz response became increasingly negative with increasing hydrophobicity [69]. This trend was rationalized in terms of polarity, where the THz response becomes more positive as molecule polarity increased. A similar study of charged amino acids found that a negative THz response correlated to an increased solvent exposed surface area, suggesting that NH_3^+ groups of charged lysine and arginine break the structure of the water network [70]. While not in agreement about the exact cause, these studies do show that the THz response is in fact correlated to the surface properties of the biomolecule surface. Theoretical predictions support this notion, where it was found that the sign of the THz response correlated with the average polarization of the protein-water surface [71].

As most experimental studies of proteins have been done in isolation, a correlation between the sign of the THz deviation and the surface properties of proteins was not obvious. A recent study of matrix metalloproteinases structural homologs, however, revealed that an increasingly positive surface electrostatic potential gives rise to a positive THz deviation [65]. Figure 8 shows the THz deviation for various proteins as a function of the net and surface charge. A clear trend emerges: as the surface charge varies from negative to positive, the THz response becomes correspondingly more positive. It is important to note that this is not a one to one correlation, i.e., a net positive surface charge does not translate to a positive THz deviation. In fact, proteins with a neutral/slightly positive charge show a small negative THz deviation, indicating that the protein surface must be strongly enriched with positive charges before the THz response changes sign. This behavior likely stems from the dominant influence of negatively charged COO^- groups on surrounding water, which appears to outweigh the effects of positively charged NH_3^+ groups. While this observation is relatively recent in the literature, it highlights an exciting possibility: THz absorption spectroscopy may provide a qualitative means to assess the surface electrostatic properties of proteins, particularly for those whose crystal structure remains unknown.

5 Dynamical Timescales of Protein Solvation Water

A distinct advantage of THz-TDS is that the full dielectric response of the sample can be obtained, as the refractive index and absorption coefficient are simultaneously measured. As measurements of aqueous solutions are typically limited to <3 THz due to higher absorption at higher frequencies [72], Debye functions are often sufficient to describe the dielectric response. In other words, the dielectric response can be modeled by the first three terms in Eq. 1, where two Debye relaxation processes are adequate. The first Debye relaxation process, τ_1, corresponds to the reorientation of the dipole of hydrogen-bonded water molecules, and dominates the dielectric response of water. The origin of the second Debye relaxation process is unclear, as various techniques find vastly different times scales, ranging from 0.1 to 2 ps [11, 73, 74]. Some studies have attributed this feature to the reorientation of weakly hydrogen-bonded water molecules [11], while others support the rapid exchange of hydrogen bonds as the origin. Most THz-TDS studies of aqueous solutions find a time scale of ~0.1–0.4 ps [75, 76], which would be consistent with a faster process such as hydrogen bond breakage and reformation.

Here, discussion of hydration water dynamics will be limited to the first Debye process, as this contributes most strongly to the dielectric response of water. In the case of aqueous solutions, the dielectric properties reflect cooperative contributions from bulk water, the solute, and the solute hydration shell. The dynamical timescale τ_1 of concentrated solutions should therefore reflect the impact of the hydrated solute

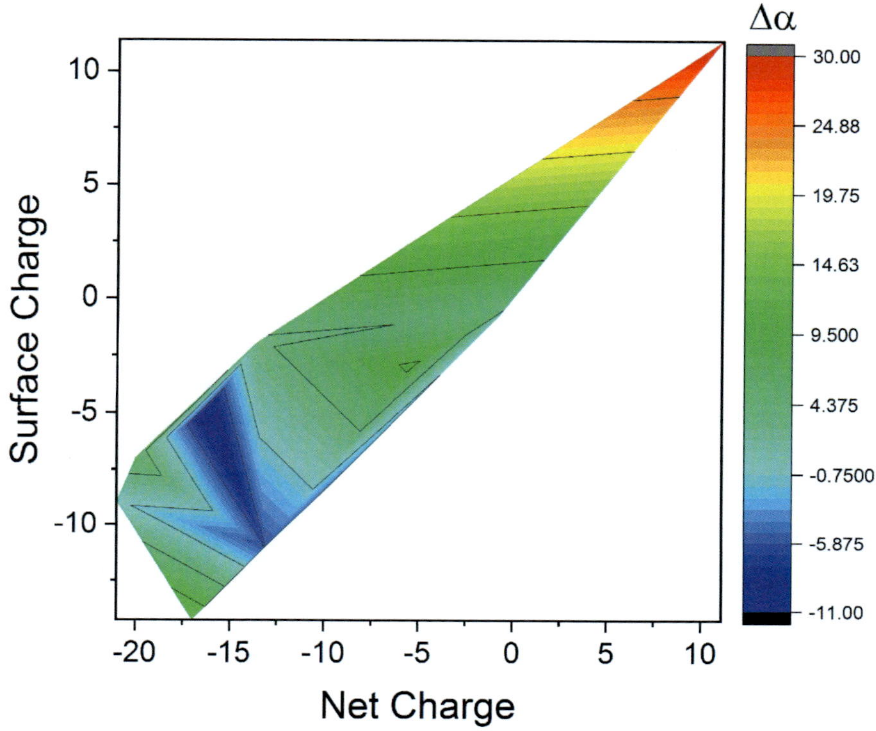

Fig. 8 THz deviation from the EM model correlates to the surface electrostatic potential of the protein. A positive deviation correlates to a more positively charge surface, while a negative deviation correlates to a more negatively charged surface

on the overall water network, where an increase corresponds to retarded dynamics in the hydration shell and a decrease indicates accelerated dynamics.

Correlating these observed dynamics with stronger or weaker hydrogen bonding of water molecules within the hydration shell remains challenging, as further deconvolution requires assumptions of the underlying origin of the timescale distributions. Figure 9 (top) shows normalized τ_1 values of various amino acids as a function of the Eisenberg hydrophobicity scale [77]. A clear trend emerges: hydrophilic amino acids have slower dynamics and dynamics accelerate as hydrophobicity increases. This could intuitively lead to the conclusion that water molecules hydrogen-bond to the hydrophilic biomolecules, and therefore have slower reorientation times due to restrained positons. However, the hydrophobic amino acid tryptophan is observed to have retarded dynamics as well [70]. This observation is rationalized as increased hydrogen bonding between neighboring water molecules above the hydrophobic surface, leading to slower reorientation times. Increased hydrogen bonding seems to correlate with slower dynamics, but cannot be explicity assigned to the hydrophobicity/hydrophilicty of the surface.

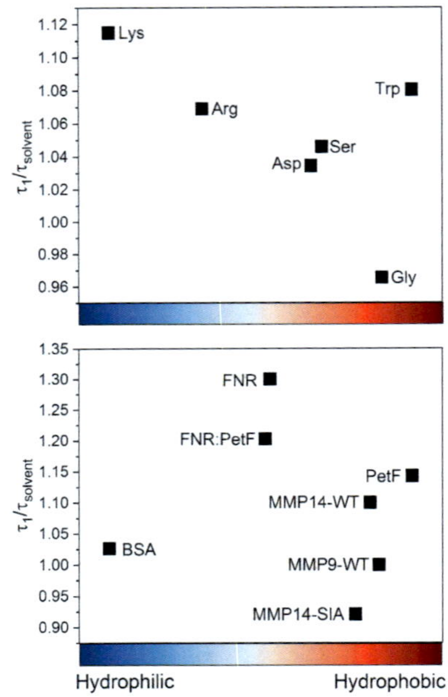

Fig. 9 Normalized Debye reorientation time, τ_1, of amino acids (top) and proteins (bottom). The x-axis represents the hydrophobicity based on the Eisenberg hydrophobicity scale. For proteins, the total hydrophobicity was determined by the sum of the solvent accessible surface area of each surface residue multiplied by the Eisenberg hydrophobicity

Figure 9 (bottom) further highlights this finding for protein solutions. While protein solutions in general have slower dynamics compared to bulk solvent, there is no direct correlation to the hydrophobicity of the surface exposed residues. It should be noted, however, that quantification of Debye relaxation time scales for aqueous proteins solutions has been sporadic, highlighting the need for more rigorous THz studies of water dynamics and protein surface properties. The protein hydration dynamics shown in Fig. 9 are limited to what is available in the current literature [65, 78, 79]. PetF, having the highest surface exposed hydrophobicity of proteins compared here, has also a highly negatively charged surface with many exposed COO^- groups [79]. These groups strongly coordinate water and could explain the retarded dynamics.

In summary, many factors must be taken into account in interpreting protein hydration dynamics. Firstly, the solvent accessible surface area of each residue determines how many water molecules can directly interact with each residue. Secondly, the surface charge distribution strongly dictates the orientational ordering of interfacial water molecules. Finally, the hydrophobic vs. hydrophilic patterning influences both water structure and hydrogen bond cooperativity. Taken together, these factors highlight how hydration water dynamics are an emergent property of the local chemical environment, not merely a sum of hydrophilic and hydrophobic effects. This reflects the heterogenous nature of the protein surface, and indicates that a future challenge in THz research is decoding how proteins perturb the delicate balance of hydrogen-bonding, sterics, and collective dynamics that govern water's dielectric response.

Acknowledgements The authors are grateful to the Havenith group. This work was in part supported by the University of Southampton (startup package); DFG (509442914, EXC 2033-390677874, EXC 2068-390729961); German Federal Ministry of Education and Research BMBF and Ministry of Culture and Research of Nord Rhine-Westphalia MKW NRW (Center for Solvation Science ZEMOS); Ruhr-Universität Bochum (Research department solvation science).

References

1. Grischkowsky D, Keiding S, van Exter M, Fattinger C (1990) Far-infrared time-domain spectroscopy with terahertz beams of dielectrics and semiconductors. J Opt Soc Am B 7:2006
2. Salén P, Basini M, Bonetti S, Hebling J, Krasilnikov M, Nikitin AY, Shamuilov G, Tibai Z, Zhaunerchyk V, Goryashko V (2019) Matter manipulation with extreme terahertz light: progress in the enabling THz technology. Phys Rep 836–837:1
3. Laage D, Hynes JT (2006) A molecular jump mechanism of water reorientation. Science 311, 832
4. Novelli F (2024) Terahertz spectroscopy of thick and diluted water solutions. Opt Express 32:11041
5. Bertie JE, Lan Z (1996) Infrared intensities of liquids XX: The intensity of the OH stretching band of liquid water revisited, and the best current values of the optical constants of H_2 O(l) at 25 °C between 15,000 and 1 cm^{-1}. Appl Spectrosc 50:1047
6. Zelsmann HR (1995) Temperature dependence of the optical constants for liquid H_2O and D_2O in the far IR region. J Mol Struct 350:95
7. Lunkenheimer P, Emmert S, Gulich R, Wolf KM, Schwab M, Loidl A (2017) Electromagnetic-radiation absorption by water. Phys Rev E 96:062607
8. Heyden M, Sun J, Funkner S, Mathias G, Forbert H, Havenith M, Marx D (2010) Dissecting the THz spectrum of liquid water from first principles via correlations in time and space. Proc Natl Acad Sci U S A 107:12068
9. Elton DC, Fernández-Serra M (2016) The hydrogen-bond network of water supports propagating optical phonon-like modes. Nat Commun 7:10193
10. Cho M, Fleming GR, Saito S, Ohmine I, Stratt RM (1994) Instantaneous normal mode analysis of liquid water. J Chem Phys 100:6672
11. Vinh NQ, Sherwin MS, Allen SJ, George DK, Rahmani AJ, Plaxco KW (2015) High-precision gigahertz-to-terahertz spectroscopy of aqueous salt solutions as a probe of the femtosecond-to-picosecond dynamics of liquid water. J Chem Phys 142:164502
12. Fukasawa T, Sato T, Watanabe J, Hama Y, Kunz W, Buchner R (2005) Relation between dielectric and low-frequency Raman spectra of hydrogen-bond liquids. Phys Rev Lett 95:197802
13. Shiraga K, Tanaka K, Arikawa T, Saito S, Ogawa Y (2018) Reconsideration of the relaxational and vibrational line shapes of liquid water based on ultrabroadband dielectric spectroscopy. Phys Chem Chem Phys 20:26200
14. Grosso G, Pastori Parravicini G (2003) Solid state physics. Academic Press
15. Millon C, Houver S, Saraceno CJ (2023) 400 kHz repetition rate THz-TDS with 24 mW of average power driven by a compact industrial Yb-laser. Opt Express 31:7922
16. Vogel T, Novelli F, Saraceno CJ (2025) Frequency-Resolved THz thermometry of nanoparticle solutions. In: 2025 conference on lasers and electro-optics Europe and European quantum electronics conference, CLEO/Europe-EQEC 2025
17. Rønne C, Åstrand P-O, Keiding SR (1999) THz spectroscopy of liquid H_2O and D_2O. Phys Rev Lett 82:2888
18. Rønne C, Keiding SR (2002) Low frequency spectroscopy of liquid water using THz-time domain spectroscopy. J Mol Liq 101:199

19. Pezzotti S, Chen W, Novelli F, Yu X, Hoberg C, Havenith M (2025) Terahertz calorimetry spotlights the role of water in biological processes. Nat Rev Chem 9:481
20. Markel VA (2016) Introduction to the Maxwell Garnett approximation: tutorial. J Opt Soc Am A 33:1244
21. Hernandez-Cardoso GG, Singh AK, Castro-Camus E (2020) Empirical comparison between effective medium theory models for the dielectric response of biological tissue at terahertz frequencies. Appl Opt 59:D6
22. He Y, Ku PI, Knab JR, Chen JY, Markelz AG (2008) Protein dynamical transition does not require protein structure. Phys Rev Lett 101:178103
23. Markelz AG, Mittleman DM (2022) Perspective on Terahertz applications in bioscience and biotechnology. ACS Photonics 9:1117
24. Jepsen PU. Phase retrieval in Terahertz time-domain measurements: a "How to" tutorial. J Infrared Millim Terahertz Waves 40:395
25. Lloyd-Hughes J, Chopra N, Deveikis J, Pandya R, Woolley J. A tutorial on THz pulse analysis: accurate retrieval of pulse arrival times, spectral energy density and absolute spectral phase. J Infrared Millimeter Terahertz Waves 46:34
26. Novelli F (2024) Terahertz transmission through a gold mirror or electrode. Materials 17:3942
27. Novelli F (2013) In search of selective excitations for studying out-of-equilibrium properties in strongly correlated electron systems and high temperature superconductors. University of Trieste, PhD Thesis
28. Leung CK, Ward-Berry JN, Wanvig i Dot E, Lee J, Zeitler JA (2025) Tutorial: accurate determination of refractive index and absorption coefficient in Terahertz time-domain spectroscopy. J Infrared Millim Terahertz Waves 46:1
29. Wynne K, Carey JJ (2005) An integrated description of terahertz generation through optical rectification, charge transfer, and current surge. Opt Commun 256:400
30. Bacon DR, Madéo J, Dani KM (2021) Photoconductive emitters for pulsed terahertz generation. J Opt 23:064001
31. Burford NM, El-Shenawee MO (2017) Review of terahertz photoconductive antenna technology. Opt Eng 56:010901
32. Singh A, Pashkin A, Winnerl S, Welsch M, Beckh C, Sulzer P, Leitenstorfer A, Helm M, Schneider H (2020) Up to 70 THz bandwidth from an implanted Ge photoconductive antenna excited by a femtosecond Er: fibre laser. Light Sci Appl 9:30
33. Kamo Y, Kitazawa S, Ohshima S, Hosoda Y (2014) Highly efficient photoconductive antennas using optimum low-temperature-grown GaAs layers and Si substrates. Jpn J Appl Phys 53:032201
34. Andreeva VA et al (2016) Ultrabroad terahertz spectrum generation from an air-based filament plasma. Phys Rev Lett 116:063902
35. Balos V, Müller P, Jakob G, Kläui M, Sajadi M (2021) Imprinting the complex dielectric permittivity of liquids into the spintronic terahertz emission. Appl Phys Lett 119:091104
36. Seifert T et al (2016) Efficient metallic spintronic emitters of ultrabroadband terahertz radiation. Nat Photonics 10:483
37. Lu X, Zhang XC (2011) Balanced terahertz wave air-biased-coherent-detection, Appl Phys Lett 98:151111
38. Wang T, Iwaszczuk K, Wrisberg EA, Denning EV, Jepsen PU (2016) Linearity of air-biased coherent detection for Terahertz time-domain spectroscopy. J Infrared Millim Terahertz Waves 37:592
39. Casalbuoni S, Schlarb H, Schmidt B, Schmüser P, Steffen B, Winter A (2008) Numerical studies on the electro-optic detection of femtosecond electron bunches. Phys Rev B 11:072802
40. Benea-Chelmus IC, Settembrini FF, Scalari G, Faist J (2019) Electric field correlation measurements on the electromagnetic vacuum state. Nature 568:202
41. Bielawski S, Veronese M, Evain C, Szwaj C, Roussel E, Di Pietro P, Cinquegrana P, Adhlakha N, Di Mitri S, Perucchi A (2023) Single-shot terahertz time-domain spectrometer using 1550 nm probe pulses and diversity electro-optic sampling. Optics Express 31(19):31072–31081

42. Couture N, Cui W, Lippl M, Ostic R, Fandio DJJ, Yalavarthi EK, Vishnuradhan A, Gamouras A, Joly NY, Ménard JM (2023) Single-pulse terahertz spectroscopy monitoring sub-millisecond time dynamics at a rate of 50 kHz. Nat Commun 14:2595
43. Gao FY, Zhang Z, Liu Z-J, Nelson KA, Nelson KA (2022) High-speed two-dimensional terahertz spectroscopy with echelon-based shot-to-shot balanced detection. Opt Lett 47(14):3479–3482
44. Hoberg C, Balzerowski P, Havenith M (2019) Integration of a rapid scanning technique into THz time-domain spectrometers for nonlinear THz spectroscopy measurements. AIP Adv 9:35348
45. Mushtaq R, Gambardella C, Miroglio R, Novelli F, Paturzo M, Rubano A, Sardo A, Balzano S, Paparo D (2025) Using diatom chain length as a bioindicator of heavy-metals contamination in marine environments. J Hazard Mater 484:136732
46. Puc U, Bach T, Günter P, Zgonik M, Jazbinsek M (2021) Ultra-Broadband and high-dynamic-range THz time-domain spectroscopy system based on organic crystal emitter and detector in transmission and reflection geometry. Adv Photonics Res 2:2000098
47. Frauenfelder H, Fenimore PW, Chen G, McMahon BH (2006) Protein folding is slaved to solvent motions. Proc Natl Acad Sci 103:15469
48. Mazur K, Heisler IA, Meech SR (2012) Water dynamics at protein interfaces: ultrafast optical kerr effect study. J Phys Chem A 116:2678–2685
49. Zaragoza JPT, Nguy A, Minnetian N, Deng Z, Iavarone AT, Offenbacher AR, Klinman JP (2019) Detecting and characterizing the kinetic activation of thermal networks in proteins: thermal transfer from a distal, solvent-exposed loop to the active site in soybean lipoxygenase. J Phys Chem B 123:8662
50. Chandler D (2005) Interfaces and the driving force of hydrophobic assembly. Nature 437:640
51. Laage D, Elsaesser T, Hynes JT (2017) Water dynamics in the hydration shells of biomolecules. Chem Rev 117:10694
52. Ebbinghaus S, Kim SJ, Heyden M, Yu X, Heugen U, Gruebele M, Leitner DM, Havenith M (2007) An extended dynamical hydration shell around proteins. PNAS 104:52
53. Heyden M, Ebbinghaus S, Havenith M (2010) Terahertz spectroscopy as a tool to study hydration dynamics. In: Meyers RA, Meyers RA (eds) Encyclopedia of analytical chemistry
54. Heyden M, Bründermann E, Heugen U, Niehues G, Leitner DM, Havenith M (2008) Long-range influence of carbohydrates on the solvation dynamics of water—answers from terahertz absorption measurements and molecular modeling simulations. J Am Chem Soc 130:5773
55. Born B, Kim SJ, Ebbinghaus S, Gruebele M, Havenith M (2009) The terahertz dance of water with the proteins: the effect of protein flexibility on the dynamical hydration shell of ubiquitin. Faraday Discuss 141:161
56. Heyden M, Havenith M (2010) Combining THz spectroscopy and MD simulations to study protein-hydration coupling. Methods 52:74
57. Bye JW, Meliga S, Ferachou D, Cinque G, Zeitler JA, Falconer RJ (2014) Analysis of the hydration water around bovine serum albumin using terahertz coherent synchrotron radiation. J Phys Chem A 118:83
58. Ding T, Li R, Zeitler JA, Huber TL, Gladden LF, Middelberg APJ, Falconer RJ (2010) Terahertz and far infrared Spectroscopy of alanine-rich peptides having variable ellipticity. Opt Express 18:27431
59. Novelli F, Ostovar Pour S, Tollerud J, Roozbeh A, Appadoo DRT, Blanch EW, Davis JA (2017) Time-Domain THz spectroscopy reveals coupled protein-hydration dielectric response in solutions of native and fibrils of human lysozyme. J Phys Chem B 121, 4810
60. Ebbinghaus S, Seung JK, Heyden M, Yu X, Gruebele M, Leitner DM, Havenith M (2008) Protein sequence- and pH-dependent hydration probed by terahertz spectroscopy. J Am Chem Soc 130:2374
61. Heugen U, Schwaab G, Bründermann E, Heyden M, Yu X, Leitner DM, Havenith M (2006) Solute-induced retardation of water dynamics probed directly by terahertz spectroscopy. Proc Natl Acad Sci 103:12301
62. Wallace VP, Ferachou D, Ke P, Day K, Uddin S, Casas-Finet J, Van Der Walle CF, Falconer RJ, Zeitler JA (2015) Modulation of the hydration water around monoclonal antibodies on addition of excipients detected by Terahertz time-domain spectroscopy. J Pharm Sci 104:4025

63. Ebbinghaus S, Meister K, Born B, DeVries AL, Gruebele M, Havenith M (2010) Antifreeze glycoprotein activity correlates with long-range protein–water dynamics. J Am Chem Soc 132:12210
64. Vondracek H, Dielmann-Gessner J, Lubitz W, Knipp M, Havenith M (2014) THz absorption spectroscopy of solvated β-lactoglobulin. J Chem Phys 141:22
65. Adams EM, Pezzotti S, Ahlers J, Rüttermann M, Levin M, Goldenzweig A, Peleg Y, Fleishman SJ, Sagi I, Havenith M (2021) Local mutations can serve as a game changer for global protein solvent interaction. JACS Au 1:1076
66. Chen X, Hua W, Huang Z, Allen HC (2010) Interfacial water structure associated with phospholipid membranes studied by phase-sensitive vibrational sum frequency generation spectroscopy. J Am Chem Soc 132:11336
67. Dreier LB, NagataY, Lutz H, Gonella G, Hunger J, Backus EHG, Bonn M (2018) Saturation of charge-induced water alignment at model membrane surfaces. Sci Adv 4:3
68. Collins K (2004) Ions from the Hofmeister series and osmolytes: effects on proteins in solution and in the crystallization process. Methods 34:300
69. Niehues G, Heyden M, Schmidt DA, Havenith M (2011) Exploring hydrophobicity by THz absorption spectroscopy of solvated amino acids. Faraday Discuss 150:193
70. Samanta N, Das Mahanta D, Choudhury S, Barman A, Kumar Mitra R (2017) Collective hydration dynamics in some amino acid solutions: a combined GHz-THz spectroscopic study. J Chem Phys 146, 125101
71. Matyushov DV (2012) Dipolar response of hydrated proteins. J Chem Phys 136:85102
72. Novelli F, Guchhait B, Havenith M (2020) Towards intense THz spectroscopy on water: characterization of optical rectification by GaP, OH1, and DSTMS at OPA wavelengths. Materials 13:1311
73. Vij JK, Simpson DRJ, Panarina OE (2004) Far infrared spectroscopy of water at different temperatures: GHz to THz dielectric spectroscopy of water. J Mol Liq 112:125
74. Rezus YLA, Bakker HJ (2005) On the orientational relaxation of HDO in liquid water. J Chem Phys 123:114502
75. Sato T, Buchner R (2004) Dielectric relaxation processes in ethanol/water mixtures. J Phys Chem A 108:5007
76. Yada H, Nagai M, Tanaka K (2009) The intermolecular stretching vibration mode in water isotopes investigated with broadband terahertz time-domain spectroscopy. Chem Phys Lett 473:279
77. Eisenberg D, Schwarz E, Komaromy M, Wall R (1984) Analysis of membrane and surface protein sequences with the hydrophobic moment plot. J Mol Biol 179:125
78. Das Mahanta D, Samanta N, Mitra RK (2017) Decisive role of hydrophobicity on the effect of alkylammonium chlorides on protein stability: a Terahertz spectroscopic finding. J Phys Chem B 121, 7777
79. Adams EM, Lampret O, König B, Happe T, Havenith M (2020) Solvent dynamics play a decisive role in the complex formation of biologically relevant redox proteins. Phys Chem Chem Phys 22:7451

Disorder at the Interface: Hydration Water and the Physics of Intrinsically Disordered Proteins

David M. Leitner, Korey M. Reid, Abhishek K. Singh, and Nguyen Q. Vinh

Abstract The combination of broadband megahertz-to-terahertz dielectric spectroscopy and molecular dynamics simulations has revealed the molecular origins of distinct structural and dynamic properties of hydration layers around intrinsically disordered proteins (IDPs) and intrinsically disordered regions (IDRs) of proteins. In this chapter, we discuss the picture of hydration water around IDRs and IDPs that has emerged from recent experimental and computational studies, the origins of the distinct properties of hydration layers around these disordered systems, as well as some of the open questions regarding the water around IDRs and IDPs, which the methods discussed in this chapter are well suited to address.

Keywords Intrinsically disordered proteins · Intrinsically disordered regions · Megahertz-to-terahertz dielectric spectroscopy · Molecular dynamics simulations

1 Introduction

Intrinsically disordered proteins (IDPs) and intrinsically disordered regions (IDRs) found in proteins that also contain structured regions play essential roles in numerous signaling and regulatory processes in the cell [1–3]. Details about how they perform their functions continue to be the subject of investigation. Physical properties of IDRs and IDPs, such as their dynamic and thermodynamic properties, are strongly influenced by their interface with surrounding water molecules, like any protein in an aqueous environment [4–14]. For this reason, there has been interest in examining

D. M. Leitner (✉)
Department of Chemistry, University of Nevada, Reno, USA
e-mail: dml@unr.edu

K. M. Reid
Department of Chemistry, Dartmouth College, Hanover, USA

A. K. Singh · N. Q. Vinh (✉)
Department of Physics and Center for Soft Matter and Biological Physics, Virginia Tech, Blacksburg, USA
e-mail: vinh@vt.edu

S. Cerveny and J. Swenson (eds.), *Dynamics of Biomolecules Studied by Dielectric Spectroscopy*, Advances in Dielectrics, https://doi.org/10.1007/978-3-032-30588-6_9

structure and dynamics of water surrounding IDPs and IDRs and comparing the nature of the hydration water around these disordered systems with hydration water around structured proteins [15–21]. Many studies have revealed that water around IDRs and IDPs exhibits more restricted dynamics than water surrounding structured proteins of similar size [15–21]. Following early work there was some debate about the origin of the difference, which was in some cases attributed to steric factors arising from structural disorder, in other cases to a greater number of charged groups in IDRs and IDPs. A combination of broadband megahertz-to-terahertz dielectric spectroscopy and molecular dynamics (MD) simulations reveals that concentration of charged groups underlies the more restrictive motion of water near the surface of the IDPs and IDRs [20, 21]. This chapter discusses these recent findings in detail.

Terahertz time-domain spectroscopy [4, 5, 22–29] and broadband megahertz-to-terahertz frequency-domain dielectric measurements [30, 31] have provided a wealth of information about the hydration layer around biomolecules. Megahertz-to-terahertz dielectric measurements can probe the number of water molecules in hydration layers of proteins and nucleic acids [30–32] that can be characterized as tightly bound and loosely bound, helping to refine the picture of "*biological water*" originally proposed by Bagchi, Zewail and their coworkers [8, 33–35]. While structural, dynamic, and thermodynamic [36–39] properties of hydration layers around structured proteins have been characterized, relatively few studies have explored hydration layers around IDPs and IDRs. The work reviewed in this chapter provides a detailed molecular picture of the hydration water properties of IDPs and IDRs and elucidates their physical origins.

We will discuss the dynamics of water around two kinds of systems, including the hydration layer and dynamics of water around three IDPs and two structured systems of similar size, all between 22 and 26 residues, five unique systems with varied ionic residue content. These systems are listed in Table 1. Distinct differences were observed in the dynamics of water near charged and neutral residues, with slower dynamics near charged groups, particularly anionic groups [20]. Moreover, a collective slowing down of water dynamics around the IDPs was detected, where the dynamics of water around neutral residues was found to be more sluggish than around neutral and even some of the charged residues in the structured systems. The results of the broadband megahertz-to-terahertz dielectric measurements indicated overall more restricted water dynamics and more tightly bound water molecules around the IDP than around the two structured systems. The companion MD simulations found consistent trends in both rotational and hydrogen bond dynamics.

To further study the role of structure and charged group density on water dynamics around IDRs and structured regions of proteins, two larger proteins and the surrounding hydration layer were studied [21]. Specifically, dynamics of hydrogen bonds between water and two proteins with both IDRs and structured regions, small ubiquitin-related modifier protein -1 (SUMO-1) and UBC9, were analyzed computationally. SUMO-1 regulates a variety of cellular proteins by covalent linkage termed sumoylation, regulating protein–protein interactions, protein–DNA interactions and protein subcellular localization [40, 41]. UBC9, a ubiquitin conjugating enzyme, participates in the cycle of sumoylation of SUMO-1. We summarize results found by

Table 1 System labels, single letter code primary sequence and sequence length

System	Sequence	Length	N^+	N^-
HP24wt	MTRSAFANLPLWKQQNLKKEKGLF	24	5	1
HP24stab	MTRSAFANLPLWKQQALMKEKGLF	24	4	1
D-myocin (DM)	RNRLRRGPLTFTTRTVRQVFRL	22	7	0
D-fibrinogen (DF)	QDGKTTFEKEGGGGRGPRILENMHES	26	4	5
D-stannin (DS)	LRLQRISQSEDEESIVGDGETKE	23	3	7

Positively charged residues are in red, while negatively charged residues are in blue. The number of cationic residues are N^+ and N^-, respectively.

MD simulations for the hydration water dynamics around these systems, the origin of differences around IDRs and structured regions and connect the results with the experimental and computational studies of hydration water around the systems listed in Table 1.

2 Methods

2.1 *Megahertz-to-Terahertz Dielectric Spectroscopy*

To advance the understanding of picosecond-scale dynamics of water and proteins in solution, we have developed a broadband megahertz-to-terahertz frequency-domain dielectric spectrometer. This instrument enables highly precise measurements of both absorbance and refractive index of aqueous samples across the microwave-to-terahertz frequency spectrum. Compared with state-of-the-art commercial devices, our spectrometer offers a substantial improvement in signal-to-noise ratio and spectral resolution. For instance, while commercial terahertz time-domain systems typically offer a dynamic range of about 10^6 and spectral resolution on the order of several gigahertz, our instrument achieves an unprecedented dynamic range of $\sim 10^{12}$ and a spectral resolution finer than 100 Hz (Fig. 1). Measurements spanning from 100 MHz to 50 GHz were conducted using a vector network analyzer (Agilent PNA N5225A) paired with an open-ended coaxial dielectric probe (HP 85070E). Frequencies from 60 GHz to 1.12 THz were accessed via terahertz frequency extenders (Virginia Diodes, VDI) integrated with the same vector network analyzer, providing continuous spectral coverage from 100 MHz to 1.12 THz with high dynamic range, suitable even for highly absorbing aqueous systems.

To maintain thermal stability of solutions and overcome the strong absorption of water, we designed a variable optical path length sample cell constructed from anodized aluminum for dielectric measurements. The cell is equipped with two transparent parallel windows at terahertz frequencies, in which one was mounted on an ultra-precise linear stage (Newport XMS series) adjustable with sub-micrometer

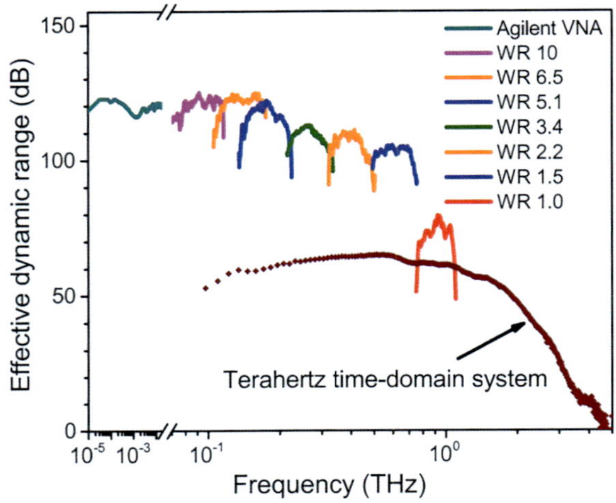

Fig. 1 Comparison of the dynamic range of our broadband terahertz frequency-domain dielectric spectrometer, comprising an Agilent vector network analyzer with frequency extenders (WR10 to WR1.0 systems), with that of a typical terahertz time-domain system [42]

precision accuracy (~0.08 μm) and the other is fixed. This design enables simultaneous determination of absorbance and refractive index from the slopes of intensity and phase as a function of sample thickness, eliminating the need for absolute thickness measurements. Temperature control was achieved by cooling the cell with Peltier coolers (Custom Thermoelectric) and heating it with power resistors embedded in the structure. The temperature was monitored and regulated within ±0.02 °C using a temperature controller (Lakeshore 336). The combination of the variable optical path-length sample cell and the high dynamic range of the terahertz spectrometer enable the most precise and accurate dielectric spectra reported to date across the megahertz-to-terahertz range of aqueous solutions.

The frequency-dependent absorption coefficient, $\alpha(\nu)$, and refractive index, $n(\nu)$, obtained from our setup were used to calculate the complex refractive index, $n^*(\nu)$, of the solution:

$$n^*(\nu) = n(\nu) + i\kappa(\nu) \tag{1}$$

where $\kappa(\nu)$ is the extinction coefficient, determined from the absorption coefficient as $\kappa(\nu) = \alpha(\nu) \cdot c/(4\pi\nu)$ with c denoting the speed of light. The complex refractive index is commonly expressed in terms of the complex dielectric function, $n^*(\nu) = \sqrt{\varepsilon^*(\nu)}$ [20, 31, 32, 43]:

$$\varepsilon^*(\nu) = \varepsilon'(\nu) + i\varepsilon''(\nu) + i\sigma/(2\pi\nu\varepsilon_0) \tag{2}$$

where $\varepsilon'(\nu)$ and $\varepsilon''(\nu)$ represent the real (dielectric dispersion) and imaginary (dielectric loss) components of the dielectric response, respectively; σ is the electrical conductivity of the solution, and ε_0 is the vacuum permittivity. The interaction between the megahertz to terahertz radiation with aqueous biological solutions provides direct access to the collective vibrational dynamics of hydrated proteins

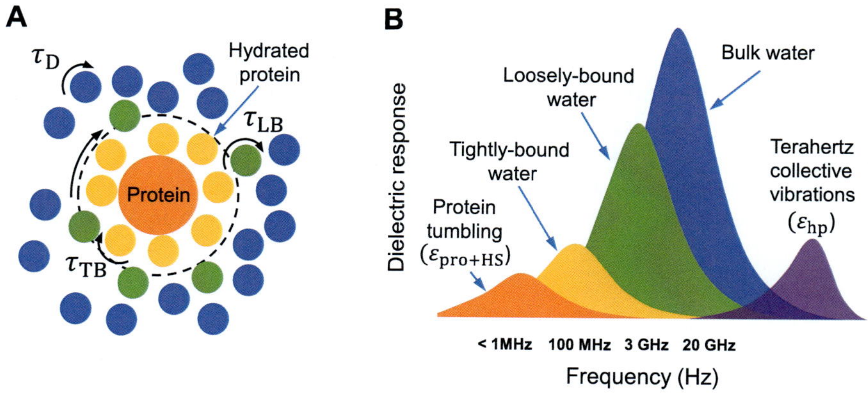

Fig. 2 Schematic representation of water molecules surrounding a protein and their dielectric responses. **a** Protein (orange) in aqueous solution with tightly bound water (yellow), loosely bound water (green), and bulk-water (blue). **b** Corresponding dielectric responses associated with relaxation dynamics, including protein-hydration tumbling, relaxations of tightly bound, loosely bound, and bulk water molecules, as well as the terahertz collective motions of proteins

and the hydration structure as well as dynamics of hydration layers, including bulk water, tightly bound and loosely bound water (Fig. 2). At each frequency, spectra were collected for multiple path lengths to extract the absorption coefficient, $\alpha(\nu)$, and refractive index, $n(\nu)$, which were then combined to obtain the complex dielectric function, $\varepsilon^*(\nu)$, with appropriate corrections for ohmic (conductive) contributions. This approach yields highly accurate, continuous dielectric spectra spanning the megahertz-to-terahertz range for aqueous systems.

2.2 *Molecular Dynamics Simulations*

Details of the MD simulations of the systems listed in Table 1 can be found in ref. [20]; we summarize some of the important steps here. The systems were constructed using PyMOL 2.1.0. Both ends were capped by an acetyl protecting group and *N*-methyl amide for the N-terminus and C-terminus, respectively. Two structured systems were also studied for comparison, specifically a portion of the villin headpiece subdomain, HP24wt (PDB: 4ZC3), and a mutant variant, HP24stab (PDB: 4ZC4). All systems were simulated with GROMACS 2016.6 and 2019.6 with the CHARMM 36 m force field including optimized parameters [44].

Computational structures were compared with experimental structures as deduced from circular dichroism (CD) spectra. The former were obtained from the MD simulation data using SESCA [45], and a favorable comparison was found [20]. The structures were furthermore characterized [20] using the GROMACS geometric clustering algorithm.

The data from the MD simulations were further examined to study the properties of water near the protein systems. Rotational time correlation functions, hydrogen bond time correlation functions, and surface area accessible to solvent were studied, as well as factors by which the dynamics of water are slowed down due to interactions with anionic, cationic and polar residues, relative to the dynamics of water near nonpolar residues, which tends to be fastest.

The first order rotational time correlation function of the water dipole is defined by

$$C_1(t) = \langle P_1(\hat{\mu}(t) \cdot \hat{\mu}(0)) \rangle, \tag{3}$$

where, $P_1(x)$ is the first order Legendre polynomial. Individual time correlations for each water molecule were averaged over the simulation time; $\hat{\mu}(t)$ is the unit vector along the dipole moment vector at time, t.

The hydrogen bond correlation function, $C_{\mathrm{HB}}(t)$, is the probability that a hydrogen bond at an initial time is found at later time, t, even if between the initial time and t the hydrogen bond may have broken and reformed multiple times. This time-correlation function was computed as

$$C_{\mathrm{HB}}(t) = \frac{\langle h_i(t) \cdot h_i(0) \rangle}{\langle h_i(0) \rangle}. \tag{4}$$

In Eq. (4), $h_i(t)$ is 1 if the ith hydrogen bond at time, t, exists if it existed at $t = 0$. A standard definition of hydrogen bond was adopted, whereby a hydrogen bond between donor, D, and acceptor, A, i.e., D–H…A, is formed when the DHA angle is greater than or equal to 150° and the distance between A and D is no greater than 3.5 Å.

Studies of larger systems with both IDRs and structured regions can provide insight into distinct dynamics of water molecules in these regions. A computational study was carried out on 2 proteins to gain insight into the distinct dynamics of water near IDRs and structured regions of proteins. Those proteins were the human small ubiquitin-related modifier 1 (SUMO-1) and ubiquitin conjugating enzyme E2I (UBC9). They were modeled with the CHARMM 36 m force field [46, 47] and simulated using GROMACS 2019.6 [48].

SUMO-1 is a protein with 101 amino acids. There are 4 IDRs ranging in length from 6 to 21 amino acids, with a total of 57 amino acids in the IDRs. UBC9 contains 158 amino acids. It has 6 IDRs ranging in length from 3 to 31 amino acids, with a total of 67 amino acids in 6 IDRs. Initial structures for SUMO-1 and UBC9 were obtained from Protein Data Bank (PDB) [49] entries 1A5R and 1A3S, respectively, and each was solvated using the TIP3P water model [50]. Full details of the MD simulation protocol used to study water dynamics around SUMO-1 and UBC9 can be found in ref. [21].

The dynamics of water molecules near the surface of proteins depends on availability of other, nearby water molecules to facilitate hydrogen bond rearrangement.

Access of additional water molecules for hydrogen bond rearrangement can be determined by computing the average solvent-accessible surface area (SASA) relative to maximum values for a particular residue, which is the relative accessible surface area (RSA). The latter is obtained relative to the average maximum accessible surface area for a specific residue found using a large data set [51]. The average SASA per residue, i, SASA_i, was computed during the simulation. The mean relative accessible surface area (RSA) per residue [51], is defined by

$$\mathrm{RSA}_i = \frac{\langle \mathrm{SASA}_i \rangle}{\mathrm{SASA}_{i,\max}}. \tag{5}$$

Finally, the retardation factor for protein-water hydrogen bonds was computed for hydrogen bond lifetimes, τ, of water with anionic, cationic, and polar residues relative to hydrogen bond lifetimes with nonpolar residues. The retardation factors are given by $\tau_{\mathrm{polar}}/\tau_{\mathrm{nonpolar}}$, $\tau_{\mathrm{cationic}}/\tau_{\mathrm{nonpolar}}$, and $\tau_{\mathrm{anionic}}/\tau_{\mathrm{nonpolar}}$ for polar, cationic, and anionic residues, respectively.

3 Results and Discussion

3.1 Characterization of Structures

To better characterize the protein systems as either IDP or structured, we clustered the MD simulation structural data. Groups of structures and frequency of occurrence are plotted in Fig. 3. The relatively rigid HP24stab displays two α-helical domains separated by a rigid turn, with one cluster containing 98% of structures and two others similar to the dominant one, and structures sampled by HP24wt indicate the center is dominated by an α-helical region. HP24wt and HP24stab exhibit far less structural variation than the IDPs. The conformational ensembles for the IDPs vary from system to system. DS exhibits a variety of extended structures, resembling the form of the disordered region of human stannin, an extended horseshoe conformation [52]. DM and DF show greater diversity as seen by the emergence of many clusters, which may facilitate binding and recognition [53, 54].

3.2 Dielectric Spectra and Rotational Relaxation

The interaction of electromagnetic waves with water molecules in aqueous solutions at megahertz-to-gigahertz frequencies provides insight into the reorientation dynamics of water. In this regime, alternating electrical fields of the electromagnetic waves couple to molecular dipole moment, inducing rotational motion. Consequently, the dielectric response of water and aqueous IDP solutions can therefore

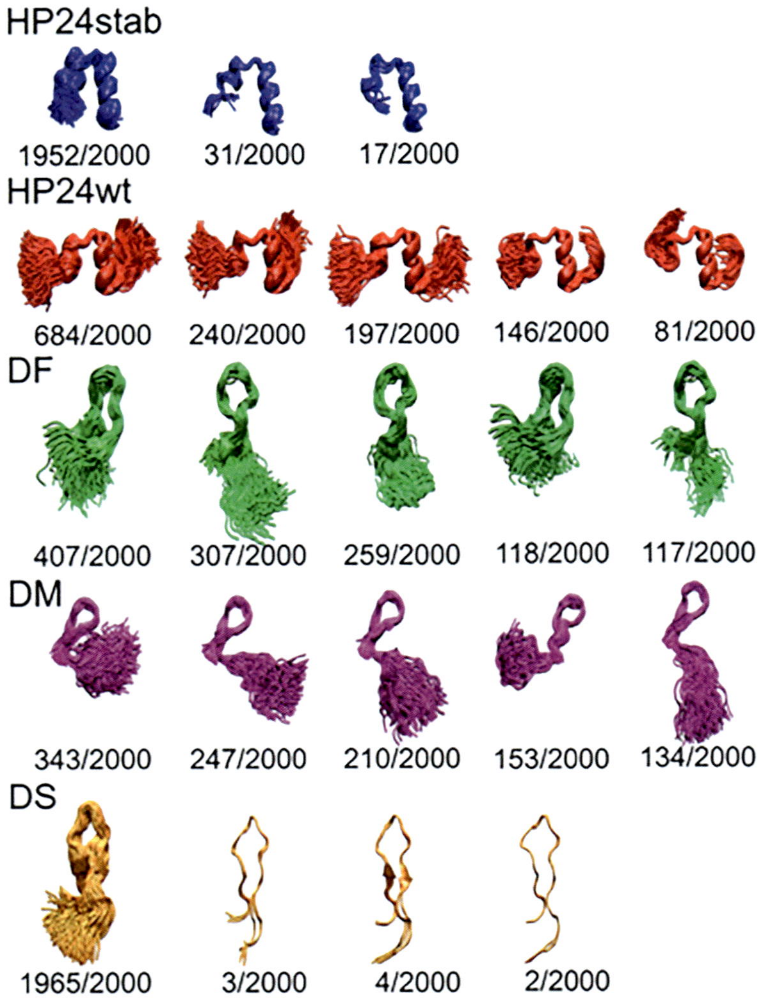

Fig. 3 Small IDPs studied by experiment and simulation, listed in Table 1. Structures and frequency of occurrence of HP24stab, HP24wt, DF, DM and DS are indicated. The 5 largest clusters of configurations with the greatest number of participants are shown. Clusters were constructed from a total of 2000 structures equidistant in time. Reprinted with permission from ref. [20], Copyright (2022), Biophysical Society

be described by Debye-type relaxation processes. At the molecular level, protein-water interactions significantly alter the dynamics of nearby water molecules, which can be categorized into three distinct types of water, including bulk water, loosely bound water and tightly bound water. Relaxation times and dielectric strengths associated with each component can be extracted from experimental spectra to quantify hydration dynamics.

The contributions of these water molecules are resolved by fitting the complex dielectric function, $\varepsilon^*_{\text{sol}}(\nu)$, with a superposition of three Debye relaxation terms, expressed as [20, 30, 55–58],

$$\varepsilon^*_{\text{sol}}(\nu) = \varepsilon_\infty + \varepsilon_{\text{pro + HS}} + \frac{\varepsilon_{\text{S}} - \varepsilon_{\text{TB}}}{1 + i\omega\tau_{\text{TB}}} + \frac{\varepsilon_1 - \varepsilon_{\text{LB}}}{1 + i\omega\tau_{\text{LB}}} + \frac{\varepsilon_2 - \varepsilon_\infty}{1 + i\omega\tau_{\text{D}}} + \varepsilon^*_{\text{hp}}, \quad (6)$$

where $\Delta\varepsilon_{\text{TB}} = \varepsilon_{\text{S}} - \varepsilon_{\text{TB}}$, $\Delta\varepsilon_{\text{LB}} = \varepsilon_1 - \varepsilon_{\text{LB}}$, and $\Delta\varepsilon_{\text{D}} = \varepsilon_2 - \varepsilon_\infty$ represent the dielectric strengths of tightly bound, loosely bound, and bulk water, respectively. The corresponding relaxation times are denoted by τ_{TB}, τ_{LB}, and τ_{D}. Here, $\omega = 2\pi\nu$ is the angular frequency of the applied electric field. Physically, τ_{TB} describes the slow reorientation of tightly bound water molecules directly interacting with the protein surface, constrained by strong electrostatic interactions, τ_{LB} characterizes the dynamics of more mobile, loosely bound water within hydration shells, while τ_{D} corresponds to the fast reorientation of bulk water governed by hydrogen-bond network rearrangements. The reorientation motion of IDP proteins and their hydration water, $\varepsilon_{\text{pro + HS}}$, occurs primarily in sub-megahertz frequencies [59], and is not investigated in this paper. Static permittivity is given by $\varepsilon_{\text{S}} = \varepsilon_\infty + \sum_{i=1}^{3} \Delta\varepsilon_i$, and ε_∞ accounts for contributions to the dielectric response from polarization modes at frequencies beyond the probed range. Finally, $\varepsilon^*_{\text{hp}}$ represents the complex dielectric response of hydrated proteins in the terahertz region, associated with terahertz collective vibrational modes of proteins in solution.

As an example, and to provide an overview of our results, the dielectric loss spectra are plotted in Fig. 4a for DF, HP24wt, and HP24stab. In Fig. 4b the rotational time correlation function, $C_1(t)$, computed with Eq. (3), is plotted for water molecules in the hydration layers.

In both panels of Fig. 4 we observe distinctly different rotational relaxation dynamics of water in the hydration layers around DF compared to that in the hydration layers around the more structured HP24stab and HP24wt. The peak in the dielectric loss spectrum for DF lies around 1 GHz, compared to ≈3 GHz for HP24stab and HP24wt. Similarly, results of the MD simulations indicate that the rotational relaxation of water molecules around the two structured HP24 systems differs from the dynamics of water molecules around the disordered DF. Water molecules in Fig. 4b are seen to relax at the same rate around HP24stab and HP24wt, while they relax 70% more slowly around DF.

Examination of the broadband dielectric spectroscopic measurements provides further information about the water molecules. Besides the ability to quantify the number of water molecules in the hydration layer, it is also possible to classify the water as tightly bound or loosely bound. Indeed, the dielectric spectra for DF, HP24wt, and HP24stab solutions can be deconvoluted into three Debye components and assigned to contributions from tightly bound, loosely bound, and bulk water orientational dynamics. The results are plotted in Fig. 5a–c.

We note that to identify the dielectric response from hydration water, the contribution of bulk water to the total dielectric response characterized by the relaxation time,

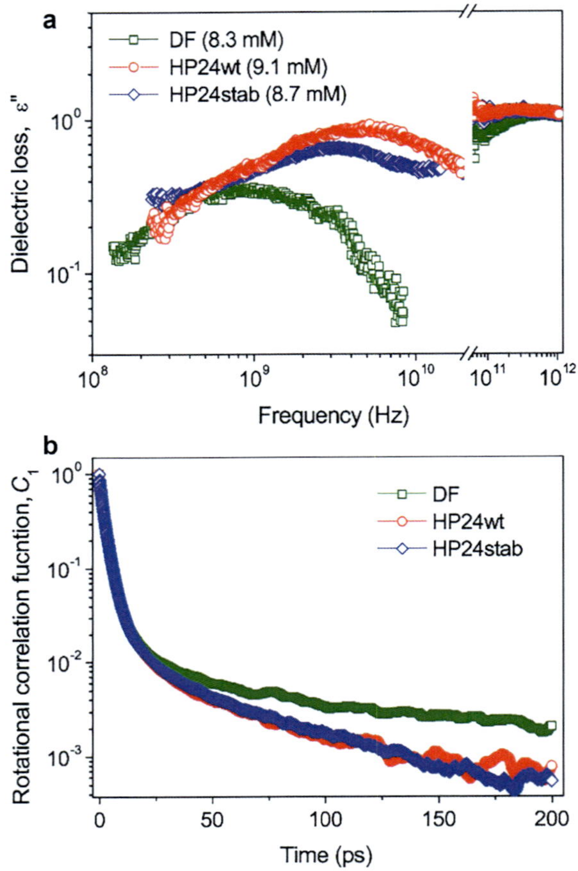

Fig. 4 Rotational relaxation of water molecules around two structured and one disordered system obtained from experiment and from molecular simulations is plotted. **a** Measured dielectric loss spectra for DF (green), HP24wt (red), and HP24stab (blue). **b** Computed rotational time correlation function, $C_1(t)$, plotted for DF (green), HP24wt (red), HP24stab (blue). Reprinted with permission from ref. [20], Copyright (2022), Biophysical Society

τ_D, and the dielectric strength, $\Delta \in_D$, needs to be subtracted from the measured dielectric spectrum. The latter is plotted in Fig. 5a–c. The complex dielectric spectra that result from this subtraction and which are plotted in Fig. 5d–f indicate contributions to the dielectric response from the hydration water molecules.

One sees in Fig. 5 that the hydration dynamics of water surrounding the intrinsically disordered DF is distinct from the hydration water dynamics around the structured HP24wt and HP24stab. Hydration water around the structured proteins exhibits a shift toward higher frequencies, dominated by loosely bound water molecules with relatively rapid orientational motion. In contrast, the majority of hydration water molecules surrounding DF undergo much slower reorientation.

Deconvolution of the dielectric spectra using the three-Debye superposition model discussed above (Eq. (6)) revealed only minor differences in the relaxation times of tightly bound water among the systems. However, the relaxation time of loosely bound water in DF was nearly four times longer than in HP24wt and HP24stab

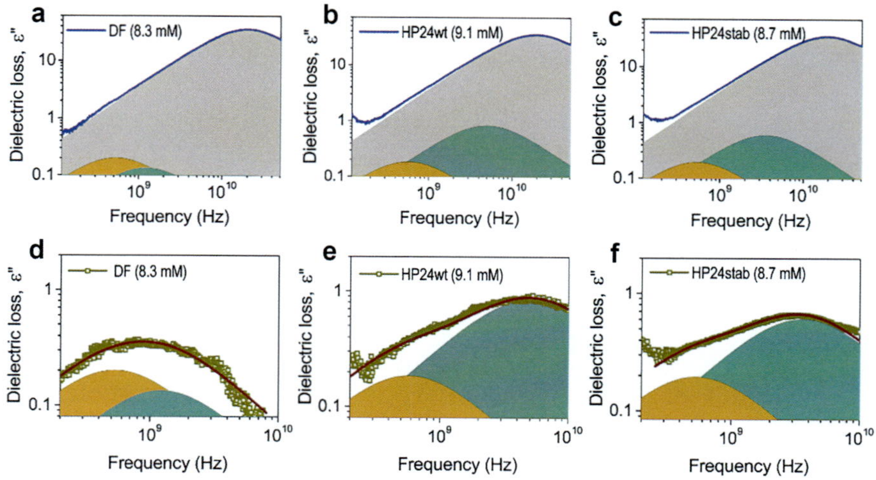

Fig. 5 Dielectric response of protein solutions is plotted. The dielectric spectra of **a** DF, **b** HP24wt, and **c** HP24stab solutions are deconvoluted into three Debye components, assigning to contributions from tightly bound water (yellow), loosely bound water (green), and bulk water orientational dynamics (gray). The dielectric spectra for protein solvation are shown for, **d** DF, **e** HP24wt, and **f** HP24stab, indicating contributions to the dielectric response from hydration water molecules only. Reprinted with permission from ref. [20], Copyright (2022), Biophysical Society

(Table 2), indicating a fundamental difference in hydration dynamics. While structured proteins maintain distinct hydration layers with clear separation between tightly and loosely bound shells, disordered proteins accumulate more water at the surface, producing blurred or merged hydration shells and a predominance of slow orientational dynamics.

To probe collective vibrational modes in the low-frequency terahertz regime, we employed an effective-medium approximation. In this framework, the protein together with its tightly bound hydration layer is treated as a single composite entity. Since the relaxation time of tightly bound water (~300 ps) is orders of magnitude slower than the sub-picosecond timescales of collective vibrational modes. Thus, the protein and its tightly bound water can be approximated as a homogeneous hydrated particle immersed in bulk water.

We employed the Bruggeman approximation, which is particularly suited for aqueous protein solutions across a wide range of concentrations and dielectric contrasts. Within this approach, the total dielectric response of the solution is expressed as a weighted combination of the dielectric contributions from hydrated proteins and pure water [11, 30]. The volumetric fraction of hydrated proteins accounts for both the protein core and the thickness of the tightly bound hydration shell. This treatment enables simultaneous estimation of the collective vibrational modes of the hydrated proteins and the number of tightly bound water molecules. Analysis using this approach revealed that approximately half the hydration water

Table 2 Parameters obtained from dielectric response spectroscopy

Solutions	$\Delta\varepsilon_{TB}$	τ_{TB} (ps)	$\Delta\varepsilon_{LB}$	τ_{LB} (ps)	$\Delta\varepsilon_{D}$	τ_{D} (ps)	σ $(S\cdot m^{-1})$	N_{hyd}	N_{TBW}
Water	–	–	–	–	73.25 ± 0.22	8.27 ± 0.20	–	–	–
DF (8.3 mM)	0.30 ± 0.05	305 ± 35	0.20 ± 0.04	131 ± 10	70.46 ± 0.22	8.23 ± 0.20	0.168	200 ± 25	90 ± 10
HP24wt (9.1 mM)	0.37 ± 0.05	280 ± 35	1.68 ± 0.20	32 ± 5	70.07 ± 0.22	8.22 ± 0.20	0.218	218 ± 25	73 ± 8
HP24stab (8.7 mM)	0.39 ± 0.05	310 ± 35	1.22 ± 0.15	45 ± 5	69.49 ± 0.22	8.25 ± 0.20	0.174	272 ± 30	65 ± 8

N_{hyd} denotes the total number of hydration water molecules, N_{TBW} is the number of tightly bound water molecules, and σ is the electrical conductivity of the solution. Relaxation times for tightly bound, loosely bound, and bulk water are τ_{TB}, τ_{LB}, and τ_{D}, respectively. The corresponding dielectric strengths are $\Delta\varepsilon_{TB}$, $\Delta\varepsilon_{LB}$, and $\Delta\varepsilon_{D}$, respectively in DF resides in the first layer, whereas only one-third to one-fourth of the hydration water occupies in the first layer of HP24wt and HP24stab (Table 2).

These results demonstrate that the water molecules around the intrinsically disordered protein are more tightly bound than those around the more structured systems. DF contains more anionic residues than HP24wt and HP24stab (Table 1). Further investigation implicates anionic residues in the tight binding of water molecules to IDRs and IDPs.

3.3 Hydrogen Bond Dynamics for Systems in *Table 1*

Other relaxation processes studied by the MD simulations also reveal more restricted dynamics of water around IDPs. The hydrogen bond lifetime is determined from the hydrogen bond correlation function, $C_{\mathrm{HB}}(t)$, defined by Eq. (4), for hydrogen bonds between water molecules and residues of the IDPs and structured proteins. The results obtained for the 5 systems listed in Table 1 are plotted in Fig. 6a. The hydrogen bonds are seen to rearrange more slowly around each of the IDPs than around the more structured HP24stab and HP24wt. The hydrogen bond lifetimes, defined as the time to decay to e^{-1}, for DS, DF, and DM are 13.1, 12.6, and 11.2 ps, respectively, while for HP24wt and HP24stab they are 9.94 ps and 9.53 ps, respectively.

In Fig. 6, lifetimes computed for hydrogen bonds between water and individual residues of each system are also plotted. In Fig. 6b–f, the average hydrogen lifetime between water molecules and each residue is indicated by a separate point, and the residues are indicated as anionic, cationic and neutral.

Information about water molecules near the surface of the protein is also provided in Fig. 6. The number of water molecules within 3.5 Å of any atom of a residue is plotted in Fig. 6b–f. There we observe a correlation between the number of water molecules close to a residue and whether the residue contains charged groups. With few exceptions, more than 20 water molecules are in the vicinity of residues with charged groups, whereas fewer than 20 water molecules are near residues that lack charged groups. For the charged groups of the 5 systems the number of nearby water molecules is found to be in the range 23.2–30.0, compared to 10.1–13.2 for neutral groups. For DF, DS, and DM, IDRs with on average more charged groups than contained in the structured systems, there are on average 16.8, 18.1, and 18.6 water molecules near each residue, respectively. For the more structured HP24wt, there are 15.4 water molecules on average near each residue, and for HP24stab, the most structured system, there are 13.0.

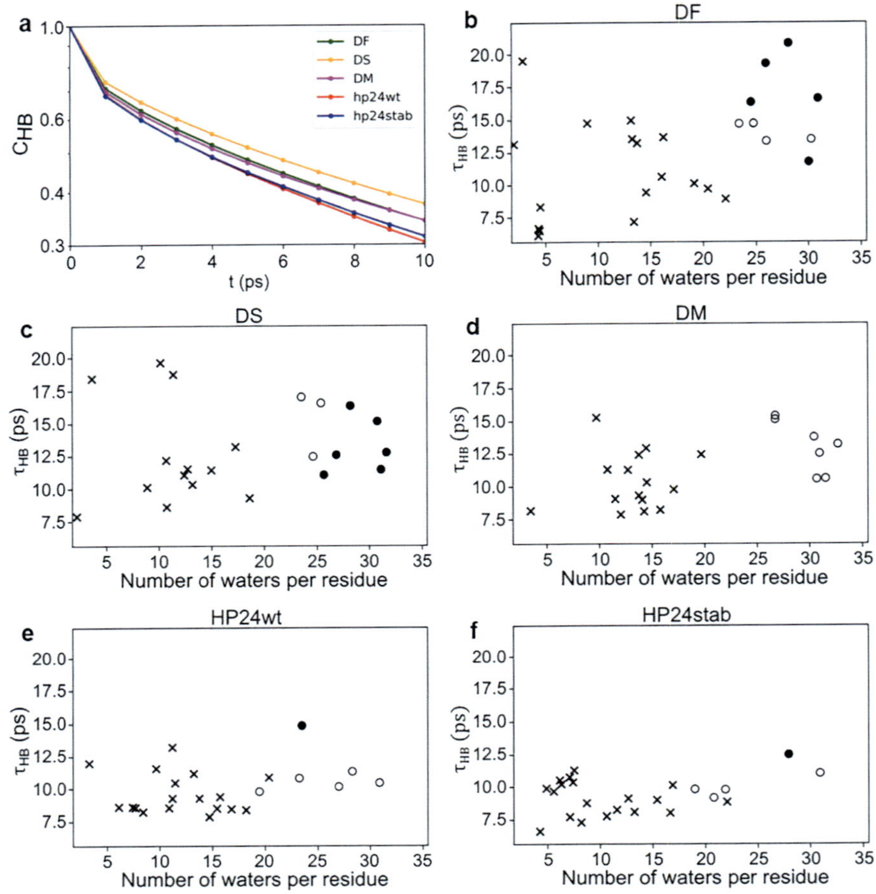

Fig. 6 Lifetimes of hydrogen bonds between water molecules and residues obtained from MD simulations and number of water molecules near each residue are plotted for three disordered and two structured systems. **a** Correlation function for hydrogen bonds, $C_{\text{HB}}(t)$, between water molecules and DF, DS, DM, HP24wt, and HP24stab (log-linear plot). Lifetimes of hydrogen bonds between water and individual residues and the corresponding mean number of water molecules in the vicinity of a residue are plotted in, **b** DF, **c** DS, **d** DM, **e** HP24wt, **f** HP24stab. Each point in **b–f** corresponds to one of the residues of the system, and all residues of each system are indicated, characterized as anionic (filled circles), cationic (open circles), and uncharged (x). Reprinted with permission from ref. [20], Copyright (2022), Biophysical Society

3.4 Hydrogen Bond Dynamics Around IDRs and Structured Regions of Proteins

Further information about the hydration layer around IDRs and IDPs can be found when studying water dynamics around IDRs and structured regions of larger proteins. For this purpose, we review recent work that has been carried out characterizing the

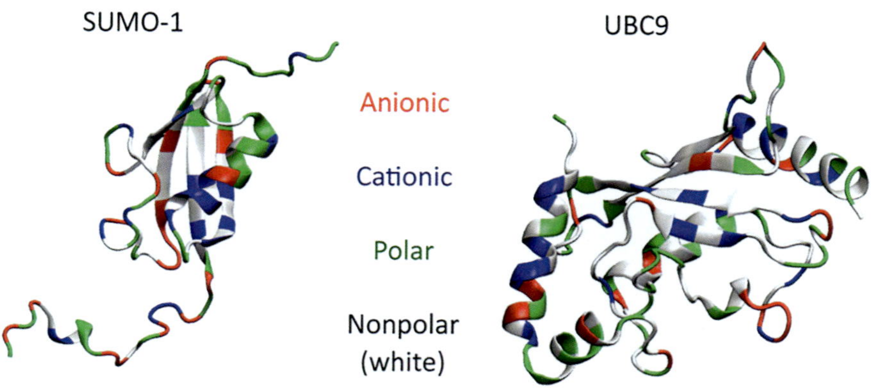

Fig. 7 Structures of SUMO-1 (left) and UBC9 (right) are colored by residue type: anionic, cationic, polar, and nonpolar. Reprinted with permission from ref. [21], Copyright (2023), American Chemical Society

water around SUMO-1 and UBC9, shown in Fig. 7. Proteins with both IDRs and structured regions provide us the opportunity to examine roles of local structure or lack thereof on water, any role of secondary structure, and the overall topology of the protein, which may or may not permit water to move into cavities near the surface of the protein, where dynamics may be different [26, 60, 61]. Larger proteins are characterized by substantial void volumes [62–70], which may permit access to water and affect dynamics at the interface, which, in addition to the nature of the residues, affects water dynamics.

It is worthwhile to first examine hydrogen bond lifetimes of water molecules with protein residues, classified as anionic, cationic, polar, and nonpolar. These are plotted for SUMO-1 and UBC9 as a function of RSA in Fig. 8. RSA quantifies accessibility of water to residues of different sizes, as given by Eq. (5). We would expect the hydrogen bond lifetimes to decrease as RSA increases, since in that case there are fewer nearby water molecules for hydrogen bond rearrangements to take place. This is indeed observed to be the trend in Fig. 8.

Besides the general decrease in the hydrogen bond lifetime with decreasing RSA, we also see that the lifetimes depend on the nature of the amino acid of the protein with which the water molecule hydrogen bonds. For a given RSA, lifetimes are generally shorter for hydrogen bonds between water and nonpolar residues. The lifetimes are longer for water molecules and polar residues, longer still for water molecules and cationic residues, and generally longest for water molecules and anionic residues. The overall trend is reasonable, as we expect stronger interactions between water molecules and ionic groups of the protein. For the cations, the water molecule can rotate around the axis of the bond between the oxygen atom and cation to form a new hydrogen bond and so is less restricted than hydrogen bonds to an anion, contributing to the longer lifetime of hydrogen bonds between water molecules and anionic residues compared to cationic.

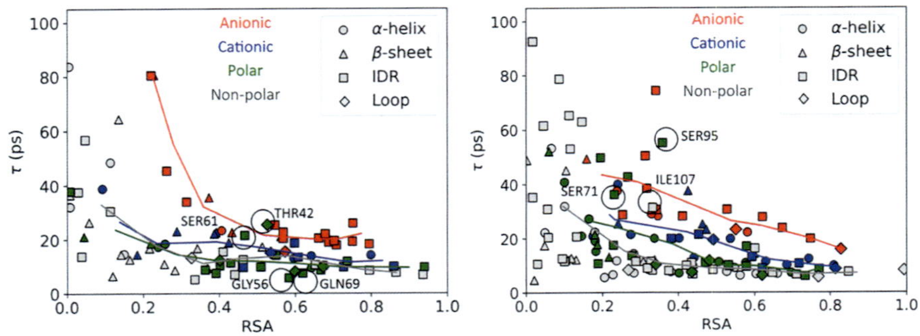

Fig. 8 Hydrogen bond lifetime, τ, versus relative solvent accessible surface area (RSA) is plotted for residues of SUMO-1 (left) and UBC9 (right). Some of the outliers discussed in the text are circled. The residues and data points are colored by residue sidechain: (red) anionic, (blue) cationic, (green) polar, and (white) nonpolar. The curves indicate running averages through the data. Reprinted with permission from ref. [21], Copyright (2023), American Chemical Society

In addition to RSA and the nature of the amino acid hydrogen bonding with water, it is of interest to examine any effects of secondary structure of the protein on dynamics of hydrogen bonds. In Fig. 9, lifetimes of hydrogen bonds are plotted as a function of RSA, where now secondary structure to which the amino acid belongs is also indicated [21]. Since anionic groups were seen to have a pronounced effect on hydrogen bond lifetime, the proportion of anionic groups to the secondary structural element is also indicated in Fig. 9. Darker points correspond to lesser anionic character of the structural element.

For a given RSA value, greater anionic character of secondary structural element, shown as lighter colors, corresponds to longer lifetimes. However, little difference is seen for anionic residues of IDRs, α-helices or β-strands. IDRs tend to have more anionic residues than structured regions, but if a structural element, such as a β-strand,

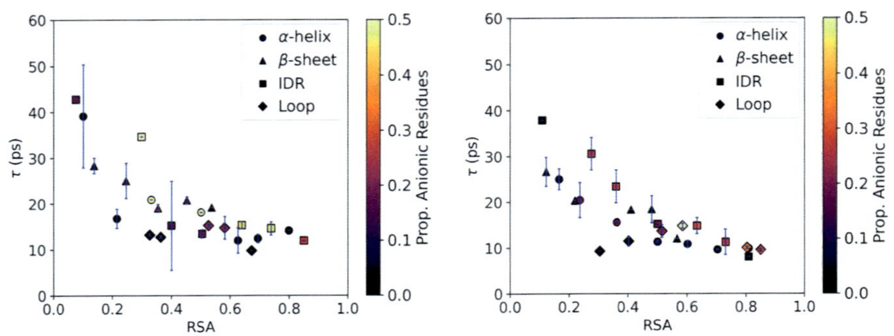

Fig. 9 Hydrogen bond lifetime, τ, is plotted as a function of relative accessible surface area (RSA), for SUMO-1 (left) and UBC9 (right). Error bars indicate one standard deviation. Coloring of each point corresponds to the proportion of anionic residues. Reprinted with permission from ref. [21], Copyright (2023), American Chemical Society

has substantial anionic character, then hydrogen bond dynamics are relatively slow near a given value of RSA. The hydrogen bond lifetime is dictated largely by the nature of the residue in contact with the water molecule and the RSA value, which determines accessibility of other water molecules that can be involved in hydrogen bond rearrangement.

While Fig. 9 reveals that hydrogen bond lifetimes become shorter with larger RSA values and with less anionic character, there are exceptions to these trends. Some of these exceptions are seen as outliers in the data plotted in Fig. 8 for SUMO-1 and UBC9, which we summarize here. Some of the deviations from the main trends reveal characteristics of interactions between proteins, their structural elements, and water.

Turning first to SUMO-1, we discuss notable outliers with exceptionally long as well as short hydrogen bond lifetimes and the reasons they are long or short. One of the outliers is Ser61, a polar residue, which is found near RSA ≈ 0.45. The average hydrogen bond lifetime between water and Ser61 is 21.5 ps. Examining the environment of Ser61, one finds several charged residues. The nearest ionic residue is Glu89, for which the minimum distance is 2.9 Å away. The minimum distance between these residues is defined as the nearest distance between an atom of Ser61 and an atom of Glu89 computed over the course of the MD simulation.[21] The average minimum distance from Ser61 to the next nearest charged residue, Arg63, is 3.4 Å. Other amino acids with ionic groups are also not far from Ser61. These include Lys48 (minimum distance 4.8 Å), Glu93 (minimum distance 6.6 Å) and Glu49 (minimum distance 6.7 Å). The large number of nearby ionic groups restricts motion of nearby water molecules, so they cannot easily facilitate rearrangement of the hydrogen bond between a water molecule and Ser61, contributing to the long lifetime at that site.

Moving to somewhat larger RSA, in the range 0.5–0.6, the residue with the longest hydrogen bond lifetime is Thr42, also polar, with a lifetime of 25.4 ps. Thr42 is found in a loop region, and, like Ser61, there are several nearby ionic residues, contributing to the long lifetime. The nearest is Asp73. The average minimum distance between Thr42 and Asp73 is found in the MD simulations to be 3.5 Å [21]. Thr42 also is found near 3 lysine residues. The average minimum distance from Thr42 to Lys46 is 3.5 Å, 5.5 Å to Lys37 and 5.9 Å to Lys39. The high concentration of nearby ionic residues again lengthens the hydrogen bond lifetime.

There is another outlier for RSA between 0.5 and 0.6, which is Gly56, this time with a short lifetime of 5.7 ps. Gly56 is found in an IDR. There are no charged residues within on average 8.3 Å to Gly56 found in the MD simulations[21]. As a result, any nearby water molecules are relatively free to move and facilitate hydrogen bond rearrangement, contributing to the short lifetime.

Consider now UBC9, for which outliers are also found. For example, for RSA ≈ 0.4 hydrogen bonds between water and Ser95, a polar residue, exhibit on average a particularly long lifetime, 55.1 ps. Ser95 is part of an IDR. Ser95 is found to be in the vicinity of several ionic residues. The nearest one, with an average minimum distance of 2.0 Å, is Asp102. The next nearest is Glu98, which is 2.3 Å away. Lys101 is an average minimum distance of 2.4 Å from Ser95 [21]. Once again,

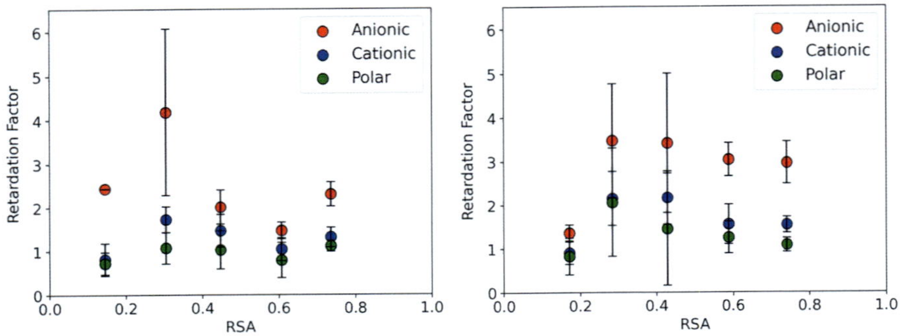

Fig. 10 Retardation factor relative to lifetimes for nonpolar residues for SUMO-1 (left), UBC9 (right) as a function of relative solvent accessible surface area (RSA). Values plotted are averaged in intervals of 0.1, and error bars indicate 1 standard deviation. Reprinted with permission from ref. [21], Copyright (2023), American Chemical Society

the presence of a high concentration of charged groups restricts the motion of nearby water molecules, providing a shortage of available waters to facilitate hydrogen bond rearrangements, yielding a relatively long hydrogen bond lifetime involving a polar residue for a particular value of RSA.

Retardation factors, which quantify the relative lifetime of hydrogen bonds between water and polar, cationic and anionic residues relative to hydrogen bond lifetimes for water and nonpolar residues, provide a summary of the role of these properties in the hydrogen bond lifetimes for different RSA values. RSA-dependent retardation factors were computed as $\tau_{polar}/\tau_{nonpolar}$, $\tau_{cationic}/\tau_{nonpolar}$, and $\tau_{anionic}/\tau_{nonpolar}$ for SUMO-1 and UBC9 [21] and are plotted in Fig. 10. We see that overall hydrogen bonds between water molecules and anionic residues exhibit the longest lifetimes, followed by cationic residues, polar and finally non-polar residues.

The enhancement of lifetimes for hydrogen bonds between water and anionic residues is particularly large for RSA around 0.3–0.4 for both SUMO-1 and UBC9. For SUMO-1, one sees in Fig. 8 that there is a high concentration of charged groups for RSA near 0.3, which explains the large average retardation factors for this protein near this value of RSA. For UBC9, there is a high concentration of charged groups for RSA between about 0.3 and 0.5, consistent with the large retardation factors in this range of RSA.

4 Summary and Outlook

In this chapter we reviewed recent studies examining the dynamics of hydration water around IDPs using broadband megahertz-to-terahertz dielectric spectroscopy together with MD simulations. The investigations addressed key aspects including hydration water dynamics, the nature and extent of the hydration layer, and the dynamic coupling between IDPs and their hydration water. The results revealed greater heterogeneity in the dynamics of water molecules surrounding IDPs compared to dynamics of water hydrating structured proteins of similar size. The megahertz-to-terahertz dielectric spectroscopy measurements quantified both the number of tightly bound and loosely bound water molecules in the hydration layer of one IDP and two structured proteins. Together with MD simulations, these results demonstrated that the higher density of charged groups contained in IDPs, particularly anionic groups, attracts more water molecules closer to them, yielding an excess of tightly bound water molecules to IDPs and the on-average more restricted orientational dynamics.

The role of charged groups in IDRs and IDPs on water dynamics was further studied by MD simulations carried out on two proteins, SUMO-1 and UBC9, which contain structured regions as well as IDRs. For these larger systems, the dynamics of the hydrogen bonds were mainly influenced by two properties. One of them is the nature of the residue to which water hydrogen bonds, specifically if the side chain is anionic, cationic, polar or nonpolar, as was also seen in the study of the smaller systems. A greater density of anionic groups was found to give rise to more restricted water dynamics, as also found for the smaller systems. The other property is the time-averaged relative solvent accessible surface area (RSA). The RSA governs access of water molecules to form new hydrogen bonds. When it is small, water molecules that might serve as new hydrogen bonding sites are less likely to be available, so that on average lifetimes are long.

While the lifetimes of hydrogen bonds between protein and water are typically longest for those between water and anionic groups for a given RSA value, in some cases we have found that they can be even longer for other kinds of interactions, such as water molecules interacting with residues with polar side chains, which occurs if there is a relatively high density of nearby charged residues, particularly anionic. In this case, there are fewer water molecules to serve as new hydrogen bond donors as their motion is hindered by relatively strong interactions with the anionic residues. This yields apparent collective effects: hydrogen bonds between water and polar or nonpolar residues are relatively long-lived compared to the case when there are fewer nearby charged residues. The collective dynamics of water around IDPs resembles that found for hydration dynamics near surfaces of antifreeze proteins found by MD simulations [71].

The results presented here provide a molecular-level picture of hydrogen bond dynamics at the surface of IDPs as well as proteins with IDRs and structured regions for dilute systems. They also raise important questions concerning the behavior of hydration layers around IDPs and IDRs in more complex environments. Liquid–liquid phase separation [72, 73] associated with regions of high and lower densities of IDPs may yield different kinds of hydrogen bond dynamics in different phases and a wider range of dynamics than those seen here. What is the nature of the hydration layer around IDRs and IDPs in the distinct phases, and how do they compare with the hydration layers characterized in the studies reviewed in this chapter? Can a firmer, more quantitative connection be made between density of charged groups and hydration water dynamics in different phases? THz spectroscopy has been characterizing the water in liquid–liquid phase separated systems [74], and it will be of interest in future work to apply the methods discussed in this chapter to examine the nature of the hydration layer around IDRs and IDPs in the distinct phases.

Acknowledgements The authors gratefully acknowledge financial support from NSF grant CHE-2245240 (DML) and Air Force Office of Scientific Research award FA9550-18-1-0263 (NQV).

References

1. Tompa P (2014) Multisteric regulation by structural disorder in modular signaling proteins: an extension of the concept of allostery. Chem Rev 114:6715–6732
2. Theillet F-X, Binolfi A, Frembgen-Kesner T, Hingorani K, Sarkar M, Kyne C, Li C, Crowley PB, Gierasch L, Pielak GJ, Elcock AH, Gerhenson A, Selenko P (2014) Physicochemical properties of cells and their effects on intrinsically disordered proteins (IDPs). Chem Rev 114:6661–6714
3. Lee R, Buljan M, Lang B, Weatheritt RJ, Daughdrill GW, Dunker AK, Fuxreiter M, Gough J, Gsponer J, Jones DT, Kim PM, Kriwacki RW, Oldfield CJ, Pappu RV, Tompa P, Uversky VN, Wright PE, Babu MM (2014) Classification of intrinsically disordered regions and proteins. Chem Rev 114:6589–6631
4. Leitner DM, Gruebele M, Havenith M (2008) Solvation dynamics of biomolecules: modeling and terahertz experiments. HFSP J 2:314–323
5. Leitner DM, Havenith M, Gruebele M (2006) Biomolecule large amplitude motion and solvation dynamics: modeling and probes from THz to X-rays. Int Rev Phys Chem 25:553–582
6. Heyden M (2019) Heterogeneity of water structure and dynamics at the protein-water interface. J Chem Phys 150:094701
7. Päslack C, Schäfer LV, Heyden M (2021) Protein flexibility reduces solvent-mediated friction barriers of ligand binding to a hydrophobic surface patch. Phys Chem Chem Phys 23:5665–5672
8. Bagchi B (2005) Water dynamics in the hydration layer around proteins and miscelles. Chem Rev 105:3197–3219
9. Laage D, Elsaesser T, Hynes JT (2017) Water dynamics in the hydration shells of biomolecules. Chem Rev 117:10694–10725
10. Sterpone F, Stirnemann G, Laage D (2012) Magnitude and molecular origin of water slowdown next to a protein. J Am Chem Soc 134:4116–4119
11. Vinh NQ, Allen SJ, Plaxco KW (2011) Dielectric spectroscopy of proteins as a quantitative experimental test of computational models of their low-frequency harmonic motions. J Am Chem Soc 133:8942–8947
12. Fogarty AC, Laage D (2014) Water dynamics in protein hydration shells: The molecular origins of the dynamical perturbation. J Phys Chem B 118:7715–7729
13. Martin DR, Matyushov DV (2017) Terahertz absorption of lysozyme in solution. J Chem Phys 147:084502
14. Gavrilov Y, Leuchter JD, Levy Y (2017) On the coupling between the dynamics of protein and water. Phys Chem Chem Phys 19:8243–8257
15. Rani P, Biswas P (2015) Local structure and dynamics of hydration water in intrinsically disordered proteins. J Phys Chem B 119:10858–10867
16. Mukhopadhyay S (2020) The dynamism of intrinsically disordered proteins: binding-induced folding, amyloid formation, and phase separation. J Phys Chem B 124:11541–11560
17. Chowdhury A, Kovalenko SA, Aramburu IV, Tan PS, Ernsting NP, Lemke EA (2019) Mechanism-dependent modulation of ultrafast interfacial water dynamics in intrinsically disordered protein complexes. Ang Chem Int Ed 58:4720–4724
18. Arya S, Singh AK, Bhasne K, Dogra P, Datta A, Das P, Mukhopadhyay S (2018) Femtosecond hydration map of intrinsically disordered α-synuclein. Biophys J 114:2540–2551

19. Gallat FX, Laganowsky A, Wood K, Gabel F, Eijck L, Wuttke J, Moulin M, Härtlein M, Eisenberg D, Colletier JP, Zaccai G, Weik M (2012) Dynamical coupling of intrinsically disordered proteins and their hydration water: comparison with folded soluble and membrane proteins. Biophys J 103:129–136
20. Reid KM, Singh AK, Bikash CR, Wei J, Tal-Gan Y, Vinh NQ, Leitner DM (2022) The origin and impact of bound water around intrinsically disordered proteins. Biophys J 121:540–551
21. Reid KM, Poudel H, Leitner DM (2023) Dynamics of hydrogen bonds between water and intrinsically disordered and structured regions of proteins. J Phys Chem B 127:7839–7847
22. Ebbinghaus S, Kim SJ, Heyden M, Yu X, Gruebele M, Leitner DM, Havenith M (2008) Protein sequence- and pH-Dependent hydration probed by terahertz spectroscopy. J Am Chem Soc 130:2374–2375
23. Ebbinghaus S, Kim SJ, Heyden M, Yu X, Heugen U, Gruebele M, Leitner DM, Havenith M (2007) An extended dynamical hydration shell around proteins. Proc Natl Acad Sci (USA) 104:20749–20752
24. Heugen U, Schwaab G, Bründermann E, Heyden M, Yu X, Leitner DM, Havenith M (2006) Solute-induced retardation of water dynamics probed directly by terahertz spectroscopy. Proc Natl Acad Sci. (USA) 103:12301–12306
25. Heyden M, Bründermann E, Heugen U, Niehues G, Leitner DM, Havenith M (2008) Long-Range influence of carbohydrates on the solvation dynamics of water: answers from erahertz absorption measurements and molecular modeling simulations. J Am Chem Soc 130:5773–5779
26. Heyden M, Havenith M (2010) Combining THz spectroscopy and MD simulations to study protein-hydration coupling. Methods 52:74–83
27. Meister K, Ebbinghaus S, Xu Y, Duman JG, DeVries A, Gruebele M, Leitner DM, Havenith M (2013) Long-range protein–water dynamics in hyperactive insect antifreeze proteins. Proc Natl Acad Sci (USA) 110:1617–1622
28. Acbas G, Niessen KA, Snell EH, Markelz AG (2014) Optical measurements of long-range protein vibrations. Nat Comm 5:3076
29. He Y, Ku PI, Knab JR, Chen JY, Markelz AG (2008) Protein dynamical transition does not require protein structure. Phys Rev Lett 101:178103
30. Charkhesht A, Regmi CK, Mitchell-Koch KR, Cheng S, Vinh NQ (2018) High-precision megahertz-to-terahertz dielectric spectroscopy of protein collective motions and hydration dynamics. J Phys Chem B 122:6341–6350
31. Doan LC, Dahanayake JN, Mitchell-Koch KR, Singh AK, Vinh NQ (2022) Probing adaptation of hydration and protein dynamics to temperature. ACS Omega 7:22020–22031
32. Singh AK, Wen C, Cheng S, Vinh NQ (2021) Long-range DNA-water interactions. Biophys J 120:4966–4979
33. Nandi N, Bagchi B (1997) Dielectric relaxation of biological water. J Phys Chem B 124:10954–10961
34. Pal SK, Peon J, Bagchi B, Zewail AH (2002) Biological water: femtosecond dynamics of macromolecular hydration. J Phys Chem B 106:12376–12395
35. Mondal S, Bagchi B (2022) From structure and dynamics to biomolecular functions: the ubiquitous role of solvent in biology. Curr Op Struct Bio 77:102462
36. Böhm F, Schwaab G, Havenith M (2017) Hydration water mapping around alcohol chains by THz calorimetry reveals local changes in heat capacity and free energy upon solvation. Angew Chem Int Ed 56:9981–9985
37. Nibali VC, Pezzotti S, Sebastiani F, Galimberti DR, Schwaab G, Heyden M, Gaigeot MP, Havenith M (2020) Wrapping up hydrophobic hydration: locality matters. J Chem Phys Lett 11:4809–4816
38. Pezzotti S, Chen W, Novelli F, Yu X, Hoberg C, Havenith M (2025) Terahertz calorimetry spotlights the role of water in biological processes. Nat Rev Chem 9:481–494
39. Buchmann A, Caminiti L, Stiewe LL, Jung S, Leitner DM, Havenith M (2025) Internal vibrational energy redistribution precedes energy dissipation into the solvent upon photoexcitation of heme proteins. Phys Chem Chem Phys 27:9470–9477

40. Sampson DA, Wang M, Matunis MJ (2001) The small ubiquitin-like modifier-1 (SUMO-1) consensus sequence mediates Ubc9 binding and is essential for SUMO-1 modification. J Biol Chem 276:21664–21669
41. Cheng X (2023) Protein SUMOylation and phase separation: partners in stress? Trends Biochem Sci 48:417–419
42. George DK, Charkhesht A, Vinh NQ (2015) New terahertz dielectric spectroscopy for the study of aqueous solutions. Rev Sci Instrum 86:123105
43. Vinh NQ, Doan LC, Hoang NLH, Cui JRR, Sindle B (2023) Correlation between macroscopic and microscopic relaxation dynamics of water: evidence for two liquid forms. J Chem Phys 158:204507
44. Abraham MJ, Murtola T, Schulz R, Páll S, Smith JC, Hess B, Lindahl E (2015) GROMACS: high performance molecular simulations through multi-level parallelism from laptops to supercomputers. SoftwareX 1–2:19–25
45. Nagy G, Igaev M, Jones NC, Hoffmann SV, Grubmuller H (2019) SESCA: predicting circular dichroism spectra from protein molecular structures. J Chem Theory Comput 15:5087–5102
46. Brooks BR, Brooks CL, Mackerell AD, Nilsson L, Petrella RJ, Roux B, Won Y, Archontis G, Bartels C, Boresch S, Caflisch A, Caves L, Cui Q, Dinner AR, Feig M, Fischer S, Gao J, Hodoscek M, Im W, Kuczera K, Lazaridis T, Ma J, Ovchinnikov V, Paci E, Pastor RW, Post CB, Pu JZ, Schaefer M, Tidor B, Venable RM, Woodcock HL, Wu X, Yang W, York DM, Karplus M (2009) CHARMM: the biomolecular simulation program. J Comput Chem 30:1545–1614
47. Huang J, Rauscher S, Nawrocki G, Ran T, Feig M, deGroot, BL Grubmuller H, MacKerell AD (2016) CHARMM36m: an improved force field for folded and intrinsically disordered proteins. Nat Methods 14;71–73
48. van der Spoel D, Lindahl E, Hess B, Groenhof G, Mark AE, Berendsen HJC (2005) GROMACS: fast, flexible, and free. J Comput Chem 26:1701–1718
49. Bernstein FC, Koetzle TF, Williams GJB, Meyer EF, Brice MD, Rodgers JR, Kennard O, Shimanouchi TM (1988) The protein data bank. A computer-based archival file for macromolecular structures. Eur J Biochem 80:319–324
50. Jorgensen WL, Chandrasekhar J, Madura JD, Impey RW, Klein ML (1983) Comparison of simple potential functions for simulating liquid water. J Chem Phys 79:926–935
51. Tien MZ, Meyer AG, Sydykova DK, Spielman SJ, Wilke CO (2013) Maximum allowed solvent accessibilites of residues in proteins. PLoS ONE 8:e80635
52. Buck-Koehntop BA, Mascioni A, Buffy JJ, Veglia G (2005) Structure, dynamics, and membrane topology of stannin: a mediator of neuronal cell apoptosis induced by trimethyltin chloride. J Mol Biol 354:652–665
53. Uversky VN (2011) Intrinsically disordered proteins from A to Z. Int J Biochem Cell Biol 43:1090–1103
54. Jakob U, Kriwacki R, Uversky VN (2014) Conditionally and transiently disordered proteins: awakening cryptic disorder to regulate protein function. Chem Rev 114:6779–6805
55. George DK, Charkhesht A, Hull OA, Mishra A, Capelluto DGS, Mitchell-Koch KR, Vinh NQ (2016) New insights into the dynamics of zwitterionic micelles and their hydration waters by gigahertz-to-terahertz dielectric spectroscopy. J Phys Chem B 120:10757–10767
56. Charkhesht A, Lou D, Sindle B, Wen CY, Cheng SF, Vinh NQ (2019) Insights into hydration dynamics and cooperative interactions in glycerol-water mixtures by terahertz dielectric spectroscopy. J Phys Chem B 123:8791–8799
57. Vinh NQ, Sherwin MS, Allen SJ, George DK, Rahmani AJ, Plaxco KW (2015) High-Precision gigahertz-to-terahertz spectroscopy of aqueous salt solutions as a probe of the femtosecond-to-picosecond dynamics of liquid water. J Chem Phys 142:164502
58. Singh AK, Doan LC, Lou D, Wen C, Vinh NQ (2022) Interfacial layers between ion and water detected by terahertz spectroscopy. J Chem Phys 157:054501
59. Oleinikova A, Sasisanker P, Weingartner H (2004) What can really be learned from dielectric spectroscopy of protein solutions? A case study of ribonuclease A. J Phys Chem B 108:8467–8474

60. Dahanayake JN, Mitchell-Koch KR (2018) Entropy connects water structure and dynamics in protein hydration layer. Phys Chem Chem Phys 20:14765–14777
61. Dahanayake JN, Shahryari E, Roberts KM, Heikes ME, Kasireddy C, Mitchell-Koch KR (2019) Protein solvent shell structure provides rapid analysis of hydration dynamics. J Chem Inf Model 59:2407–2422
62. Chen CR, Makhatadze GI (2015) Protein volume: calculating molecular van Der Waals and void volumes in proteins. BMC Bioinf 16:101
63. Enright MB, Leitner DM (2005) Mass fractal dimension and the compactness of proteins. Phys Rev E 71:011912
64. Enright MB, Yu X, Leitner DM (2006) Hydration dependence of the mass fractal dimension and anomalous diffusion of vibrational energy in proteins. Phys Rev E 73:051905
65. Reid KM, Yu X, Leitner DM (2021) Change in vibrational entropy with change in protein volume estimated with mode Grüneisen parameters. J Chem Phys 154:055012
66. Leitner DM, Hyeon C, Reid KM (2020) Water-mediated biomolecular dynamics and allostery. J Chem Phys 152:240901
67. Leitner DM, Pandey HD, Reid KM (2019) Energy transport across interfaces in biomolecular systems. J Phys Chem B 123:9507–9524
68. Liang J, Dill KA (2001) Are proteins well-packed? Biophys J 81:751–766
69. Yu X, Leitner DM (2003) Anomalous diffusion of vibrational energy in proteins. J Chem Phys 119:12673–12679
70. Roche J, Royer CA (2018) Lessons from pressure denaturation of proteins. J Royal Soc Interface 15:20180244
71. Xu Y, Gnanasekaran R, Leitner DM (2012) Analysis of hydrogen bond dynamics at the surface of an antifreeze protein. J Atomic Mol Phys 125071
72. McCarty J, Delaney KT, Danielsen SPO, Fredrickson GH, Shea J-E (2019) Complete phase diagram for liquid-liquid phase separation of intrinsically disordered proteins. J Phys Chem Lett 10:1644–1652
73. Pezzotti S, König B, Ramos S, Schwaab G, Havenith M (2023) Liquid-liquid phase separation? Ask the water! J Phys Chem Lett 14:1556–1563
74. Mukherjee S, Ramos S, Pezzotti S, Kalarikkal A, Prass TM, Galazzo L, Gendreizig D, Barbosa N, Bordignon E, Havenith M, Schäfer LV (2024) Entropy tug-of-war determines solvent effects in the liquid-liquid phase separation of a gobular protein. J Phys Chem Lett 15:4047–4055

Index

S. Cerveny and J. Swenson (eds.), *Dynamics of Biomolecules Studied by Dielectric Spectroscopy*, Advances in Dielectrics, https://doi.org/10.1007/978-3-032-30588-6